# Petroleum Geology of the Southern North Sea:
# Future Potential

Geological Society Special Publications
*Series Editor* A. J. FLEET

# Petroleum Geology of the Southern North Sea: Future Potential

EDITED BY

**KAREN ZIEGLER**
University of Western Ontario, Canada

**PETER TURNER**
University of Birmingham, UK

AND

**STEPHEN R. DAINES**
Conoco (UK) Ltd, Aberdeen

1997

Published by

The Geological Society

London

# THE GEOLOGICAL SOCIETY

The Society was founded in 1807 as The Geological Society of London and is the oldest geological society in the world. It received its Royal Charter in 1825 for the purpose of 'investigating the mineral structure of the Earth'. The Society is Britain's national society for geology with a membership of around 8000. It has countrywide coverage and approximately 1000 members reside overseas. The Society is responsible for all aspects of the geological sciences including professional matters. The Society has its own publishing house, which produces the Society's international journals, books and maps, and which acts as the European distributor for publications of the American Association of Petroleum Geologists, SEPM and the Geological Society of America.

Fellowship is open to those holding a recognized honours degree in geology or cognate subject and who have at least two years' relevant postgraduate experience, or who have not less than six years' relevant experience in geology or a cognate subject. A Fellow who has not less than five years' relevant postgraduate experience in the practice of geology may apply for validation and, subject to approval, may be able to use the designatory letters C. Geol. (Chartered Geologist).

Further information about the Society is available from the Membership Manager, The Geological Society, Burlington House, Piccadilly, London W1V 0JU, UK. The Society is a Registered Charity, No. 210161.


Published by The Geological Society from:
The Geological Society Publishing House
Unit 7, Brassmill Enterprise Centre
Brassmill Lane
Bath BA1 3JN
UK
(*Orders*: Tel. 01225 445046
Fax 01225 442836)

First published 1997


The publishers make no representation, express or implied, with regard to the accuracy of the information contained in this book and cannot accept any legal responsibility for any errors or omissions that may be made.




**British Library Cataloguing in Publication Data**
A catalogue record for this book is available from the British Library.

ISBN 1-897799-82-9
ISSN 0305-8719

Typeset by Aarontype Ltd, Unit 47, Easton Business Centre, Felix Road, Easton, Bristol, UK

Printed by The Alden Press,
Osney Mead, Oxford, UK


**Distributors**

*USA*
AAPG Bookstore
PO Box 979
Tulsa
OK 74101-0979
USA
(*Orders*: Tel. (918) 584-2555
Fax (918) 560-2652)

*Australia*
Australian Mineral Foundation
63 Conyngham Street
Glenside
South Australia 5065
Australia
(*Orders*: Tel. (08) 379-0444
Fax (08) 379-4634)

*India*
Affiliated East-West Press PVT Ltd
G-1/16 Ansari Road
New Delhi 110 002
India
(*Orders:* Tel. (11) 327-9113
Fax (11) 326-0538)

*Japan*
Kanda Book Trading Co.
Tanikawa Building
3-2 Kanda Surugadai
Chiyoda-Ku
Tokyo 101
Japan
(*Orders*: Tel. (03) 3255-3497
Fax (03) 3255-3495)

# Contents

# Introduction

KAREN ZIEGLER[1], PETER TURNER[2] & STEPHEN DAINES[3]

[1] *Department of Earth Sciences, University of Western Ontario, London, Canada N6A 5B7*
[2] *School of Earth Sciences, University of Birmingham, Edgbaston,
Birmingham B15 2TT, UK*
[3] *Conoco UK, North Anderson Drive, Aberdeen AB15 6FZ, UK*

This volume follows similarly themed ones such as those on the Brent Group (Morton *et al.* 1992), and on the Central Graben and the Moray Firth (Hurst *et al.* 1996). It is a natural consequence that a conference on the southern North Sea complements this group of North Sea volumes that are concerned with linking knowledge growth with technical advances so that hydrocarbon resources can be exploited with increasing facility. As the southern North Sea is one of the oldest hydrocarbon provinces in Europe where exploration for hydrocarbons began as early as 1918 (see Glennie, this volume), it is timely that this important region is examined by its students who consider what the future may bring.

The early exploration of the petroleum potential of the southern North Sea concentrated mainly on the Lower Permian Rotliegendes aeolian dune sandstones. Whilst in the 1970s focus was on the sedimentological and environmental aspects, in the 1980s on diagenesis, and in the early 1990s on reservoir geochemistry, interest has now proceeded to stratigraphic re-assessment using correlation concepts such as sequence stratigraphy, climatic cyclicity (e.g. Howell & Aitken 1996) and chemo-stratigraphy. Exploration now aims for subtle traps (ten years from now authors will still be making the same statement) and satellite structures. Explorers are re-evaluating the economic value of marginal and existing fields, and they are continually searching for and applying new techniques and technologies (e.g. Turner & Turner 1995; De Haan 1995) for enhanced productivity. Detailed and advanced techniques in diagenesis also play an important role (e.g. Ashton 1993; Cubitt & England 1995).

Early success also occurred in the Bunter Formation, but this has been markedly short-lived compared to the continuing success in the Rotliegend. However, more recent discoveries in the Carboniferous Westphalian and Namurian reservoirs provide an expanded focus on the basin's wealth.

This volume presents new aspects and methodologies relating to both exploration and production. Much attention is paid to the structural and tectonic characterization of the basin, including reference to outcrop analogues. Other topics were discussed at the meeting from which this volume arose, but are not presented here, including the potential of the Cretaceous, new concepts and results on subsalt imaging, magnetostratigraphy and geostatistical techniques. Also, one of the key technological challenges facing explorers is obtaining a clear view of trap geometry, is conspicuously absent from this volume. The editors hope that relevant contributions on these topics are presented elsewhere in the near future.

**Ken Glennie**, as much a pioneer of recent and fossil desert environments as of the Rotliegend of the southern North Sea, provides the keynote papers for this volume in which he captures the 'History of exploration in the southern North Sea' and the 'Recent advances in understanding the southern North Sea'. His two papers give an excellent overview and introduction to the area and the historical evolution of its commercial exploitation. He comments on the changing attitude of oil companies from carefully guarding and selectively trading hard-won information in the early phases of exploration, to the realization that benefits are possible from sharing results in conferences and the literature.

**George & Berry's** contribution describes the aeolian deposition of the Rotliegend as a response to complex synsedimentary tectonics and cyclical climatic changes ('Permian (Upper Rotliegend) synsedimentary tectonics, basin development and palaeogeography of the southern North Sea'). Their detailed facies mapping results in a series of palaeogeographic maps showing lateral displacements along synsedimentary faults and progressive migration of marginal alluvial fans.

Building on this aspect of sedimentation, **Howell & Mountney's** paper on the 'Climatic cyclicity and accommodation space in arid to semi-arid depositional systems: an example from the Rotliegend Group of the UK southern North Sea' expands this concept to the question of accommodation space created within a

*From* Ziegler, K., Turner, P. & Daines, S. R. (eds), 1997, *Petroleum Geology of the Southern North Sea: Future Potential*, Geological Society Special Publication No. 123, pp. 1–3.

dominantly continental basin. The concept of time-based stratigraphic correlation decoupled from sea-level changes is being tested. The authors define 12 climatically driven depositional units which they use for an improved palaeogeographic reconstruction.

The need for a good understanding of fault development and facies correlations in the Rotliegend becomes clear in the two contributions by **Leveille *et al.***: 'Compartmentalization of Rotliegendes gas reservoirs by sealing faults, Jupiter Fields area, southern North Sea' and 'Diagenetic controls on reservoir quality in Rotliegendes sandstones, Jupiter Fields area, southern North Sea'. These case studies clearly describe how fluid flow and diagenesis, and hence production, is directly controlled by the structural setting, i.e. fault-related and burial-related geochemical aspects. It is shown that 3D-seismic, detailed fault analysis, and modelling of Mesozoic structural geometries can greatly contribute to better predictions of the reservoir quality in undrilled fault blocks.

The stratigraphic expansion of the Rotliegend to the deeper Carboniferous strata is investigated by **Cameron & Ziegler** in 'Probing the lower limits of a fairway: further pre-Permian potential in the southern North Sea'. The authors present a conceptual template needed to begin the systematic and methodical exploration of new opportunities found in the lower tier of the southern North Sea fairway. They identify, through different approaches, probable Carboniferous to Devonian source rocks, and inspect regional seismic for structural and stratigraphic traps. One main factor for the more detailed pre-Permian interpretation has been the advance in seismic acquisition and processing.

**Quirk & Aitken** use their detailed mapping and description of the tectonic framework of the Westphalian ('The structure of the Westphalian in the northern part of the southern North Sea') for improving seismic interpretation, leading to the possibility of finding new targets within already drilled structures.

In the following, complementary paper on 'Sequence stratigraphy of the Westphalian in the northern part of the southern North Sea', **Quirk** provides a low-resolution sequence-stratigraphic interpretation for the Westphalian. The study is based on a large data-set of wells, 2D and 3D seismic. It is an important contribution because detailed lithostratigraphic schemes are required to allow the precise mapping and correlation of known intervals.

One of the first developments of a gas field in the Carboniferous Barren Red Measures Group

in the southern North Sea is described by **Mijnssen** ('Modelling of sand body connectivity in the Schooner Field'). This paper clearly presents the need for detailed understanding of facies and sedimentology, stratigraphy and correlation, presence and type of faults, and also the appropriate choice of the drilling methods. In that sense it is a perfect case study to demonstrate the need for the understanding of many of the aspects discussed in the previous papers. However, it also shows that the Barren Red Group has potential as a future hydrocarbon play in the southern North Sea.

**Yang & Baumfalk** present the 'Application of high-frequency cycle analysis in high-resolution sequence stratigraphy'. First, this paper is an excellent overview and introduction to the methods used in sequence stratigraphy, and shows how they are integrated with traditional methods. Secondly, it illustrates how such techniques can be applied to some southern North Sea examples (Neogene and Carboniferous) in order to improve understanding and prediction of reservoir and source rock distribution.

Overall, this volume presents some highlights in the most recent developments that will allow delivery of the future potential of the southern North Sea. By their nature, such volumes contain the knowledge and interpretions current at the time that they are being produced. By our nature, we students are never satisfied with the limitations imposed by these circumstances. A parallel observation can be made about our incomplete knowledge of the mechanisms that have created one of the world's richest gas provinces, no matter how well it is argued and presented. Of course, here lies the basis for the future discoveries of knowledge and wealth that will occur in this area. Most contributions to the volume have clearly demonstrated that there is significant future potential for the southern North Sea, within and beyond the traditional Rotliegend targets. Owing to the research and development invested in new geological, geophysical and engineering techniques, we will be able to explore and exploit these new targets.

The volume arose from a conference on 'The Petroleum Geology of the southern North Sea: Future Potential' that was organized through the Petroleum Group of the Geological Society in April 1995.

We would like to acknowledge the staff of the Geological Society and the Petroleum Group for their help in organizing and running the conference.

We thank all authors for their efforts, and all referees for their invaluable help: A. Robinson,

S. Knott, R. Wrigley, J. Howell, J. Aitken, C. Fielding, S. Stewart, M. Sweet, K. Glennie, D. Quirk, G. George, C. North, M. Hounslow, E. Hailwood, P. Sansom, J. Maynard, N. Cameron, R. Gaupp, N. Jones and G. Leveille. Thanks also to C. Palmer, G. Kocurek, and G. Plint for their help.

K.Z. would like to thank both the Postgraduate Research Institute for Sedimentology at Reading University (UK) and the Department of Earth Sciences at the University of Western Ontario in London/Ontario (Canada) for the access to facilities and the time necessary to prepare this volume.

The editors also gratefully acknowledge Conoco (UK) Ltd, Scott Pickford Group Ltd and BP Exploration Ltd for the contributions to cover the cost of printing the colour figures.

# References

ASHTON, M. (ed.) 1993. *Advances in Reservoir Geology*. Geological Society, London, Special Publications, **69**.

CUBITT, J. M. & ENGLAND, W. A. (eds) 1995. *The Geochemistry of Reservoirs*. Geological Society, London, Special Publications, **86**.

DE HAAN, H. J. (ed.) 1995. *New Developments in Improved Oil Recovery*. Geological Society, London, Special Publications, **84**.

HOWELL, J. & AITKEN J. (eds) 1996. *High Resolution Sequence Stratigraphy: Innovations and Applications*. Geological Society, London, Special Publications, **104**.

HURST, A., JOHNSON, H., BURLEY, S. D., CANHAM, A. C. & MACKERTICH, D. S. (eds) 1996. *Geology of the Humber Group: Central Graben and Moray Firth, UKCS*. Geological Society, London, Special Publication, **114**.

MORTON, A. C., HASZELDINE, R. S., GILES, M. R. & BROWN, S. (eds) 1992. *Geology of the Brent Group*. Geological Society, London, Special Publications, **61**.

TURNER, P. & TURNER, A. (eds) 1995. *Palaeomagnetic Applications in Hydrocarbon Exploration and Production*. Geological Society, London, Special Publications, **98**.

# History of exploration in the southern North Sea

K. W. GLENNIE

*Department of Geology & Petroleum Geology, King's College,
University of Aberdeen AB9 2UE, UK
Correspondence address: 4 Morven Way, Ballater, Grampian AB35 5SF, UK*

**Abstract:** Exploration history in the Southern North Sea Basin extends back to 1858 in Germany, which was the centre of Europe's oil industry for about a century, with a resurgence in oil discoveries from 1934 to 1945. This changed dramatically in 1963 with the realization of the immense size of the Dutch Groningen gas field. Core data indicated that the Rotliegend reservoir extended westward into the southern North Sea. Between December 1965 and October 1966, five major gas discoveries totalled reserves of almost 20 tcf, sufficient to saturate the UK monopoly market. By 1969, with cheaper seismic ships and a billion barrel oil field in Norwegian waters, UK exploration activity shifted to the central and northern North Sea. The UK southern North Sea became a backwater until about 1981 when, with a pending shortage, British Gas encouraged renewed activity by increasing the price for new gas, the Carboniferous soon becoming an exploration target. Meanwhile, exploration in the Dutch land and offshore, and German land areas resulted in a steady stream of gas discoveries and some oil derived from Liassic source rocks, both enhanced by the later use of 3D seismic data.

The southern North Sea is the main gas-producing area of NW Europe's continental shelf. The gas, mainly in Late Permian (Tatarian) Rotliegend reservoir rocks, is derived from a Carboniferous source in what, structurally, is the Southern Permian Basin, which extends eastward across The Netherlands and Germany to Poland. Although gas fields occur along virtually the whole length of this 1200 km long basin, almost two thirds (97 trillion cubic feet (tcf); Petroleum Geological Circle, 1993) of the ultimate recoverable reserves (>150 tcf) are trapped in just one field, Groningen. Much smaller reserves of oil occur in Zechstein and Lower Cretaceous reservoirs, mostly on land east of the North Sea, although some oil is present in the Dutch offshore area.

The history of the exploration that led to the discovery of these reserves is the subject of this paper.

## Geological outline of the Southern Permian Basin

The Southern Permian Basin (Fig. 1) is framed by the deformed rocks of the Variscan Highlands and the more stable London–Brabant Platform in the south, and to the north by the Mid North Sea–Ringkøbing–Fyn High, and its continuation eastward into the ancient Tornquist–Teisseyre shear zone, which flanks the southern edge of the Fennoscandian Shield.

The basin peters out to the southeast within the Variscan fold belt, and to the west it abuts the Pennine High (Ziegler 1990).

The basin became an important gas province because basin subsidence coincided with the deposition of a vertically stacked sequence of sediments whose facies were controlled by a changing climate as the area slowly drifted northward from the equator (Glennie 1990*a*). The Carboniferous source rocks comprise mostly Westphalian and some Namurian coals (Fig. 2), the products of burial of the equatorial swamps of a foreland basin. These swamps were repeatedly flooded by geologically short-lived marine transgressions, nineteen of which are known within the British Isles (Anderton *et al.* 1979). Comparison of the known distribution and cyclicity of Permo-Carboniferous glaciations (e.g. Crowell 1995) with those of the late Quaternary (e.g. Boulton 1993) suggests that the 'marine bands' probably resulted from interglacial melting of ice and an associated rise in sea level.

Towards the end of the Carboniferous, the humid coal-swamp climate changed to one of increasing aridity under wholly continental conditions as the rising Variscan Highlands created a barrier to easterly trade winds. In addition, the Southern Permian Basin was located somewhere between about 10° and 30° north of the Equator, the Permian latitudinal equivalent of the modern Sahara Desert (Glennie 1990*b*). Thus dune sands, which were to form the best reservoirs for gas of Carboniferous origin, eventually came

*From* Ziegler, K., Turner, P. & Daines, S. R. (eds), 1997, *Petroleum Geology of the Southern North Sea: Future Potential*, Geological Society Special Publication No. 123, pp. 5–16.

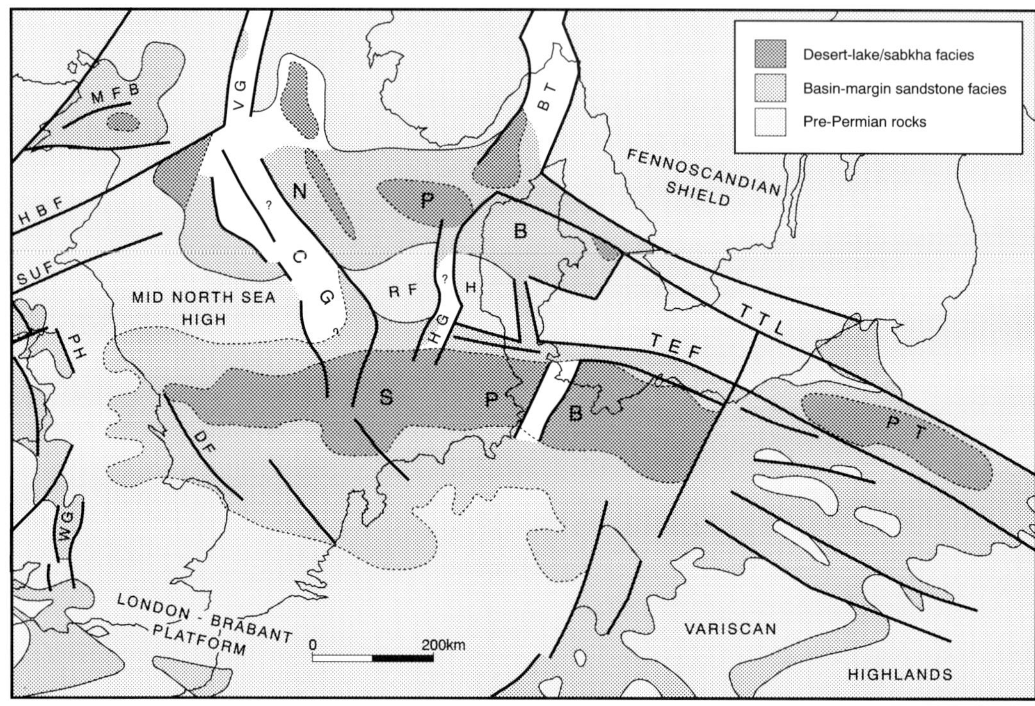



**Fig. 1.** Simplified map of the Southern Permian Basin (SPB) showing basin-centre desert lake/sabkha and basin-margin sandstone facies. BT, Bamble Trough; CG, Central Graben; DF, Dowsing Fault; HBF, Highland Boundary Fault; HD, Hessian Depression; HG, Horn Graben; MFB, Moray Forth Basin; NPB, Northern Permian Basin; OG, Oslo Graben; PH, Pennine High; PT, Polish Trough; RFH, Ringkøbing–Fyn High; SUF, Southern Uplands Fault; TEF, Trans European Fault; TTL, Tornquist–Teisseyre Line; WG, Worcester Graben; VG, Viking Graben.

to occupy much of the southern margin of the basin. Prior to this, however, the most rapidly subsiding part of the basin, centred over northern Germany, became the site of very extensive volcanic activity (Lower Rotliegend volcanics; Ziegler 1990; Plein 1993). The volcanic episode was followed by the long-lived (20–30 Ma) period of erosion marked by the Saalian Unconformity (Fig. 2), above which were deposited the Upper Rotliegend fluvial and dune sands. Because basin-centre subsidence proceeded more rapidly than sedimentation, a major saline desert lake occupied the basin centre (Fig. 1). Climatic changes, controlled by changes in the size of the Gondwana ice cap (Crowell 1995; Parrish 1995), caused cyclicity in sedimentation which, in addition to repeated halite sequences, can also be recognised in basin-margin sequences (e.g. George & Berry 1993, 1994).

Progressive melting of the Gondwana ice cap led to a global rise in sea level and the eventual flooding of the Permian continental basin by marine waters of the Zechstein Sea. Still in an essentially hyper-arid environment, early basin-margin carbonates and anhydrites were replaced, especially towards the basin centre, by the deposition of thick sequences of halite (Taylor 1990). Desiccation to probable dryness was interspersed with several floods induced by glacial melting, which controlled deposition of the various Zechstein cycles. The Zechstein halite forms a most efficient seal for gas trapped in the Rotliegend reservoirs, while the later diapiric movement of this salt was to cause deformation and the creation of potential dip-closed traps for hydrocarbons in overlying reservoirs. East of the North Sea, the Zechstein also contains small accumulations of commercial oil derived from intra-Zechstein sources (Taylor 1990; Müller et al. 1993).

Whatever the reason, the cyclic supply of oceanic water to the Zechstein Sea eventually ceased, and the area reverted to the continental conditions under which the Lower Triassic Bunter sandstones were deposited. Carboniferous

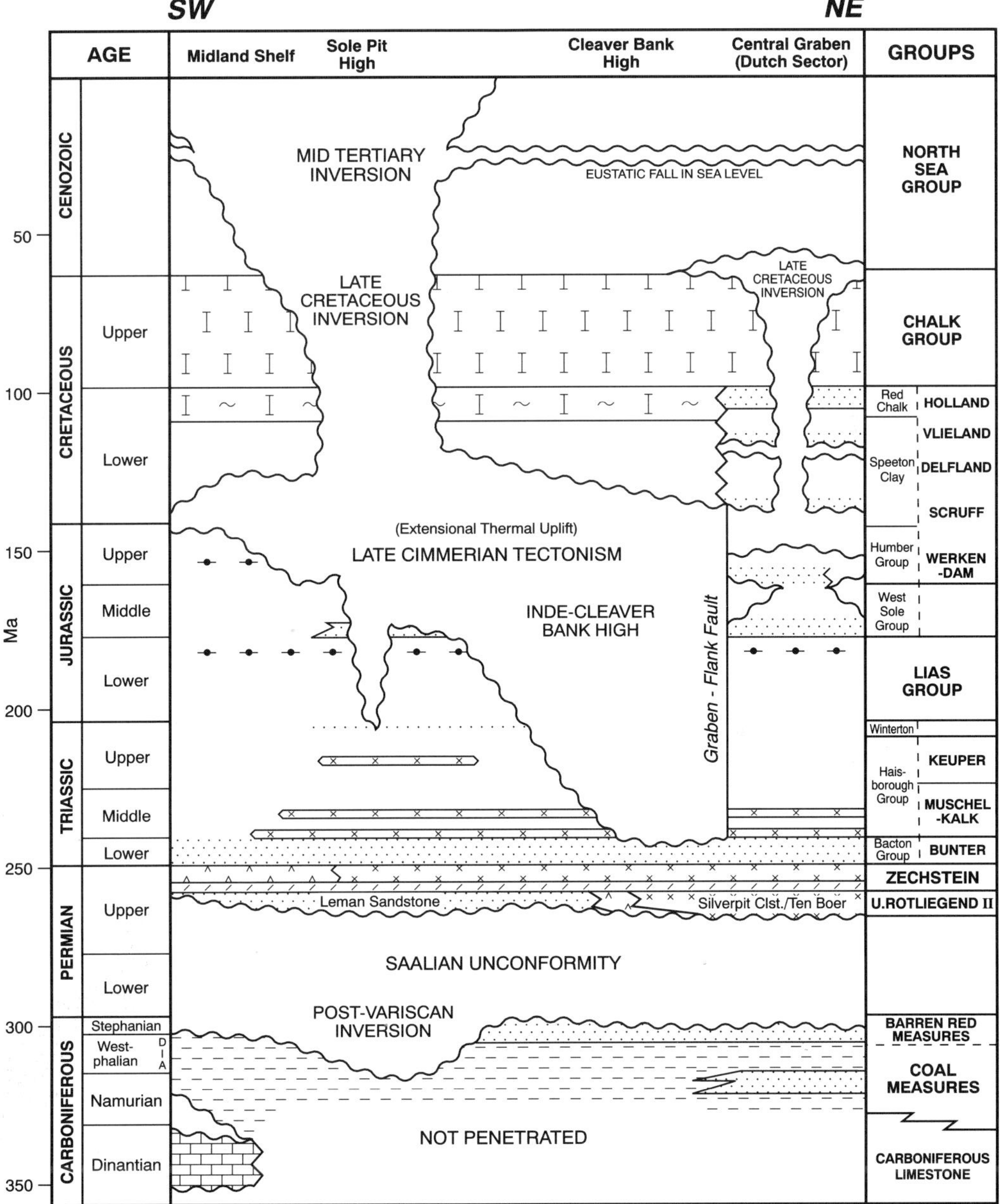



**Fig. 2.** Simplified stratigraphic chart for the Southern Permian Basin showing changes between the East Midland Shelf and the Dutch Central Graben. The main source rocks, reservoirs and seals are shown by standard symbols. Note the effect that several stages of inversion have had on rock-sequence preservation.

gas was later to be trapped in Bunter sandstone reservoirs of fluvial, lacustrine and aeolian origins. During much of the ensuing Triassic, marine transgressions of Tethyan origin extended westward across Poland and Germany to the North Sea area; associated halites (e.g.

Röt) provided a top seal for gas trapped in Bunter sandstones. The whole of the North Sea area was eventually transgressed at the end of the Triassic.

Anoxic conditions over much of Germany and Poland during the Early Jurassic resulted in deposition of an upper Liassic oil source rock (Posidonia Shale; Fig. 2) (Fleet *et al.* 1987). Oil from this source was later trapped in Lower Cretaceous sandstones in Germany and The Netherlands, including some Dutch offshore fields. Following the mid-Jurassic uplift associated with the Rattray volcanic centre of the northern Central Graben, the Late Jurassic Kimmeridge Clay was developed in a source-rock facies over much of the western North Sea (Cornford 1990). In the southern North Sea Basin, however, where preserved this source rock was never buried sufficiently deeply to reach maturity, and over continental Europe it is either absent (Fig. 2) or its time equivalent sediments were not deposited in a source-rock facies (Ziegler 1990).

Major Late Jurassic to earliest Cretaceous transtensional movements, possibly associated with regional thermal uplift especially of the flank areas of the North Sea graben system, resulted in the widespread Cimmerian Unconformity. In Dutch waters these movements were associated with, and followed by, deposition of the continental sands that form important oil and gas reservoirs (Fig. 2). In this context, it is of interest to note that towards the UK–Dutch median line in UK waters, Late Cimmerian erosion down to the top Zechstein seems to have been compensated for isostatically by deposition of a post-Jurassic sequence of equal thickness, although the degree of compensation is considered by Alberts & Underhill (1991) to be nearer 500 m rather than the possible 2000 m suggested by Glennie (1986, fig. 10).

The Sole Pit, Broad Fourteens and West Netherlands structural highs contain important reserves of Rotliegend gas, while the Broad Fourteens and West Netherlands highs have additional reserves of Upper Jurassic–Lower Cretaceous oil.

There was a two-stage inversion of the Sole Pit Basin; limited Late Cretaceous (Turonian–Coniacian) inversion resulted in progressive thinning and then onlap of the Chalk onto the Sole Pit High (Van Hoorn 1987; Alberts & Underhill 1991, fig. 12.8), and was followed by non-deposition or erosion of early Tertiary sequences during the Miocene (Badley *et al.* 1989). To the east, in the Broad Fourteens and West Netherlands areas, Late Cretaceous inversion was much stronger than the Tertiary event (Oele *et al.* 1981; Van Wijhe 1987; Glennie 1990*a*, fig. 2.21). This difference was attributed by Ziegler (1990) to the early 'locking' of the structures in Dutch waters. Knott *et al.* (1993) suggest that, kinematically and temporally, both these inversion phases are more likely to have resulted from the transmission of stresses associated with the Pyrenean Orogeny than with the Alpine (Helvetic) Orogeny.

## History of exploration in the southern North Sea and adjacent areas

### Pre-Groningen

Seep-derived Tertiary oil at Pechelbronn (France) has been exploited since the Middle Ages, while the Scottish oil-shale industry at Broxburn began in 1851. Ignoring these mine-related activities, one could reasonably claim that Europe's conventional hydrocarbon exploration began at Wietze, in the Lower Saxony Basin of North Germany, in 1858 (the year before Drake discovered oil at Titusville in Pennsylvania), when a well was drilled at a geologically chosen site. With the sort of serendipity that one is used to in the oil industry, the well was drilled to locate coal and discovered oil that flowed at the rate of $1\frac{1}{2}$ buckets a day for the next 12 months from a depth of 35.5 m (Plein 1978; Glennie *et al.* 1987). This was the start of the German oil industry. Between then and 1890, a further 100 wells were drilled, 60 of which were producers that gave an annual production between 1908 and 1910 of 11 000 tonnes. Together with four other fields elsewhere in Germany, the country had a total yearly production of 50 000 tonnes (Schröder *et al.* 1991).

No new fields were discovered between 1910 and 1934. In that latter year all mineral rights in Germany were nationalized, thereby bringing order to what had been a virtually unworkable concession situation. Nine hundred wildcats were drilled between 1935 and 1945, resulting in the discovery of 28 new fields that had mostly Jurassic and Cretaceous reservoirs. By the end of World War II, 26 million barrels of oil had been produced from these fields. Even today, over 99% of Germany's gas reserves and almost 93% of its oil have been found in the Northwest German Basin (Schröder *et al.* 1991).

Exploration in eastern England began in 1918. Gas was discovered at Eskdale in Yorkshire in 1938, and the following year Eakring was the start of a series of small oil discoveries in the East Midlands, Kelham Hills in 1940 and, in 1943, Caunton and Dukes Wood. Despite drilling over 400 wells between 1936 and 1954, and another 150 exploration wells between 1955 and 1972, with the discovery of 17 oil fields, all with Carboniferous reservoirs, production did not exceed 1 million barrels of oil per year. The cumulative production from the Eakring field amounted to about $6.5 \times 10^6$ bbl from an estimated $25.6 \times 10^6$ bbl in place (Storey & Nash 1993); the main source rocks for oil are probably Dinantian and lower Namurian shales deposited in the Widmerpool and Gainsborough gulfs. Apart from Eskdale, Zechstein gas was also discovered at Lockton and Malton in North Yorkshire.

In the years preceding the German invasion in 1940, only 12 dry holes had been drilled in The Netherlands, although, during the 1938 World Petroleum Congress, an oil show was noted in basal Tertiary sands when a demonstration well was drilled at Mient, near The Hague. Because of German pressure, exploration was resumed in 1941, and the Schoonebeek oilfield was discovered close to the German border in 1943; it did not get into production until the war was over (Visser & Sung 1958). Gas was discovered in the same area at Coevorden in 1948 and went into production in 1951. By 1954, there were six gas and oil fields in the east and three oilfields in the west, including the Ijsselmonde–Ridderkerk field beneath Rotterdam (Glennie *et al.* 1987).

A wildcat drilled in the Groningen area in 1952 found 180 m of the Lower Permian Rotliegend Sandstone to be water bearing. In 1956, a little gas was discovered in Zechstein dolomites in the Ten Boer well. Three years later, however, the Slochteren-1 well found the Rotliegend fully gas bearing. At that time, when almost every village had its own plant for manufacturing coal gas, another gas discovery was not very welcome. Though exploration for oil continued, with the Zechstein still the main target, 180 m or so of gas-bearing reservoir implies a potentially large field. Following further seismic interpretation and the drilling of four appraisal wells, it was realised that the gas had a common water table, and thus probably belonged to one major structure (Stäuble & Milius 1970). This was confirmed when the Ten Boer well was deepened in 1963 to find the gas–water contact at a depth of 2950 m. The recoverable reserves were then estimated to be 58 tcf (but, as mentioned above, are now known to be nearer 97 tcf; Petroleum Geological Circle 1993), which was considered sufficient to alter the fuel economy of The Netherlands and adjoining land areas. The source of this gas was believed to be the Coal Measures of the underlying Carboniferous.

Once the great size of the Groningen gas field was realized late in 1963, two pressing exploration questions had to be answered. What was the environment of deposition of the reservoir rock, and could this information lead to finding more 'Groningens'? The reservoir rock was thought to be a desert sandstone, probably at least in part of aeolian dune facies. Because the foresets of these sands dipped essentially to the west, attention was immediately focussed on the possibility of finding more gas fields under the North Sea. No one in Shell (operator of Groningen), however, was quite certain of the sedimentary interpretation. Although many geologists had worked in desert areas, few, if any, had the time to study the surface sediments. Groningen was discussed the week before I was due to transfer to Shell Research (KSEPL) to take charge of the turbidite research team. When I reported for duty, I was told that the budget for turbidite research had been cancelled and the money put into desert research instead. I was automatically classified 'the desert expert', and told to 'find out what makes deserts tick' in order to improve our interpretation of the Rotliegend (Glennie 1970, 1972).

## Post-Groningen

The foresets of Rotliegend dune sand in the Slochteren well indicated that the southern North Sea would be a major exploration area. The Continental Shelf Convention of 1958 granted national sovereignty over mineral rights out to the median line (Brennand *et al.* 1990). With the size of Groningen known and already in production in 1963, the Dutch government had no need to hurry the finalization of its offshore licensing regulations, so its offshore exploration began only in 1968. Thus the first offshore wells were drilled in German and British waters. The early German North Sea wells found gas, but with a high, non-commercial proportion (45–65%) of nitrogen. Despite this disappointment, the wells provided useful information on the geographical limits of the Rotliegend play. On land in Germany, Rotliegend gas was discovered in 1968 with the Wustrow Z1 well and, despite a high (40–60%) nitrogen content was, developed across the border in the former East Germany as the 8.5 tcf Salzwedel field. A string of small Rotliegend discoveries had an acceptable content of nitrogen in the west, but eastward this increases to values of up to 80%. In the east, the Zechstein is the reservoir for a string of small oil and condensate finds, and a little gas also occurs in Bunter reservoirs (Müller *et al.* 1993).

Offshore drilling in UK waters began in December 1964, and it is interesting to see where the early exploration wells were drilled (Fig. 3). The first two (Amoseas and Shell) were located on the southern flank of the largest

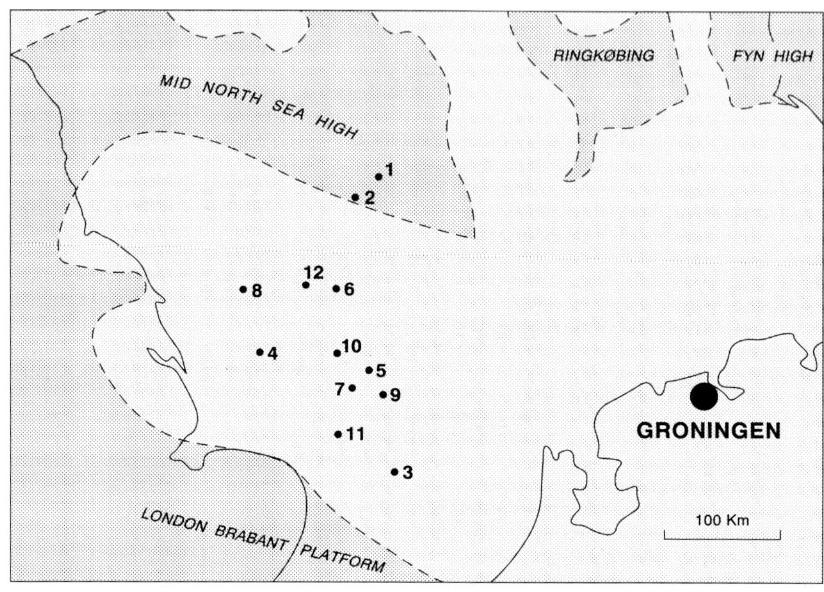



| DRILLING SEQUENCE NUMBER | WELL NUMBER | OPERATOR | FIELD | |
|---|---|---|---|---|
| 1 | 38/29-1 | AMOSEAS | | D & A |
| 2 | 44/2-1 | SHELL | | D & A |
| 3 | 53/10-1 | GULF | | D & A |
| 4 | 48/6-1 | BP | WEST SOLE | |
| 5 | 49/13-1 | GULF | | SHOW |
| 6 | 44/21-1 | TOTAL | | D & A |
| 7 | 49/17-1 | CONOCO | S. VIKING | |
| 8 | 41/20-1 | SIGNAL | | D & A |
| 9 | 49/19-1 | SHELL | | D & A |
| 10 | 49/6-1 | PHILLIPS | ANN | |
| 11 | 49/26-1 | SHELL | LEMAN | |
| 12 | 42/23-1 | BURMAH | | D & A |

**Fig. 3.** The drilling sequence, operators and discoveries made by the first 12 offshore wells in the UK southern North Sea.

structure in the North Sea, the Mid North Sea High, a palaeo-high over which the Rotliegend was absent. The third (Gulf) found the Rotliegend but was south of the Zechstein salt seal and was water-wet. The fourth well (BP) discovered the West Sole gas field in December 1965. The first dozen wells found three commercial fields (West Sole, Leman and South Viking), one apparently non-commercial field (Ann) and one gas show (49/13-1), and by October 1966, Indefatigable and Hewett had been added to the list of successes. Thus, within less than one year, the recoverable reserves from the first five major discoveries totalled almost 20 tcf of gas.

These early discoveries reflected the importance of acquiring the right acreage as much as any real geometrical knowledge of the prospects. At that time, using analog seismic data, it was generally not possible to recognise structures below the top of the Zechstein salt. That development would not come until the use of digital recording and processing which, with the continued enormous improvement in computer technology, have revolutionised structural interpretation in recent years.

The Hewett discovery was unusual for the time in having gas in two Lower Triassic reservoirs, the Bunter and Hewett sandstones, as well as in the Hauptdolomite of the Zechstein. Migration of gas through the relatively thin Zechstein salt seal into Triassic reservoirs was probably assisted by faulting.

The high hopes associated with the large Triassic structures overlying Zechstein salt swells in quadrants 43 and 44 were soon dashed. Well 43/20-1, later known as the Gordon field, drilled in 1969, encountered only 249 ft of gas-bearing Bunter sandstone in a reservoir 477 ft thick; the reservoir has a top seal of Röt halite (Bifani 1986). Unlike the Rotliegend fields, all of which seemed to be filled to spill/leak point, this unfaulted structure was only partly filled. A similar situation was found in the nearby Esmond and Forbes structures, the horizontal gas-water contact being clearly visible on seismic in the former field (Bifani 1986, fig. 4; Ketter 1991). Seismic data indicate the lateral flow of salt into the pillows (see Taylor 1990, fig. 6.21 for an example), which allowed gas from the underlying Rotliegend reservoir to break through the adjacent thinned salt where it had been disrupted by faulting; the volume of gas drained from the Rotliegend at that time was insufficient to fill the Bunter reservoir. This interpretation was virtually confirmed by the gas content of four small closely spaced structures in Dutch block K/13 (Roos & Smits 1983). The Bunter reservoir of two of them were gas bearing, the underlying Rotliegend being water-wet, with the reverse situation in the other two structures.

Once the fields had been developed, the UK North Sea gas was brought ashore by pipeline to Easington in Yorkshire (West Sole), Theddlethorpe in Lincolnshire (Viking) and Bacton in Norfolk (Hewett, Leman and Indefatigable), where it was fed into a newly developed national gas grid system.

After the early successes in finding gas, exploration in UK waters almost ground to a halt within about four years, although with obligation wells, the peak of drilling activity did not occur until 1969. Britain had a state-owned monopoly market (Gas Council, now British Gas) for the gas from these first five economic fields, and their combined recoverable reserves of almost 20 tcf had, at least temporarily, saturated the market. By the end of the decade, exploration activity was moving to the central North Sea, where the billion barrel Ekofisk oil field (Chalk reservoir) was discovered in Norwegian waters in 1969, to be followed a year later by Forties (Palaeocene sandstones), another billion barrel field. With a rapid development of the commercially more valuable oil, especially following the 1971 discovery of yet another billion barrel field, Brent (Mid-Jurassic), the southern North Sea was to become an exploration backwater for the next decade, although field development continued into the early 1970s. The development and purchase of gas from both the UK and Norwegian sectors of the giant Frigg field further delayed any renewal of exploration activity in the southern North Sea.

Although exploration in the Dutch offshore did not start until 1968, it was then continuous. There was a steady stream of mostly Rotliegend gas discoveries (the first in 1968 with Mobil's well P/6-1) over the years because there was a free market and the price of gas was related to the thermal value of oil. Apart from the known Schoonebeek field with its Lower Cretaceous Bentheim Sandstone reservoir, oil exploration occurred in two other areas with reservoirs of the same general age (Lower Cretaceous Vlieland Sandstone): the southern Central Graben (F/18 and F/3 fields; the former being the first offshore discovery of oil); and the landward Rijswijk province (Bodenhausen & Ott 1981), which extends offshore to blocks Q/1 (Haven, Helder, Helm and Hoorn fields; Hastings *et al.* 1991; Roelofsen & de Boer 1991), P/15 (Rijn), K/18 and L/16 (now the Kotter and Logger fields). The major source rock for this oil is the Liassic Posidonia Shale. Oil trapped in Early Cretaceous structures remigrated during Late Cretaceous inversion and suffered from biodegradation (Roelofsen & de Boer 1991).

Exploration on the German part of the North Sea had been disappointing right from the beginning because of a non-commercial content of nitrogen and carbon dioxide in a reservoir that was rich in shale and beds of halite. There has been much more success in the adjacent land area of the Lower Saxony Basin (Betz *et al.* 1987), gas being discovered in reservoirs ranging in age from the Carboniferous to the Jurassic, while oil is produced from Cretaceous reservoirs. In the former East Germany, however, the Salzwedel field was already under development in the late 1960s despite its high nitrogen content. Since the Salzwedel find in 1968 , the rate of discovery has been fairly constant, with another 5 tcf of gas being added to the reserves in a series of mostly small fields (Burri *et al.* 1993).

The ongoing German oil production in the Lower Saxony Basin reached a peak around 1968 with an annual production of about 7 million tonnes, whereas gas production in the former West Germany did not peak until 1978, with a production of $20 \times 10^6 \, \text{m}^3$ Schröder *et al.* 1991).

Even though the incentive to find more gas in UK waters had been removed in the late 1960s by a saturated market, in 1981, with a pending shortage, British Gas encouraged renewed activity by increasing the price for new gas (Fig. 4). This activity has continued fairly steadily until the present, with cost-saving unmanned platforms allowing small accumulations to be developed economically. Additionally, the widespread use of 3D seismic methods as an exploration tool as well as for development, has enabled gas accumulations to be defined with greater precision, thereby reducing the number of dry holes. Also, fields that formerly were considered to be uneconomical because the permeability of their reservoirs had been damaged by the deep-burial growth of diagenetic illite, could now be brought into production by the use of horizontal drilling; this technique greatly increased the drainage area in contact with the production well.

Some operators had long realized that at the base of the Rotliegend sequence were local anomalous developments of clean sand (e.g. 49/24-J1000) which, because they were gas-bearing, had good porosities and permeabilities and lacked coals, were thought also to be Rotliegend. Because of their limited development (not found in adjacent wells), such rare occurrences are now believed to be 'shoe-string' fluvial sands of late Westphalian or Stephanian age. The realization that gas could be trapped more widely if such sands had a top seal of Silverpit claystones and evaporites led inevitably in the 1980s to the Carboniferous play (Fig. 4) beneath the Rotliegend of the basin centre.

In the early days of Rotliegend exploration, the Carboniferous had been treated as 'economic basement', with well penetrations commonly not exceeding three or four metres. The contact was generally recognized by a change in drilling rate, the first cuttings of coal, or the typical purplish coating on Carboniferous sands – palynology failed, except in a negative sense, because all spores were generally oxidized to carbon dioxide for the first 20 m below the Rotliegend. Although oil occurs in Carboniferous reservoirs in the East Midlands, the Westphalian source rocks of the Silverpit Basin are gas prone and have yet to reach peak maturation. The early difficulties of interpreting 2D seismic data below the Zechstein salt have been largely overcome with the widespread use of 3D surveys, which allow the Barren Red Measures to be mapped in some detail.

The estimated ultimate recoverable reserves for over 40 UK gas fields of the southern North Sea, derived from a variety of sources, can be

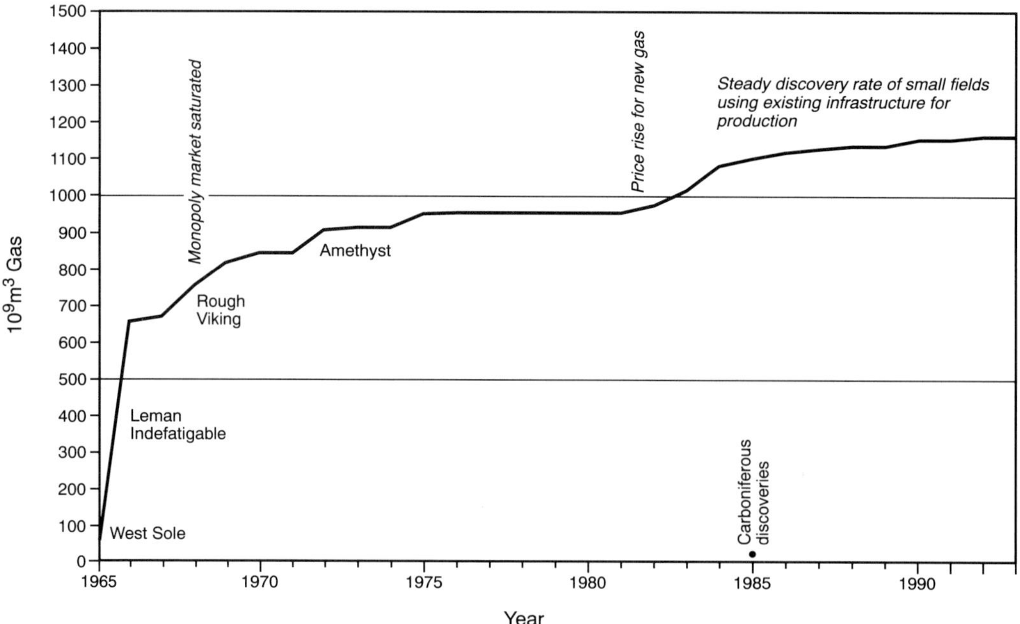


**Fig. 4.** Annual accumulative increase in discovery of additional reserves of ultimately recoverable gas 1965–1993, UK southern North Sea. Note the lack of discoveries from 1975 until a rise in price for new gas in 1981.

found in Table 1. The table emphasizes the great importance of the Rotliegend gas discoveries, and the relatively minor contribution made by the Carboniferous and Bunter reservoirs Some of these fields have either already gone out of production or are nearing the end of their economic life, while others are still in the early stages of development.

## Future potential of the southern North Sea Basin

Exploration for hydrocarbons is driven by the current working hypothesis, which at the beginning of any new play ought to be kept as simple as possible. Foresets in the Groningen gas-bearing sands, for example, indicated that reservoir rock could continue westward under the North Sea, where the Zechstein caprock was probably also present – no more geological sophistication was needed to select where the initial offshore exploration should be centred; the spur to action was the huge economic incentive of possibly finding another 'Groningen'. Once hydrocarbons have been found in any particular play, then the working hypotheses naturally become more complex as additional facts become known.

All the obvious structures and stratigraphic horizons beneath the southern North Sea have been tested by the drill, and were found either to be gas bearing or were dry. The more distant future of the basin lies in the less-popular 'not-so-obvious' targets.

The Carboniferous play of the 1980s shows that there is still an element of truth in the old driller's expression that 'oil (or gas) is where you find it', and not necessarily where you think it should be. That play was the outcome of a realization that Carboniferous strata need not equate with 'economic basement'; the improved quality of seismic data permitted potential reservoir horizons to be mapped; but the necessary work and initial drilling was driven by the economic incentive of an improved gas price. Economics will be the driving force behind any future discoveries. But hydrocarbons are not found by economics; that needs the addition of human imagination and logical deduction, coupled with an ability to persuade management to take a financial risk on unusual, non-standard prospects. Where could hydrocarbons still be lurking within the Southern Permian Basin?

The explored parts of the Southern Permian basin are underlain by an abundance of gas-prone Carboniferous source rock (Cornford 1990). We need to find reservoir rocks and seals with a trapping configuration. Since most

**Table 1.** *Estimated ultimate recoverable reserves of gas in UK southern North Sea gasfields*

| Gasfield | Ultimate reserves | |
|---|---|---|
| | $(10^9\,\mathrm{m}^3)$ | Reservoir |
| Amethyst | 24 | Rot |
| Anglia | 7 | Rot |
| Ann | 4 | Rot |
| Audrey | 3 | Rot |
| Baird | 2 | Rot |
| Barque | 38 | Rot |
| Caister | 9 | Cbn |
| Camelot | 6 | Rot |
| Cleeton | 8 | Rot |
| Clipper | 16 | Rot |
| Della | 2 | Rot |
| Esmond | | |
| Forbes | 19 | Bnt |
| Gordon | | |
| Excalibur | 7 | Rot |
| Galleon | 40 | Rot |
| Guinivere | 2 | Rot |
| Hewett | | |
| Deborah | 115 | Bnt |
| Dotty | | |
| Hyde | 5 | Rot |
| Indefatigable | 128 | Rot |
| Indefatigable SW | 2 | Rot |
| Johnston | 5 | Rot |
| Lancelot | 6 | Rot |
| Leman | 343 | Rot |
| Markham | 20 | Rot |
| Murdoch | 9 | Cbn |
| Orwell | 7 | Bnt |
| Pickerill | 17 | Rot |
| Ravenspurn N | 35 | Rot |
| Ravenspurn S | 18 | Rot |
| Rough | 11 | Rot |
| Sean N | 5 | Rot |
| Sean S | 7 | Rot |
| Thames | | |
| Yare | 11 | Rot |
| Bure | | |
| Tristan | 1 | Rot |
| V-fields | 33 | Rot |
| Victor | 27 | Rot |
| Viking | 80 | Rot |
| Welland | 8 | Rot |
| West Sole | 57 | Rot |

Sources: Shell UK, Abbotts (1991); Parker (1993); Woodland (1975), Dept. of Energy (unpublished). Reservoirs: Bnt, Bunter; Cbn, Carboniferous; Rot, Rotliegend.

of the obvious structural targets have already been drilled, the less obvious ones must be sought. Within the Carboniferous of the Dutch–German border area, there may still be potential for oil generation from Namurian source rocks, similar to that of the Midlands of England (e.g. Glennie 1986, fig. 4). Could such a source rock give potential to a Carboniferous Limestone play, perhaps with a karstified reservoir, along the southern flank of basin? In such a location, both oil and gas could migrate up-dip into structural and stratigraphic traps. Deeper still, the older Palaeozoic strata are mature for gas generation in the Midlands of England (gas shows at Cooles Farm; Smith 1993) and similar rocks could be gas bearing beneath East Anglia. Or across the other side of the basin, adjacent to the Mid North Sea–Ringkøbing–Fyn High, there has been far too little drilling to rule out the possibility of stratigraphic trapping of gas within the Yoredale Series (see Besly 1990); similarly, do we really know whether there are no Rotliegend reservoir sands in the same general area?

Have all the Rotliegend structures been tested? I think not. Almost every viable structure seems to contain gas trapped beneath that most efficient of seals, the Zechstein halite. With an inevitable long-term increase in the price of gas, many small structures will become economically viable even if only for one-well developments; they will require cheap wells, with the possibility of retrievable, flexible, production tubing. How many potential fields have been written off when the drill found a tight reservoir because it was located too close to the Dowsing or South Hewett faults? Would a sidetrack have found producible gas? And perhaps we need not confine our thinking only to very small accumulations. Rotliegend gas was generated during subsidence of the Sole Pit Basin. With a cover of that excellent Zechstein salt, and with lateral fault seals, some of this gas could still be trapped in synclinal or other down-dip positions; this possibility is supported by the 460 ft difference across the dividing fault between the gas–water contacts of the A and F reservoirs of the Viking Field (Morgan 1993); careful reconstruction of the Sole Pit inversion on a regional scale might lead to the recognition of some sizeable traps.

Lake-margin sands contain 60% of Germany's reserves of Rotliegend gas (Gast 1991). Although there are marked palaeogeographic differences with Germany, the UK southern lake margin is under-explored, while the northern shore is untouched; until tested by the drill, all hypotheses about facies distribution are of a working nature only, and must be considered fallible.

Following early disappointments (e.g. Eskdale and Lockton, on land), the Zechstein has always been down-graded as an exploration target in UK waters. But is that fair? The Zechstein was the prime target when the first Groningen well was drilled (Slochteren-1), and ever since it has been overshadowed by the discovery of prolific volumes of Rotliegend gas. In Germany and Poland, however, the Zechstein is the reservoir for many small self-sourced oil accumulations as well as gas (Taylor 1990), It may well be worth studying the Zechstein in these eastern parts of the Southern Permian Basin to find out why the reservoir there is a viable exploration target and apparently is much less so in UK waters; perhaps differences in the style of basin development at the time of deposition was an important factor (see e.g. Kiersnowski *et al.* 1995). Since writing the above, the Zechstein of central Germany is reported to be an exploration target again following inactivity for about 20 years while oil and Rotliegend gas was developed further north (Karnin *et al.* 1996).

Post-Permian sequences are less likely than others to contain undiscovered accumulations of hydrocarbons. Even here, however, small gas fields have possibly been overlooked within Triassic sandstones and dolomites in proximity to thin or breached Zechstein salt; and oil of Liassic and Kimmeridgian origin could still be present near the UK–Dutch border.

Some of the foregoing, perhaps seen from a different viewpoint, can also be found in a very readable anecdotal book compiled and edited by Richard Moreton (1995), entitled 'Tales from Early UK Oil Exploration 1960–1979'

## Conclusions

(1) Although the hydrocarbon industry has a history extending back over 130 years in Germany and about 60 years in Britain, in the mid-1960s there was almost explosive activity in the Southern North Sea following realization of the size of the giant Groningen gas field in The Netherlands, with its Rotliegend reservoir. The first five offshore discoveries saturated the UK monopoly market and exploration interest moved to the Central and Northern North Sea where oil had been discovered. Exploration for Rotliegend gas in Dutch waters has continued steadily since 1968.

(2) A pending shortage of gas for the UK market led to a price rise and renewed development of gas fields, including those with 'Barren Red' reservoirs, from 1981 onward. Improved seismic enabled structures beneath the Zechstein salt to be better defined, especially once 3D seismic surveys came to be used for exploration as well as development. The latter, combined with the use of horizontal wells, has enabled the development of many formerly unproducible fields.

(3) Zechstein oil and condensate discoveries are confined largely to Dutch, German and Polish land areas. Liassic oil in mostly Cretaceous reservoirs extends offshore west of The Netherlands and occurs in the Dutch part of the Central Graben, but cannot be expected in adjacent UK waters because of a lack of maturity of the source rock.

(4) Long-cherished working hypotheses on hydrocarbon generation and entrapment should not deter geologists from looking for oil or gas in unconventional locations such as the older Palaeozoic or in synclines.

# References


ABBOTS, I. L. (ed.) 1991. *UK Oil and Gas Fields, 25 Years Commerative Volume.* Geological Society, London, Memoirs, **14.**

ALBERTS, M. A. & UNDERHILL, J. R. 1991. The effect of Tertiary structuration on Permian gas prospectivity, Cleaver Bank area, southern North Sea, UK. *In*: SPENCER, A. M. (ed.) *Generation, Accumulation and Production of Europe's hydrocarbons.* Oxford University Press, 161–173.

ANDERTON, R., BRIDGES, P. H., LEEDER, M. R. & SELLWOOD, B. W. 1979. *A Dynamic Stratigraphy of the British Isles.* George Allen & Unwin, London.

BADLEY, M. E., PRICE, J. D. & BACKSHALL, L. C. 1989. Inversion, reactivated faults and related structures: seismic examples from the southern North Sea. *In*: COOPER, M. A. & WILLIAMS, G. D. (eds) *Inversion Tectonics.* Geological Society, London, Special Publications, **44,** 201–219.

BESLY, B. M. 1990. Carboniferous. *In*: GLENNIE, K. W. (ed.) *Introduction to the Petroleum Geology of the North Sea* (3rd edn). Blackwell Scientific Publications, 90–119.

BETZ, D., FÜHRER, F., GREINER, G. & PLEIN, E. 1987. Evolution of the Lower Saxony Basin. *Tectonophysics,* **137,** 127–170.

BIFANI, R. 1986. Esmond gas complex. *In*: BROOKS, J., GOFF, J. & VAN HOORN, B. (eds) *Habitat of Palaeozoic Gas in north west Europe.* Geological Society, London, Special Publications, **23,** 209–222.

BODENHAUSEN, J. W. A. & OTT, W. F. 1981. Habitat of the Rijswijk oil province, onshore The Netherlands. *In*: ILLING, L. V. & HOBSON, G. D. (eds) *Petroleum Geology of the continental shelf of North-west Europe.* Heyden, London, 301–309.

BOULTON, G. S. 1993. Ice Ages and Climatic Change. *In*: DUFF, P. McL.D. (ed.) *Holmes' Principles of Physical Geology* (4th edn). Chapman & Hall, London, 439–469.

BRENNAND, T. P., VAN HOORN, B. & JAMES, K. H. 1990. Historical Review of North Sea Exploration. *In*: GLENNIE, K. W. (ed.) *Introduction to Petroleum Geology of the North Sea.* 3rd Edition. Blackwell Scientific Publications, 1–33.

BURRI, P., FAUPEL, J. & KOOPMANN, B. 1993. The Rotliegend in northwest Germany, from frontier to fairway. *In*: PARKER, J. R. (ed.) *Petroleum Geology of Northwest Europe: Proceedings of the 4th Conference.* Geological Society, London, 741–748.

CORNFORD, C. 1990. Source rocks and hydrocarbons of the North Sea. *In*: GLENNIE, K. W. (ed.) *Introduction to the Petroleum Geology of the North Sea.* Blackwell Scientific Publications, 294–361.

CROWELL, J. C. 1995. The Ending of the Late Paleozoic Ice Age During the Permian Period. *In*: SCHOLLE, P. A., PERYT, T. M. & ULMER-SCHOLLE, D. S. (eds) *The Permian of Northern Pangea,* **1.** Springer-Verlag, Berlin, 62–74.

FLEET, A. J., CLAYTON, C. J., JENKYNS, H. C. & PARKINSON, D. N. 1987. Liassic source-rock deposition in Western Europe. *In*: BROOKS, J. & GLENNIE, K. (eds) *Petroleum Geology of North West Europe.* Graham & Trotman, London, 59–70

GAST, R. E. 1991. The Perrenial Rotliegend Saline Lake in NW Europe. *Geologisches Jahrbuch,* **A119,** 25–59.

GEORGE, G. T. & BERRY, J. K. 1993. A new lithostratigraphy and depositional model for the Upper Rotliegend of the UK Sector of the Southern North Sea. *In*: NORTH, C. P. & PROSSER, D. J. (eds) *Characterization of Fluvial and Aeolian Reservoirs.* Geological Society, London, Special Publications, **73,** 291–319.

—— & ——1994. A new palaeogeographic and depositional model for the Upper Rotliegend, offshore The Netherlands. *First Break,* **12**(3), 147–158.

GLENNIE, K. W. 1970. *Desert Sedimentary Environments.* Developments in Sedimentology, **14.** Elsevier, Amsterdam.

——1972. Permian Rotliegendes of northwest Europe interpreted in light of modern desert sedimentation studies. *American Association Petroleum Geologists Bulletin,* **56,** 1048–1071.

——1986. Development of NW Europe's Southern Permian Gas Basin. *In*: BROOKS, J., GOFF, J. & VAN HOORN, B. (eds) *Habitat of Palaeozoic Gas in NW Europe.* Geological Society, London, Special Publications, **23,** 3–22.

——1990a. Outline of North Sea History & Structural Framework. *In*: GLENNIE, K. W. (ed.) *Introduction to the Petroleum Geology of the North Sea.* Blackwell Scientific Publications, 34–77.

——1990b. Lower Permian – Rotliegend. *In*: GLENNIE, K. W. (ed.) *Introduction to the Petroleum Geology of the North Sea.* Blackwell Scientific Publications, 120–152.

——, BROOKS, J. & BROOKS, J. R. V. 1987. Hydrocarbon exploration and geological history of North West Europe. *In*: BROOKS, J. & GLENNIE, K. (eds) *Petroleum Geology of North West Europe,* Graham & Trotman, 1–10.

HASTINGS, A., MURPHY, P & STEWART, L. 1991. A multidisciplinary approach to reservoir characterization: Helm field, Dutch North Sea. *In*: SPENCER, A. M. (ed.) *Generation, Accumulation and Production of Europe's Hydrocarbons.* Oxford University Press, 193–202.

KARNIN, W.-D., IDIZ, E., MERKEL, D. & RUPRECHT, E. 1996. The Zechstein Stassfurt Carbonate hydrocarbon system of the Thuringian Basin, Germany. *Petroleum Geoscience*, **2**, 52–58.

KETTER, F. J. 1991. The Esmond, Forbes and Gordon Fields, Blocks 43/8a, 43/13a, 43/15a, 43/20a, UK North Sea. *In*: ABBOTTS, I. L. (ed.) *United Kingdom Oil and Gas Fields, 25 Years Commemorative Volume*. Geological Society, London, Memoirs, **14**, 425–432

KIERSNOWSKI, H., PAUL, J., PERYT, T. M. & SMITH, D. B. 1995. Facies, paleogeography, and sedimentary history of the Southern Permian Basin in Europe. *In*: SCHOLLE, P. A., PERYT, T. M. & ULMER-SCHOLLE, D. J. (eds) *The Permian of Northern Pangea*. **2**, *Sedimentary Basins and Economic Resources*. Springer-Verlag, Berlin, 119–136.

KNOTT, S. D., BURCHELL, M. T., JOLLEY, E. J. & FRASER, A. J. 1993. Mesozoic to Cenozoic plate reconstructions of the North Atlantic and hydrocarbon plays of the Atlantic margins. *In*: PARKER, J. R. (ed.) *Petroleum Geology of Northwest Europe: Proceedings of the 4th Conference*. Geological Society, London, 953–974.

MORGAN, C. P. 1993. The Viking Complex Field, Blocks 49/12a, 49/16, 49/17, UK North Sea. *In*: ABBOTTS, I. L. (ed.) *United Kingdom Oil and Gas Fields, 25 Years Commemorative Volume*. Geological Society, London, Memoirs, **14**, 509–515.

MORETON, R. 1995. *Tales from early UK Oil Exploration 1960–1979*. Petroleum Exploration Society of Great Britain, 30th Anniversary Book.

MÜLLER, E. P., DUBSLAFF, H, EISERBECK, W. & SALLUM, R. 1993. Zur Entwicklung der Erdöl- und Erdgasexploration zwischen Ostsee und Thüringer Wald. *Geologisches Jahrbuch*, **A131**, 5–30.

OELE, J. A., HOL, A. C. P. J. & TIEMANS, J. 1981. Some Rotliegend gas fields of the K and L blocks, Netherlands offshore (1968–1978) – a case history. *In*: ILLING, L. V. & HOBSON, G. D. (eds) *The Petroleum Geology of the Continental Shelf of North-west Europe*. Heyden, 289–300.

PARKER, J. R. (ed.) 1993. *Petroleum Geology of Northwest Europe: Proceedings of the 4th Conference*. Geological Society, London.

PARRISH, J. T. 1995. Geologic Evidence of Permian Climate. *In*: SCHOLLE, P. A., PERYT, T. M. & ULMER-SCHOLLE, D. S. (eds) *The Permian of Northern Pangea*, 1. Springer-Verlag, Berlin, 53–61.

PETROLEUM GEOLOGICAL CIRCLE 1993. *Synopsis: Petroleum Geology of the Netherlands – 1993*. Royal Geological & Mining Society of the Netherlands.

PLEIN, E. 1978. Rotliegend-Ablagerungen im Norddeutschen Becken. *Zeitschrift der Deutschen Geologischen Gesellschaft*, **129**, 71–97.

——1993. Bemerkungen zum Ablauf der paläogeographischen Entwicklung im Stefan und Rotliegend des Norddeutschen Beckens. *Geologisches Jahrbuch*, **A131**, 99–116.

ROELOFSEN, J. W. & DE BOER, W. D. 1991. Geology of the Lower Cretaceous Q/1 oil-fields, Broad Fourteens Basin, The Netherlands. *In*: SPENCER, A. M. (ed.) *Generation, Accumulation and Production of Europe's Hydrocarbons*. Oxford University Press, 203–216

ROOS, B. M. & SMITS, B. J. 1983. Rotliegend and Main Buntsandstein gas fields in block K/13. A case history. *Geologie en Mijnbouw*, **62**, 75–83.

SCHRÖDER, L., LÖSCH, J., SCHÖNEICH, H., STANCU-KRISTOFF, G. & TAFEL, W.-D. 1991. Oil and gas in the north-west German basin. *In*: SPENCER, A. M. (ed.) *Generation, Accumulation and Production of Europe's Hydrocarbons*. Oxford University Press, 139–148.

SMITH, N. J. P. 1993. The case for exploration of deep plays in the Variscan fold belt and its foreland. *In*: PARKER, J. R. (ed.) *Petroleum Geology of Northwest Europe: Proceedings of the 4th Conference*. Geological Society, London, 667–675.

STÄUBLE, A. J. & MILIUS, G. 1970. *Geology of the Groningen gas field*. American Association Petroleum Geologists, Memoirs, **14**, 359–369.

STOREY, M. W. & NASH, D. F. 1993. The Eakring–Dukes Wood oil field: an unconventional technique to describe a field's geology. *In*: PARKER, J. R. (ed.) *Petroleum Geology of Northwest Europe: Proceedings of the 4th Conference*. Geological Society, London, 1527–1537.

TAYLOR, J. C. M. 1990. Upper Permian–Zechstein. *In*: GLENNIE, K. W. (ed.) *Introduction to Petroleum Geology of the North Sea*. Blackwell Scientific Publications, 153–190.

VAN HOORN, B. 1987. Structural evolution, timing and tectonic style of the Sole Pit inversion. *Tectonophysics*, **137**, 239–284.

VAN WIJHE, D. H. 1987. The structural evolution of the Broad Fourteens Basin. *In*: BROOKS, J. & GLENNIE, K. (eds) *Petroleum Geology of North West Europe*. Graham & Trotman, 315–323.

VISSER, W. A. & SUNG, G. C. L. 1958. Oil and natural gas in northeastern Netherlands. *In*: WEEKS, L. G. (ed.) *Habitat of Oil*. American Association of Petroleum Geologists Memoirs, 1067–1090.

WOODLAND, A. W. (ed.) 1975. *Petroleum and the Continental Shelf of North West Europe: Vol 1. Geology*. Applied Sciene Publishers, Barking.

ZIEGLER, P. A. 1990. *Geological Atlas of Western and Central Europe*. 2nd Edition. Shell International Petroleum Maatschappij.

# Recent advances in understanding the
# southern North Sea Basin: a summary


K. W. GLENNIE

*Department of Geology & Petroleum Geology,*
*King's College, University of Aberdeen AB9 2U, UK*
*Correspondence address: 4 Morven Way, Ballater, Grampian AB35 5SF, UK*



**Abstract:** After a short-lived but very successful phase of exploration in the UK North Sea, the need for extreme confidentiality ceased to exist for those fields that had already been discovered, including the gas fields of the southern North Sea; this led to a widespread dissemination of data from 1974 onwards, to the great benefit of the industry as a whole. Early difficulties in producing gas from reservoirs whose permeability had been reduced by the growth of diagenetic illite led to studies in depth- and facies-related diagenesis, which continues today in different parts of the Southern Permian Basin. A change from analogue to digital recording of seismic data coincided with improvements in computer technology, and transformed seismic from a tool with limited depth penetration, which required the skill of a dedicated specialist to interpret, to one that today is used universally by all exploration and production geologists; these advances led, in turn, to much improved stratigraphic and structural resolution.

Recent studies, especially in the former East Germany, have led to a better definition of the Permian. In Central Germany, Lower Rotliegend volcanics are overlain by sedimentary sequences of the Müritz sub-group (Upper Rotliegend 1), which are coeval with part of the combined Altmark and Saalian Unconformity. In northern Germany, this major hiatus is overlain by the Havel and Elbe subgroups (Upper Rotliegend 2) which, together with the succeeding Zechstein sequences, were all deposited within the youngest stages of the Permian, covering a probable time span of no more than 16 Ma. Deposition of up to 4000 m or so of Late Permian strata in such a short time indicates structurally controlled subsidence in addition to any of a thermal nature.


## The growth of geological knowledge and hypotheses

Oil companies are supposed to be notorious for the way they guard their geological secrets, and where their commercial life is at stake, this should be so. For the first few years of North Sea activity, the exploration companies involved were no exception. Geological data were exchanged between individual companies, to their mutual benefit but to the exclusion of everyone else, on a well-by-well basis, discovery for discovery and dry hole for dry hole (initially, even counting the footage drilled).

The above situation more or less prevailed until the first of the major conferences (Bloomsbury Conference of 1974; see Woodland, 1975) on the Petroleum Geology of NW Europe, when the results of regional studies by oil companies and, more significantly, descriptions of the fields already discovered, were made available to the geological community at large. Since then, the enormous amount of geological knowledge gained through hydrocarbon exploration and production in the North Sea area has reached a much wider audience via numerous other conferences and their published transactions. In addition, a stream of publications from both oil-company and academic sources on topics ranging from basin development and plate tectonics, the sedimentology and diagenesis of reservoir rocks, and the generation, migration and entrapment of hydrocarbons, has grown almost exponentially since 1959. In the process, the geology of NW Europe in general and of the North Sea area in particular has probably become better known than any other comparable area of the world. A little of that knowledge, and some of the related working hypotheses relevant to the southern North Sea Basin, are discussed below. To do these topics justice, several aspects of the Southern Permian Basin that lie beyond their strict geographical limits must also be considered.

## Reservoir quality

With most attention in the Southern Permian Basin being paid to the Rotliegend (Fig. 1), it soon became apparent that some fields had much poorer reservoir quality than others, and this seemed to have nothing to do with either


*From* Ziegler, K., Turner, P. & Daines, S. R. (eds), 1997, *Petroleum Geology of the Southern North Sea: Future Potential*, Geological Society Special Publication No. 123, pp. 17–29.

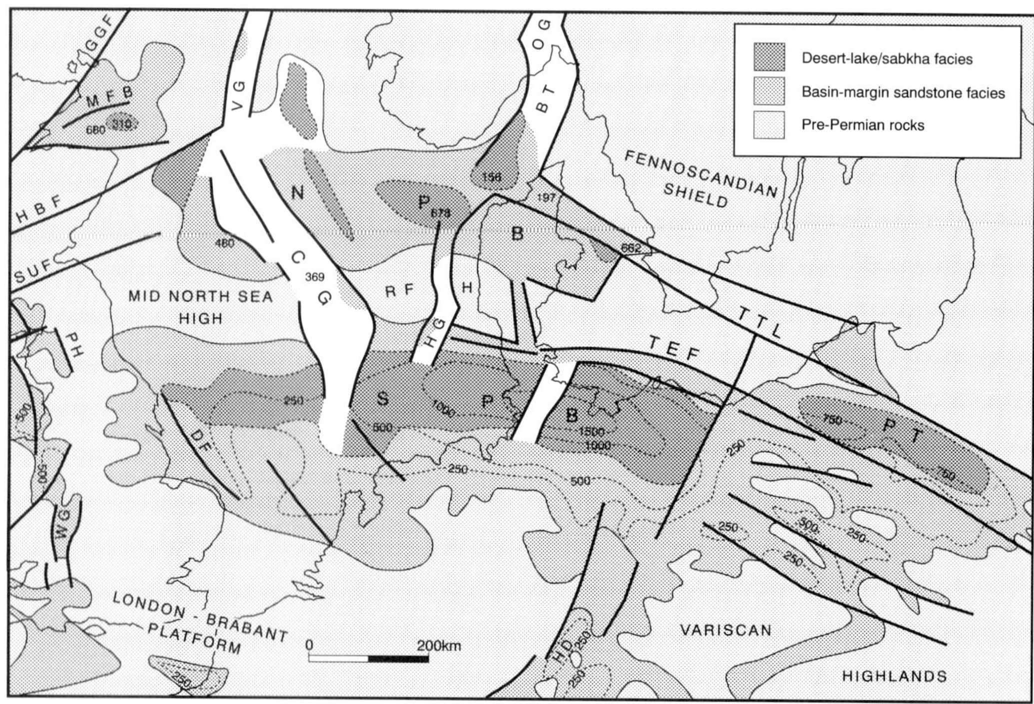



**Fig. 1.** The Southern Permian Basin (SPB) with showing basin-centre desert lake/sabkha and basin-margin sandstone facies. BT, Bamble Trough; CG, Central Graben; DF, Dowsing Fault; GGF, Great Glen Fault; HBF, Highland Boundary Fault; HD, Hessian Depression; HG, Horn Graben; OG, Oslo Graben; PH, Pennine High; PT, Polish Trough; RFH, Ringkøbing–Fyn High; SUF, Southern Uplands Fault; TEF, Trans European Fault; TTL, Tornquist–Teisseyre Line; WG, Worcester Graben; VG, Viking Graben. Contours in metres. In the Northern Permian Basin (NPB) and Moray Firth (MFB) there are insufficient well penetrations to isopach the Rotliegend; isolated thicknesses in metres derived from well data.

depositional facies or compaction related to present depth of burial. In the Leman Field, for instance, the maximum reservoir depth is about 2100 m (Table 1), the average porosity is 12.9% and air permeability in the range 0.5–15 mD, although the latter could be less than 0.1 mD in some horizons. In Groningen, on the other hand, at the greater depth of 2950 m the average porosity is 17% whereas in West Sole, at the same depth, the average porosity is only 12% even though in the best reservoir horizon, aeolian sandstone, it is 15%; permeability, however, is generally less than 3 mD (Winter & King 1991). In between these two extremes of depth, the porosity in several fields is better than 15% and permeability can exceed 1000 mD. The main culprit for loss of permeability seemed to be the presence of cotton-wool-like illite whiskers, which increase the tortuosity in the pore throats so that in some Sole Pit wells, the rate of gas production on test was much less than $1 \times 10^6 \, \text{ft}^3/$ day, when a rate of at least $15 \times 10^6 \, \text{ft}^3/$day was

**Table 1.** *Rotliegend reservoir quality v. depth*

| Field | Reservoir GWC (m ss) | Average $\phi$ (%) | Perm. (mD) |
|---|---|---|---|
| Leman | 2100 | 12.9 | 0.5–15 |
| Sean S | 2620 | 17.1 | 190–420 |
| Sean N | 2650 | 17.5 | 130–400 |
| Victor | 2670 | 16 | 52 |
| Barque | 2680 | 11 | 0.02–100 |
| Clipper | 2710 | 11 | 0.02–100 |
| Indefatigable | 2730 | 15 | 0.09–1000 |
| Amethyst | 2760 | 11–25 | 1–1000 |
| Cleeton | 2850 | 18 | 40–100 |
| Rough | 2910 | 8.2–18.2 | 2–291 |
| Groningen | 2950 | 17 | 0.1–1000 |
| West Sole | 2950 | 12 | <1 |

Derived from many sources.
Although not always directly comparable, there is clearly no straight correlation between present depth of the reservoir and the average porosity and permeability of similar reservoir rock (mostly aeolian sandstone).

essential for the well to be economically viable. Similar pore-destroying illite has been described from Germany by Hancock (1978).

It took some years to realise that the poor quality of Rotliegend reservoirs in Sole Pit and elsewhere was associated with a depth of burial that, prior to inversion in the Late Cretaceous and mid-Tertiary, could have been some 1500 m greater than at present (Marie 1975; Glennie *et al.* 1978). Early estimates were based on compaction of a basal Bunter shale as deduced from sonic logs, and has since been supported on seismic and well evidence by Van Hoorn (1987). Similar inversion occurred in what is now the Cleveland Hills (Kent 1980), where Cope (1986) based the amount of uplift on vitrinite reflectance values, and in the Broad Fourteens Basin where poor Rotliegend permeability prevented development of NAM's first offshore exploration well in block K/17 (Oele *et al.* 1981). Fibrous illite seemed to form at palaeodepths in excess of about 3000 m (Glennie 1990*b*, fig. 5.27).

Illite does not necessarily form at such depths, however. In Germany, dune sands at a depth of around 5000 m in the Söhlingen field had a better poroperm quality in its Schneverdingen Sandstone reservoir (Drong *et al.* 1982) than in the shallower Hauptsandstein (now Havel and Dethlingen sandstones respectively (Schröder *et al.* 1995). Drong *et al.* (1982) ascribed the difference in quality to differences in depositional climate, but the Schneverdingen sandstones were deposited in a graben environment and average twice the grain diameter of the younger dune sands; thus both grainsize and environment of deposition relative to structural location within the graben probably influenced diagenesis and thus the reservoir quality.

The twin effects of diagenesis and depositional environment in Germany have been discussed at some length by Gaupp *et al.* (1993), with a shorter treatment in Burri *et al.* (1993). They found a close relationship between environment of deposition of Rotliegend sandstones and the type of grain-rimming clay cement.

Most of the North German gas fields occur in a lake-margin shoreline facies (Gast 1991) that developed across a broad embayment along the southern side of the Rotliegend desert lake; the sands are interbedded with the equivalent of the Silverpit Claystone of UK waters. Within the shoreline facies the pore-lining clays consist of chlorite, which reduces porosity and permeability only marginally. Gaupp *et al.* (1993) also found that illite was prevalent in a zone up to about 1 km wide when the Rotliegend was juxtaposed with Westphalian Coal Measures. In the present horst and graben sub-surface environment, illite cementation lessens northwards as throws diminish and the Rotliegend reservoirs are separated from the Coal Measures by an increasing thickness of Lower Rotliegend volcanics and Barren Red Measures. In UK waters, there may be diagenetic parallels. In Cleeton (Table 1), where the main authigenic clay is kaolinite, porosity in the dune sands typically reaches 22% and permeability averages 95 mD (Heinrich 1993). It is possible that the better quality sands in the northern part of Ravenspurn North also result from the distribution of shoreline sands or the underlying presence of Namurian sandstones rather than Coal Measures. Alternatively, Turner *et al.* (1993) suggest that the early presence of methane within the reservoir would have inhibited the growth of illite.

The effect of the illite cementation on productivity was to seriously delay the development of some fields in the Sole Pit and Broad Fourteens areas. With an economic incentive in the early 1980s to explore for new gas and to develop known gas fields, the problem had to be overcome. The initial target was to seek zones of natural open fractures, such as had been encountered in the West Sole field (see Winter & King 1991). With no visible offsets, however, such fractures cannot be located on seismic and, as we have seen (Gaupp *et al.* 1993), proximity to actual faults had the opposite effect to that sought. The problem was resolved with the development of horizontal drilling and the ability to tap many high-permeability fractures in order to achieve an economic rate of production.

## Ages and Rotliegend basin development

In recent years, German geologists have been responsible for some important new geological data involving the Rotliegend, especially with respect to age and rock correlation. This has implications for processes of basin development, and rates of basin subsidence and sedimentation in the North Sea area (see e.g. the revised Rotliegend stratigraphy in Schröder *et al.* 1995, and Schneider *et al.* 1995, and a variety of earlier applications of this stratigraphy by Gebhardt *et al.* 1991; Schneider & Gebhardt 1993; Plein 1993, 1995 and Gast 1993).

Traditionally, the Rotliegend has been divided into the Upper and Lower Rotliegend. It seems likely that both the Upper Rotliegend and Zechstein in the North Sea–North Germany area were deposited within the youngest Permian stages, which many authors consider to be the Tatarian (as shown in Fig. 2). It should be noted, however, that the uppermost Tartarian (as shown in Fig. 2) is now represented by the Dzulfian and Changsingian stages (see e.g. Schneider *et al.* 1995, table 2). Menning (1995, fig. 1) gives the Tatarian stage (as in Fig. 2) a time span of 16 Ma but limits Zechstein and Upper Rotliegend deposition to only 15 Ma, divided about equally between them. If true, this would imply that up to about 4 km of Tatarian post-compaction subsidence over parts of the North German Plain took place at a rate exceeding 250 m Ma$^{-1}$. The problem with these time spans is that they are poorly constrained both radiometrically and biostratigraphically, and have been estimated in part by comparison with rates of deposition deduced in other areas of Europe.

The Lower Rotliegend, which possibly had a maximum duration of 10 or 15 Ma, is dominated by the volcanic episode that followed the Variscan Orogeny (Fig. 3); associated sediments contain fauna and flora indicative of an Autunian age (Menning *et al.* 1988). This volcanism,

in turn, was succeeded in North Germany and the North Sea area by the long (10–30 Ma) combined Saalian and Altmark unconformities, which occupied much of the time span of the Upper Rotliegend (Fig. 2). Some sedimentation within the Variscan orogenic belt of Central Germany was coeval with the Saalian Unconformity (e.g. Saar–Nahe Basin). These latter sediments (Müritz sub-group) are referred to by Gebhardt *et al.* (1991), Schneider & Gebhardt (1993) and Schneider *et al.* (1995) as Upper Rotliegend 1 (UR1; Fig. 2), and the sequence overlying the Saalian Unconformity (Havel and Elbe sub-groups) as Upper Rotliegend 2 (UR2). The terms UR1 and UR2 may eventually be dropped in favour of the Müritz sub-group and Havel and Elbe sub-groups respectively, but in the context of this discussion, UR1 and UR2 are distinctly useful as basin-wide terms and will be retained.

Because it is so poorly constrained, the time span of UR2 sedimentation in North Germany, where its thickness locally reaches 2000 m seems to vary, according to author, from 4 to 5 Ma

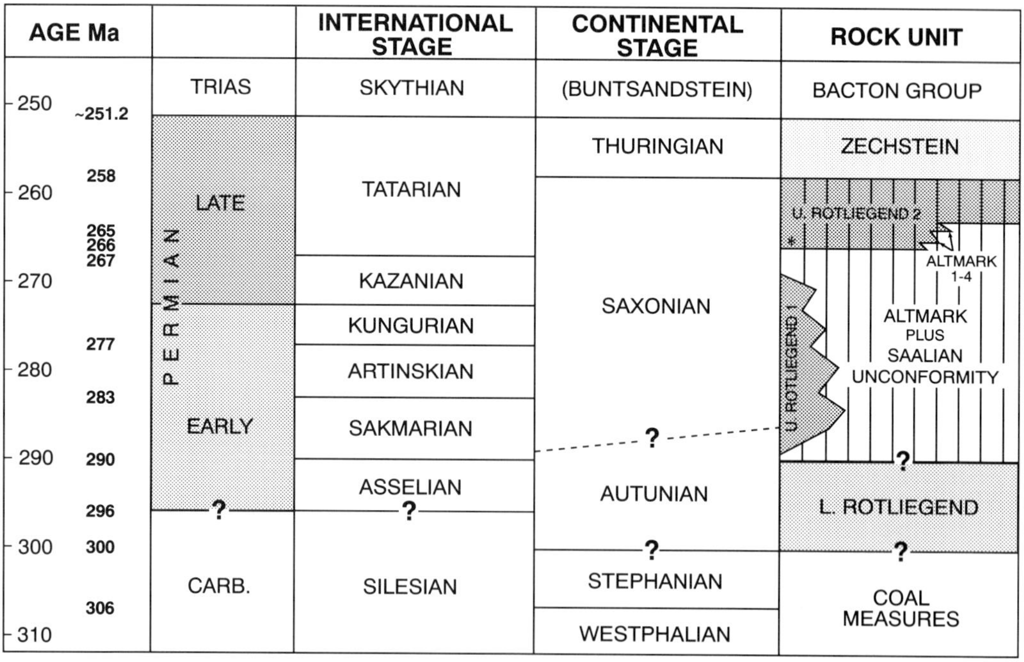

| AGE Ma | | INTERNATIONAL STAGE | CONTINENTAL STAGE | ROCK UNIT |
|---|---|---|---|---|
| −250   ~251.2 | TRIAS | SKYTHIAN | (BUNTSANDSTEIN) | BACTON GROUP |
| | | | THURINGIAN | ZECHSTEIN |
| −260   258 | LATE | TATARIAN | | U. ROTLIEGEND 2 / ALTMARK 1-4 |
| 265 266 267 | | | SAXONIAN | ALTMARK PLUS SAALIAN UNCONFORMITY |
| −270 | | KAZANIAN | | |
| | | KUNGURIAN | | |
| 277 | | ARTINSKIAN | | |
| −280   283 | | SAKMARIAN | | |
| EARLY | | | | |
| −290   290 | | ? | ? | ? |
| | | ASSELIAN | AUTUNIAN | L. ROTLIEGEND |
| 296 | ? | ? | | |
| −300   300 | | | ? | ? |
| | CARB. | SILESIAN | STEPHANIAN | COAL MEASURES |
| 306 | | | WESTPHALIAN | |
| −310 | | | | |

(The "AGE Ma" scale carries the tick values: 250, ~251.2, 258, 260, 265, 266, 267, 270, 277, 280, 283, 290, 290, 296, 300, 300, 306, 310; the second column is labelled vertically PERMIAN, divided into LATE and EARLY; in the ROCK UNIT column "U. ROTLIEGEND 1" is labelled vertically.)

**Fig. 2.** The Permian period seen in its possible time frame. Between the Lower Rotliegend and units of the Upper Rotliegend 2 (UR2) is the long-lived Saalian Unconformity, which is partly coeval with Upper Rotliegend 1 (UR1). In NE Germany, UR2 was affected by the four tectonically induced Altmark unconformities. Based on Haq & Van Eysinga (1987); Gast (1988); Leeder (1988); Taylor (1990); Menning *et al.* (1988); Menning (1991, 1995); Gebhardt *et al.* (1991); Schneider *et al.* (1995). Asterisk marks possible age of Illawarra magnetic reversal (265 Ma BP).

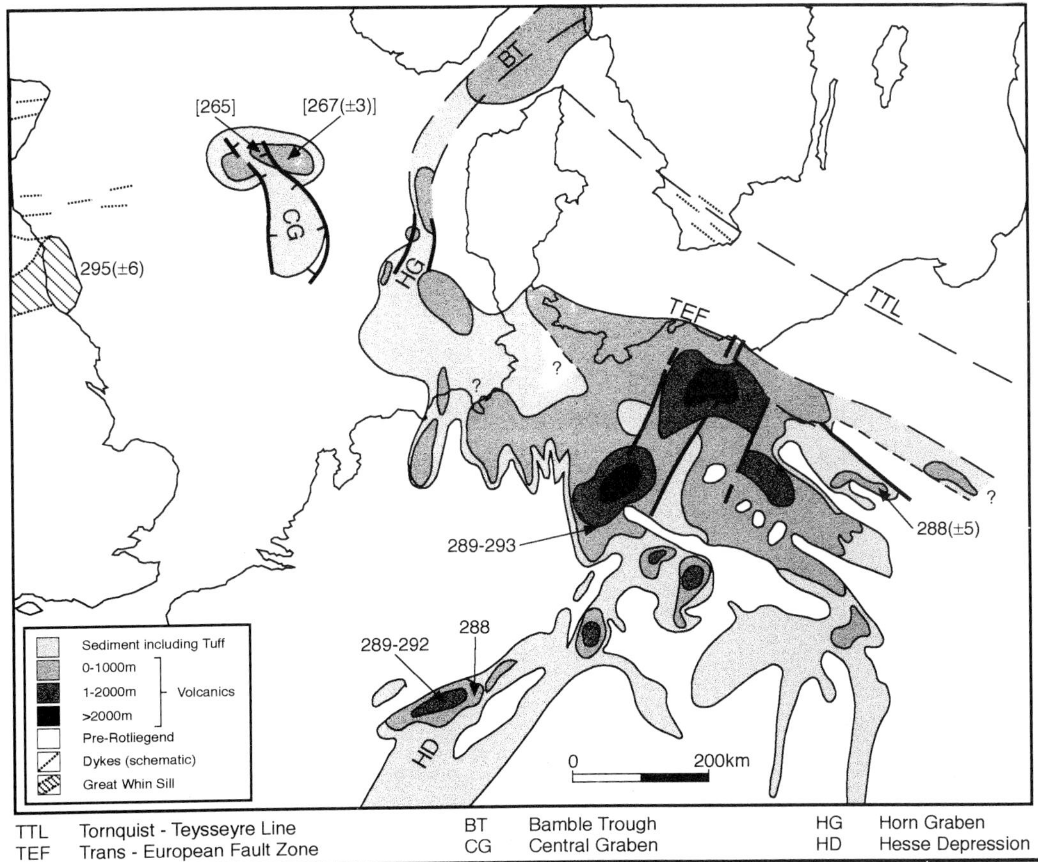


**Fig. 3.** Distribution and thickness of the Lower Rotliegend volcanics. The areas of thickest volcanics were probably subjected to caldera subsidence. Numbers indicate isotopic age. Modified from Plein (1993) by addition of some early Permian dykes, the Great Whin Sill (with age from Fitch & Miller 1967), and ages in vicinity of the Central Graben (in square brackets) from Ineson (1993). Trans- European Fault zone from Coward (1990).

(Gebhardt *et al.* 1991) for the Elbe subgroup (Fig. 4) to about 10 Ma for the complete UR2 sequence (Gast 1993), but in both cases is still well within the Late Permian time span. The 10 Ma time span implies UR2 basin subsidence beneath the North German Plain at a rate of $200\,\mathrm{m\,Ma^{-1}}$.

The key to correlation between the basins of North and Central Germany is the recognition of the Illawarra Magnetic Reversal (Menning *et al.* 1988). It occurs within the hiatus between the UR1 and UR2 sequences of the intra-montane basins of central Germany and in the lower part of the Havel Subgroup of the Southern Permian Basin; there, it is overlain by more than 1000 m of UR2 sediments in the offshore well J/5–1 (Fig. 4). The numerical age of the Illawarra Reversal is still very imprecise.

Menning *et al.* (1988, figs 6 & 7) placed it at 255 Ma, or 10 Ma below their then presumed date for the base Triassic. This may soon be revised to 9 or even 8 Ma below the base Zechstein (R. Gast pers comm. 1995), but will still not be a reliable age.

It is clear from the above analysis that although knowledge of precise ages is important for our understanding of events, they are mostly poorly constrained. With improvement in dating techniques, older radiometric ages are revised, new ones are added and, as their implications are realised, even the time spans of geological periods have changed. Van Eysinga (1975) for instance shows the upper and lower limits of the Permian period as 230 and 280 Ma respectively; those dates were revised by Haq & Van Eysinga (1987) to 250 and 290 Ma. Based on the age of

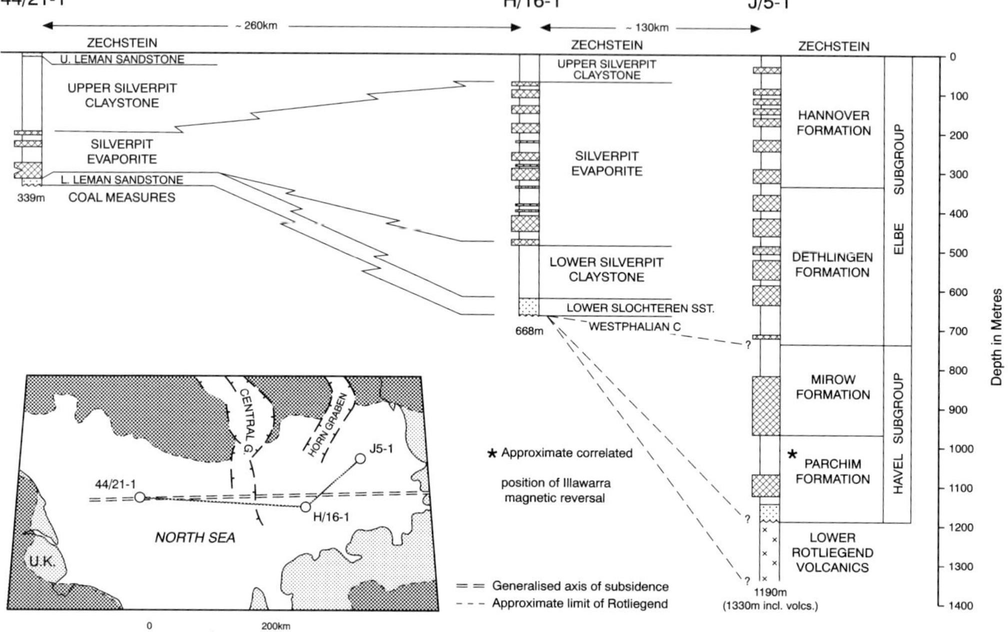


**Fig. 4.** Comparison of the thickness and distribution of bedded halites from German well J/5–1, Dutch well H/16–1 and UK well 44/21–1. Inset map suggests that J/5–1 is not in an axial location within the basin; Rotliegend beneath present land in half tone. The distribution of halites implies that each would change laterally to a sabkha and then sandy facies, which onlaps the basin floor. The approximate position of the Illawarra Magnetic Reversal in J/5–1 is marked by the asterisk (*) within the Parchim Formation. The Havel Subgroup is probably not present in the Dutch and UK North Sea.

late Westphalian tuffs in Germany (Lippolt *et al.* 1984), it was suggested that the Carboniferous–Permian boundary should be revised to around 300 Ma BP (see Leeder 1988; Menning 1991, fig. 1); Menning (1992*a,b*), however, quotes 295 ± 5 Ma and 296 Ma for this boundary, the latter being used in Menning (1995).

With a revision of their data on the Lower Rotliegend volcanics, Lippolt & Hess (1989) now consider their ages to cluster around 293–288 Ma (Fig. 4). If these figures are correct, most, if not all the Lower Rotliegend volcanics of Germany will be confined to the Early Permian rather than extending into the Stephanian as was previously the case. In this context, the Great Whin Sill of NE England (Fig. 3) is possibly now of Permian rather than Carboniferous age. The age of the Inge Volcanics (Cameron 1993) of the Central Graben (Fig. 3; 265–267 Ma, Ineson 1993), on the other hand, seems to coincide more nearly with the possible start of UR2 sedimentation (266 Ma?) rather than with the Lower Rotliegend. Similarly, in NE Germany, at least two of the four Altmark unconformities (Fig. 2) is interspersed with

basaltic volcanism (Gebhardt *et al.* 1991, fig. 7). It is clear that the presence of volcanics can no longer be taken as an indication of an early Permian age or of Lower Rotliegend affinities.

Based on a continuous sequence in China, the Permian–Triassic boundary is now placed at 251.2 ± 3 Ma (Claoue-Long *et al.* 1991), but how closely this coincides with the Zechstein–Bunter depositional boundary in NW Europe is not known.

The time spans of the UR2 sequences in the Southern Permian Basin, and the variation in their thickness and facies, have direct implications for our understanding of basin development and associated sedimentary facies. In North Germany the following sequence of events took place: short-term Lower Rotliegend volcanism; deep erosion during the time span represented by the Saalian Unconformity (Müritz sub-group); short-term deposition of a thick UR2 sequence (Havel and Elbe subgroups). This history indicates that the combined Saalian–Altmark Unconformity was probably the result of thermal uplift of the crust following intense Early Permian volcanism, and then, some 20 Ma later, cooling-related crustal subsidence reaching the point where sedimentation began to replace erosion.

## Implications of German Permian stratigraphy for structural development of the greater North Sea

Because North Germany was the centre of Lower Rotliegend volcanism, both uplift and later subsidence would have been driven from that area. It is likely that, at least in part, German volcanicity also influenced uplift and erosion along the axial trend in UK waters which, being distant from the thermal centre, would explain why the Rotliegend sequence there is only one third or less of that in Germany (Fig. 4). This interpretation has a further implication. UR2 sedimentation in UK waters may not have begun until two thirds of the deposional time span in Germany had already elapsed (see stages in development proposed by Plein 1993). A cover of thick Lower Rotliegend volcanics is probably why Stephanian sediments

were preserved over a large area despite the thermal uplift of North Germany (Fig. 5), while preservation of the same sequence in the North Sea, where the Lower Rotliegend volcanics are absent, may have resulted from coeval differential uplift (sub-Rotliegend Namurian in Ravenspurn North; Turner *et al.* 1993) and subsidence (Westphalian C, D of Barren Red Measures in the Carboniferous gas play area to the northeast; Besly *et al.* 1993; see also Glennie & Boegner 1981, fig. 2b).

Differential uplift and subsidence! It seems unlikely that thermal cooling was solely responsible for differential subsidence in the UK North Sea or for 2000 m of UR2 subsidence within ten million years in Germany (4000 m in perhaps 15–16 Ma if the Zechstein is included). For comparison, post-Cimmerian subsidence of some 5 km in the Central North Sea took about 140 m Ma to achieve (36 m Ma$^{-1}$). Other evidence indicates that crustal stretching should also be invoked. Perhaps both the Southern and Northern Permian basins (or at least sub-basins within them) were giant pull-apart basins (e.g. Sole Pit Basin, Fig. 1; see also Rotliegend isopachs in Glennie & Boegner 1981, fig. 2c).

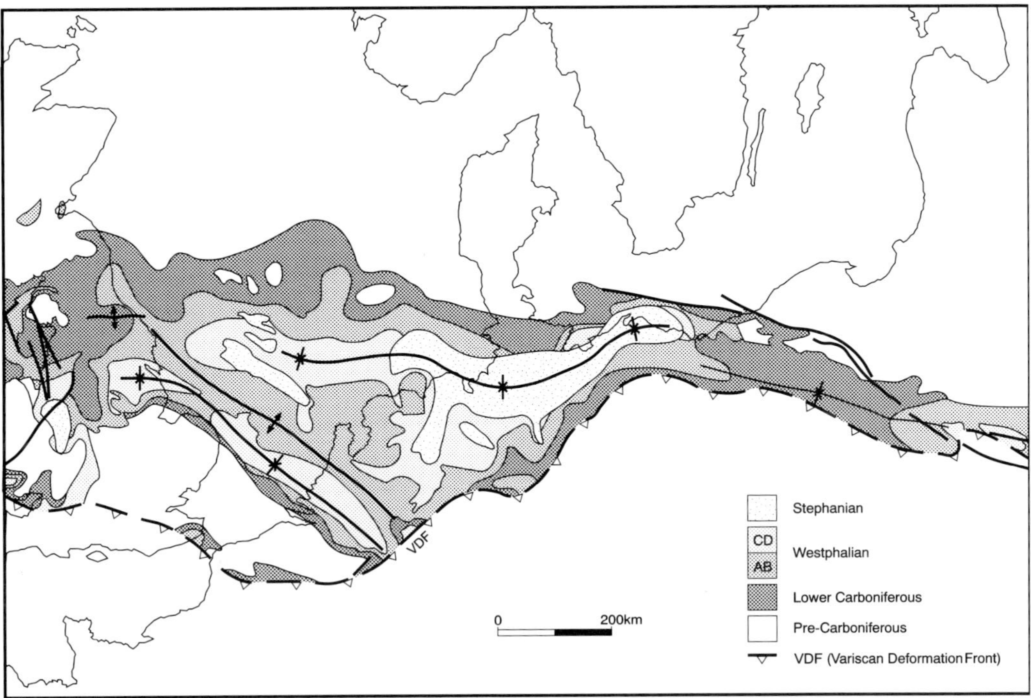


**Fig. 5.** Carboniferous stratigraphic units of the Southern Permian Basin. Simplified from Ziegler (1990), with axes of post-Variscan basin subsidence and inversion added. Note that the Stephanian sequence in N Germany was preserved beneath the Lower Rotliegend Volcanics. For simplicity, this figure is framed by Fig. 6.

The Tornquist–Teisseyre line had been an area of active transcurrent faulting since the Ordovician, and seems to have been associated with 'pop-up' structures of Devonian rocks at the margin of the Rotliegend basin (Fig. 6). Indeed, much of the North Sea area had probably been subjected to alternating transtensional and transpressive movements from the Late Carboniferous onwards (e.g. Sole Pit, Cleveland Hills, Broad Fourteens, West Netherlands basins and post-Variscan highs; Glennie this volume, fig. 2), resulting in both basin subsidence and inversion (Fig. 5). The NW–SE-trending fault pattern recognized over the areas of Rotliegend deposition in Central Europe have right-lateral displacements across most of them (Fig. 7), and had already provided accommodation for up to 2000 m of Lower Rotliegend volcanics and associated sediments in Central Germany.

The trends of the North Sea graben systems (Viking–Central, Horn–Bamble–Oslo) and the Hessian Depression are strongly oblique to the axes of the two Permian basins. It is suggested that the pattern of grabens resulted when the Variscan orocline acted as a wedge driven into the re-entrant between the Laurentian and Baltic shields, causing these more rigid blocks to move apart along the line of the old Iapetus suture (approx. Viking Graben; see Glennie 1990c, fig. 1). Such a separation may have begun already early in the Carboniferous, providing a point source for the Yoredale Series of the central North Sea (Besly 1990). Southward propagation of this crustal separation may have created the Central Graben at, or slightly before, the start of UR2 deposition. This could account for the late date of volcanism reported from the Danish Central Graben (Fig. 3). The creation of associated pull-apart basins would also account for some of the thick Rotliegend sequences in the Northern Permian Basin close to the Tornquist–Teisseyre fault system (Fig. 1).

Figure 1 indicates that the Polish Trough is not on trend with the main Southern Permian Basin, but is separated from it by the Trans-European Fault marking the NE limit of Lower Rotliegend volcanic activity, and by a continuation of the NE trending fault bounding the Hessian Depression. Although the presence of a

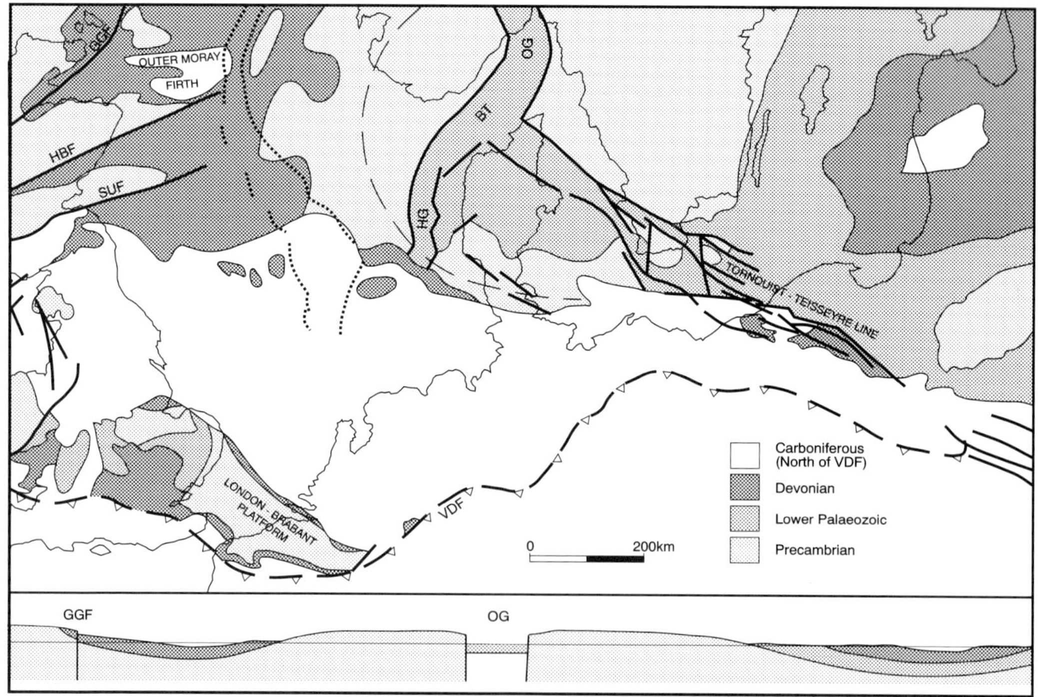



**Fig. 6.** Pre-Permian geology of area framing the Carboniferous rocks of the Southern Permian Basin. Simplified from Ziegler (1990), with outline of Central–Viking graben system dotted. Note the pop-up nature of the Precambrian and Devonian rocks in vicinity of Tornquist–Teisseyre line. Legend to letters in Fig. 1. W–E cross-section through the Oslo Graben indicates the gently warped nature of the basement rocks.

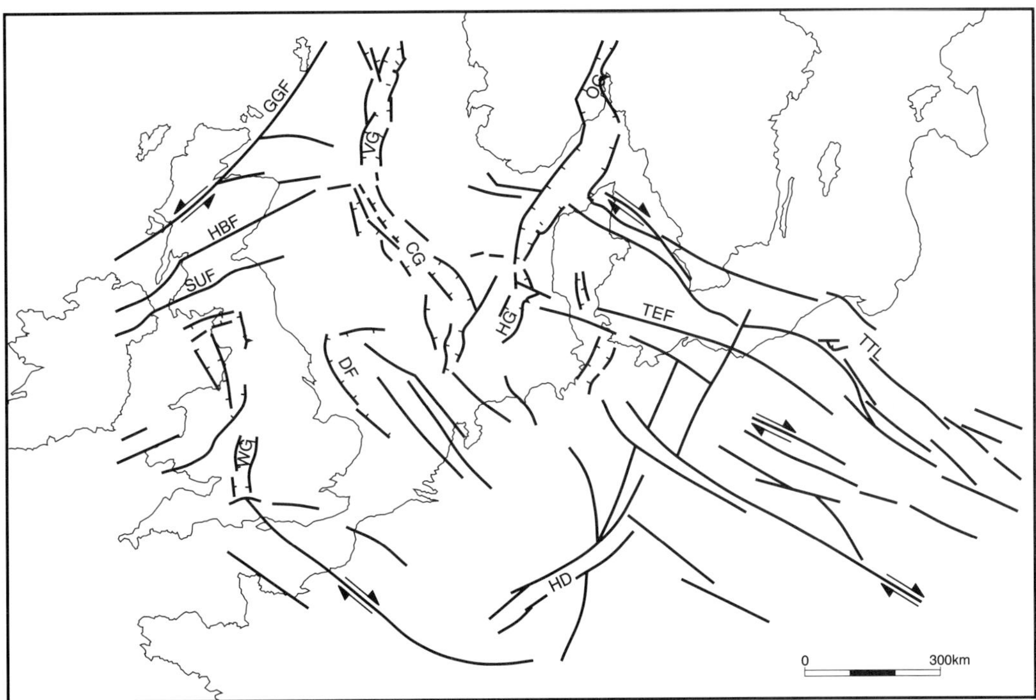


**Fig. 7.** Pattern of major faults believed to have been active during the Permian. Note the horizontal sense of movement across some of the faults. Simplified mostly from Ziegler (1990). CG, Central Graben; DF, Dowsing Fault; GGF, Great Glen Fault; HBF, Highland Boundary Fault; HD, Hesse Depression; HG, Horn Graben; OG, Oslo Graben; SUF, Southern Uplands Fault; TTL, Tornquist–Teisseyre Line; WG, Worcester Graben; VG, Viking Graben.

sabkha facies is known, bedded halite does not occur in the Polish Trough (Pokorski 1989; Kiersnowski *et al.* 1995, fig. 5), perhaps because sedimentation there kept up with a lower rate of subsidence. The Southern Permian Basin and Polish Trough can be considered as one basin only when the outline of their respective basin-margin sandstones is considered. As implied by Fig. 5 and depicted by Plein (1993, figs 4–6), the east–west development of the Southern Permian Basin possibly went in steps controlled by the cross-cutting system of North Sea grabens.

The isometric sketch Fig. 8 emphasises that, rather than being on trend with the Southern Permian Basin, the Polish Trough aligns with the Northern Permian Basin, the two being separated by the Ringkøbing–Fyn structural high, which may have resulted from transpression between the Trans-European and Tornquist–Teisseyre fault zones. The southern margin of this high also marks a fundamental boundary (Trans-European Fault) between the Caledonian 'front' of Lower Palaeozoic rocks that cross the North Sea region between Britain and Germany, and the

southern margin of the Svecofennian basement (Coward 1990, fig. 8, 1993).

Rotliegend basin subsidence progressed more rapidly than sediment infill of the Southern Permian Basin, so that by the time of the Zechstein Transgression, the surface of the Rotliegend desert lake was some 250–300 m below global sea level (Smith 1979; Ziegler 1990). This could explain why no evidence of sub-aerial exposure is known from the central part of the lake, as the level of the regional water table was probably always higher than the lake floor. It would also explain why the greatest number of halite horizons occurs in the German part of the basin where, unlike the marginal sabkhas, the water was always sufficiently deep to permit the precipitation of bedded halite. It is pertinent that for the UK and Dutch wells shown on Fig. 4 free water was either absent (i.e. sabkha within the capillary fringe of the water table) or was too shallow to permit bedded halite to be precipitated in the Silverpit Clays-tone sequences. The shown boundaries between claystone and evaporite members represent

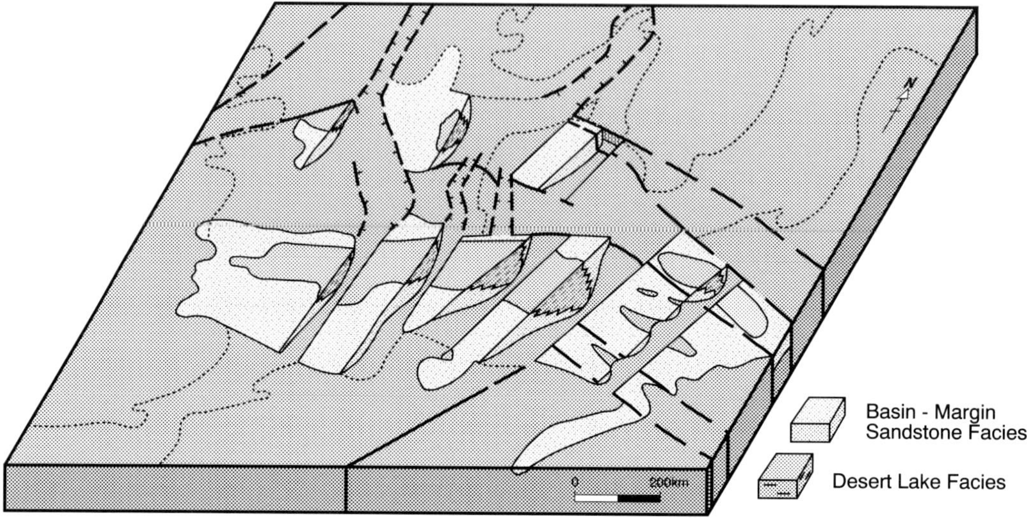



**Fig. 8.** Simple isometric sketch of Southern and Northern Permian basins to emphasize the major 'wet' and 'dry' facies of the Upper Rotliegend 2 the en-echelon relationship of the Southern Permian Basin to the Polish Trough, and its alignment with the Northern Permian Basin. The cross-cuts are designed to show the eastward thickening desert-lake facies, and the abrupt thinning across the NNE-trending fault extension of the Hesse Depression. Compare with Fig. 1.

facies changes and are not time lines; time lines will be recognized only if horizons of bedded halite can be matched from well to well. As the halites represent periods of maximum aridity and desiccation, they should coincide with aeolian activity in basin-margin areas.

Provided the regional water table maintained its level as the floor of the Rotliegend basin subsided, the desert lake would have extended westwards from North Germany, and time-bounded units of the basin centre would onlap the basin floor to its west (see e.g. Gast 1993, fig. 3), the flanking basin-margin areas above the water table being the sites of fluvial and aeolian activity, with shore-line sedimentation near the water's edge.

To judge from the repetition of bedded halites in German well J/5–1 (Fig. 4), basin subsidence, rather than any change to a more humid climate, was possibly responsible for extension of the Ten Boer lake facies over the Slochteren dune sands in the Groningen field (e.g. Glennie 1990*b*, fig. 5.24) and for the earlier incursion of the Ameland Claystone between the Upper and Lower Slochteren sandstones. Nevertheless, the halite horizons in J/5–1 possess a rhythmic pattern of halite precipitation, which could have been induced by the Gondwana glaciations in much the same way that Pleistocene glaciations affected the cycles of dune and sabkha activity in deserts (Glennie 1994; see also Boulton 1993;

Crowell 1995). Helmuth & Süssmuth (1993) suggest seventeen cycles in the UR2 sequence of NE Germany following the Illawarra Reversal. The duration of each cycle depends on the UR2 time span, but in any case seems to be longer than the Pleistocene cycles (perhaps 500 ka versus 100 ka).

In the context of glacially driven cyclicity, Gebhardt (1994) suggests that the normal state of desert-lake sedimentation was one of continuous precipitation of halite. She suggests that the cyclicity was the outcome of interglacial melting of the icecap, which resulted in periodic incursions of less saline marine water together with some marine fauna.

Because of a lack of direct evidence of the presence of ice sheets, many publications imply that the Permo-Carboniferous glaciations of Gondwana had either ceased by the early Late Permian (Kazanian; Crowell 1995), or were confined to mountain glaciers that involved only small volumes of ice. The indirect evidence of the cyclic flooding of the Rotliegend desert-lake and later Zechstein basin suggests that, although perhaps dwindling in size, the Gondwana glaciations (and interglacials) continued until at least the end of the Permian (Glennie 1990*a,b*).

Rapid basin subsidence continued after the Zechstein Transgression, allowing the accommodation of up to 2000 m or so of mostly halite in the centre of the Southern Basin, and an equal thickness in the Norwegian–Danish Basin (Taylor 1990). Whether thermal cooling still played a part in the latter basin is not known, but transtensional movements must be invoked to account for most of the 3000 m of Triassic basin fill in that area. Right-lateral transtensional stresses during the Jurassic and Cretaceous are also invoked for the development of sub-basins such as Sole Pit (Glennie & Boegner 1981) and Broad Fourteens (Van Wijhe 1987). Transpression, probably acting along the same system of faults, led to Late Cretaceous and Tertiary inversion and the traps that now form oil and gas fields. For an outline of areas affected by strong Mesozoic subsidence and inversion see Ziegler (1990, encl. 4).

## Conclusions

(1) The Permian period probably lasted from 296 to 251 Ma. The short-lived Lower Rotliegend volcanism straddled 290 Ma. Following the 10–30 Ma duration of the Saalian–Altmark Unconformity, both the Upper Rotliegend and Zechstein of the North Sea area were deposited within the Tatarian stage, which had a time span of possibly no more than 16 Ma. Volcanism in the Central Graben possibly coincided (267–265 Ma BP) with early graben development and the start of UR2 sedimentation.

(2) The combined Saalian–Altmark Unconformity in northern Germany and, by inference of basin shape, also in the southern North Sea area, was caused by thermal uplift. The rapidity of the ensuing Late Permian subsidence probably resulted from transtensional movements as well as thermal cooling. Bedded halite is confined to the basin-centre areas where the Silverpit desert lake was a constant feature because subsidence exceeded the rate of sedimentation.

(3) A history of local deep burial led to the destruction of reservoir properties, especially in dune sands and close to faults, by the creation of fibrous illite. Reservoirs damaged in this way now have their contained gas produced at economic rates by using horizontal drilling techniques. Rotliegend lake-margin sands were cemented with chlorite, whose crystal habit is far less damaging to the reservoir; such sands reservoir about 60% of German Rotliegend gas.

Shell Expro are thanked for draughting all except Fig. 2 which was prepared by B. Fulton of Aberdeen University. M. Epting is thanked for data on the NAM well H/16-1 used in Fig. 5 and J. Ineson for the ages of volcanics in the Central Graben. U. Gebhardt, R. Gast and R. Gaupp provided invaluable discussion and data on the German Rotliegend


## References

BESLY, B. M. 1990. Carboniferous. *In*: GLENNIE, K. W. (ed.) *Petroleum Geology of the North Sea* (3rd Edition). Blackwell Science, 90–119.

——, BURLEY, S. D. & TURNER, P. 1993. The late Carboniferous 'Barren Red Bed' play of the Silver Pit area, Southern North Sea. *In*: PARKER, J. R. (ed.) *Petroleum Geology of Northwest Europe: Proceedings of the 4th Conference*. Geological Society, London, 727–740.

BOULTON, G. S. 1993. Ice Ages and Climatic Change. *In*: D. DUFF, P. McL. (ed.) *Holmes' Principles of Physical Geology* (4th edn). Chapman & Hall, London, 439–469.

BURRI, P., FAUPEL, J. & KOOPMANN, B. 1993. The Rotliegend in northwest Germany, from frontier to fairway. *In*: PARKER, J. R. (ed.) *Petroleum Geology of Northwest Europe: Proceedings of the 4th Conference*. Geological Society, London, 741–748.

CAMERON, T. D. J. 1993. Triassic, Permian and Pre-Permian of the Central and Northern North Sea. *In*: KNOX, R. W. O'B & CORDEY, W. G. (eds) *Lithostratigraphic Nomenclature of the UK North Sea*. British Geological Survey.

CLAOUE-LONG, J. C., ZHANG, Z., MA, G. & DU, S. 1991. The age of the Permo-Triassic boundary. *Earth and Planetary Science Letters*, **105**, 182.

COPE, M. J. 1986. An interpretation of Vitrinite Reflectance Data from the Southern North Sea Basin. *In*: BROOKS, J., GOFF, J. & VAN HOORN (eds) *Habitat of Palaeozoic Gas in NE Europe*. Geological Society, London, Special Publications, **23**, 85–98.

COWARD, M. P. 1990. The Precambrian, Caledonian and Variscan Framework to NW Europe. *In*: HARDMAN, R. F. P. & BROOKS, J. (eds) *Tectonic Events Responsible for Britain's Oil and Gas Reserves*. Geological Society, London, Special Publications, **55**, 1–34.

——1993. The effect of Late Caledonian and Variscan continental escape tectonics on basement structure, Paleozoic basin kinematics and subsequent Mesozoic basin development in NW Europe. *In*: PARKER, J. R. (ed.) *Petroleum Geology of Northwest Europe: Proceedings of the 4th Conference*. Geological Society, London, 1195–1108.

CROWELL, J. C. 1995. The Ending of the Late Paleozoic Ice Age During the Permian Period. *In*: SCHOLLE, P. A., PERYT, T. M. & ULMER-SCHOLLE, D. S. (eds) *The Permian of Northern Pangea*, **1**, Springer-Verlag, Berlin, 62–74.

DRONG, H. J., PLEIN, E., SANNEMANN, D., SCHUEPBACH, M. A. & ZIMDARS, J. 1982. Der Scheveningen-Sandstein des Rotliegenden – eine äolische Sedimentfüllung alter Grabenstrukturen. *Zeitschrift der deutschen geologischen Gesellschaft*, **133**, 699–725.

FITCH, F. J. & MILLER, J. 1967. The age of the Whin Sill. *Geological Journal*, **5**, 233–250.

GAST, R. E. 1988. Rifting im Rotliegenden Niedersachsens. *Die Geowissenschaften*, **6**, 115–122.

——1991. The Perrenial Rotliegend Saline Lake in NW Europe. *Geologisches Jahrbuch*, **A119**, 25–59.

——1993. Sequenzanalyse von äolischen Abfolgen im Rotliegenden und deren Verzahnung mit Küstensedimenten. *Geologisches Jahrbuch*, **A131**, 117–139.

GAUPP, R., MATTER, A., PLATT, J., RAMSEYER, K & WALZEBUCK, J. 1993. Diagenesis and Fluid Evolution of Deeply Buried Permian (Rotliegende) Gas Reservoirs, Northwest Germany. *American Association of Petroleum Geologists Bulletin*, **77**, 1111–1128.

GEBHARDT, U. 1994. Zur Genese der Rotliegend-Salinare in der Norddeutschen Senke (Oberrotliegend II, Perm). *Freiberger Forschungsheft*, **C452**, 3–22.

——, SCHNEIDER, J. & HOFFMANN, N. 1991. Modelle zur Stratigraphie und Beckenentwicklung im Rotliegenden der Norddeutschen Senke. *Geologisches Jahrbuch*, **A127**, 405–427.

GLENNIE, K. W. 1990a. Outline of North Sea History & Structural Framework. *In*: GLENNIE, K. W. (ed.) *Introduction to the Petroleum Geology of the North Sea*. Blackwell Scientific Publications, 34–77.

——1990b. Lower Permian – Rotliegend. *In*: GLENNIE, K. W. (ed.) *Introduction to the Petroleum Geology of the North Sea*. Blackwell Scientific Publications, 120–152.

——1990c. Rotliegend sediment distribution: a result of late Carboniferous movements. *In*: HARDMAN, R. F. P. & BROOKS, J. (eds) *Tectonic Events Responsible for Britain's Oil and Gas Reserves*, Geological Society, London, Special Publications, **55**, 127–138.

——1994. Quaternary dunes of SE Arabia and Permian (Rotliegend) dunes of NW Europe: some comparisons. *Zentralblatt für Geologie und Paläontologie*. Teil 1(11/12): 1199–1215.

——1997. History of Exploration in the Southern North Sea. *This volume*.

—— & BOEGNER, P. 1981. Sole Pit inversion tectonics. *In*: ILLING, L. V. & HOBSON, G. D. (eds) *Petroleum Geology of the Continental Shelf of North-West Europe*. Heyden, London, 110–120.

GLENNIE, K. W., MUDD, G. C. & NAGTEGAAL, P. J. C. 1978. Depositional environment and diagenesis of Permian Rotliegendes sandstones in Leman Bank and Sole Pit areas of the UK southern North Sea. *Journal of the Geological Society, London*, **135**, 25–34.

HANCOCK, J. M. 1978. Possible causes of Rotliegend sandstone diagenesis in northern West Germany. *Journal of the Geological Society, London*, **135**, 35–40.

HAQ, B. U. & VAN EYSINGA, F. W. B. 1987. *Geological Time Table* (4th Edition). Elsevier, Amsterdam.

HEINRICH, R. D. 1993. The Cleeton Field, Block 42/29, UK North Sea. *In*: ABBOTTS, I. L. (ed.) *United Kingdom Oil and Gas Fields, 25 Years Commemorative Volume*. Geological Society, London, Memoirs, **14**, 409–415.

HELMUTH, H.-J. & SÜSSMUTH, S. 1993. Die lithostratigraphische Gliederung des jüngeren Oberrotliegenden (Oberrotliegendes II) in Nordostdeutschland. *Geologisches Jahrbuch*, **A131**, 31–55.

INESON, J. 1993. Rotliegend (Lower Permian) of Denmark. *In*: *Dynamisk/Stratigrafisk analyse af Palaeozoicum I Danmark. Kunderapport 33, EFP-89; Område 1*. Olie og Naturgas, **1**. Danmarks Geologiske Undersøgelse, 1–29.

KENT, P. E. 1980. Subsidence and uplift in East Yorkshire and Lincolnshire: a double inversion. *Proceedings of the Yorkshire Geological Society*, **42**, 502–524.

KIERSNOWSKI, H., PAUL, J., PERYT, T. M. & SMITH, D. B. 1995. Facies, Paleogeography, and Sedimentary History of the Southern Permian Basin in Europe. *In*: SCHOLLE, P. A., PERYT, T. M. & ULMER-SCHOLLE, D. J. (eds) *The Permian of Northern Pangea. Vol. 2. Sedimentary Basins and Economic Resources*. Springer-Verlag, Berlin, 119–136.

LEEDER, M. R. 1988. Recent developments in Carboniferous geology; a critical review with implications for the British Isles and NW Europe. *Proceedings of the Geologists' Association*, **99**, 74–100.

LIPPOLT, H. J. & HESS, J. C. 1989. Isotopic evidence for the stratigraphic position of the Saar-Nahe Rotliegend volcanism III. Synthesis of results and geological implications. *Neues Jahrbuch für Geologie und Paläontologie*, **9**, 553–559.

——, —— & BURGER, K. 1984. Isotopische Alter von pyroklastischen Sanidinen aus Kaolin-kohlensteinen als Korrelationsmarken für das mitteleuropäische Oberkarbon. *Fortschritte Geologie Rheinland und Westfalen*, **32**, 119–150.

MARIE, J. P. P. 1975. Rotliegendes stratigraphy and diagenesis. *In*: WOODLAND, A. W. (ed.) *Petroleum and the Continental Shelf of North-west Europe*. Applied Science Publishers, Barking, 205–210.

MENNING, M. 1991. Rapid subsidence in the Central European Basin during the initial development (Permian-Triassic boundary sequences, 258–240 Ma). *Zentralblatt für Geologie und Paläontologie*, **1**, 809–824.

——1992a. Stratigraphic working chart for the Pangea project. *Abstracts of 13th IAS Regional Meeting of Sedimentology*, Friedrich-Schiller Universität, Jena 1992, 175.

——1992b. Numerical Time Scale for the Permian and Triassic Lithostratigraphic Units of Central Europe. *Abstracts of 13th IAS Regional Meeting of Sedimentology*, Friedrich-Schiller Universität, Jena 1992, 176.

——1995. A numerical time scale for the Permian and Triassic Periods: an integrated time analysis. *In*: SCHOLLE, P. A., PERYT, T. M. & ULMER-SCHOLLE, D. S. (eds) *The Permian of Northern Pangea. 1*. Springer-Verlag, Berlin, 77–97.

——, KATZUNG, G. & LÜTZNER, H. 1988. Magnetostratigraphic investigations in the Rotliegendes (300–252 Ma) of central Europe. *Zeitschrift für geologische Wissenschaften*, Berlin, **16**, 1045–1063.

OELE, J. A., HOL, A. C. P. J. & TIEMANS, J. 1981. Some Rotliegend gas fields of the K and L blocks, Netherlands offshore (1968–1978 – a case history. *In*: ILLING, L. V. & HOBSON, G. D. (eds) *The Petroleum Geology of the Continental Shelf of North-west Europe*. Heyden, 289–300.

PLEIN, E. 1993. Bemerkungen zum Ablauf der paläogeographischen Entwicklung im Stefan und Rotliegend des Norddeutschen Beckens. *Geologisches Jahrbuch*, **A131**, 99–116.

——(ed.) 1995. *Norddeutsches Rotliegendbecken Rotliegend-Monographie Teil II*. Courier Forschungsinstitut Senckenberg, Frankfurt.

POKORSKI, J. 1989. Evolution of the Rotliegendes Basin in Poland. *Bulletin of the Polish Academy of Sciences, Earth Sciences*, **37**, 49–55.

SCHNEIDER, J. & GEBHARDT, U. 1993. Litho- und Biofaziesmuster in intra- und extramontanen Senken des Rotliegend (Perm, Nord- und Ostdeutschland). *Geologisches Jahrbuch*, **A131**, 57–98.

——, RÖSSLER, R. & GAITSCH, B. 1995. Time lines of the Late Variscan volcanism – a holostratigraphic synthesis. *Zentralblatt für Geologie und Paläontologie*, **1**, 477–490.

SCHRÖDER, L., PLEIN, E., BACHMANN, G. H., GAST, R. E., GEBHARDT, U., GRAF, R., HELMUTH, H.-J., PASTERNAK, M., PORTH, H. & SÜSSMUTH, S. 1995. Stratigraphische Neugliederung des Rotliegend im Norddeutschen Becken. *Geologisches Jahrbuch*, **A148**.

SMITH, D. B. 1979. Rapid marine transgression and regression of the Upper Permian Zechstein Sea. *Journal of the Geological Society, London*, **136**, 155–156

TAYLOR, J. C. M. 1990. Upper Permian–Zechstein. *In*: GLENNIE, K. W. (ed.) *Introduction to Petroleum Geology of the North Sea*. Blackwell Scientific Publications, 153–190.

TURNER, P., JONES, M., PROSSER, J., WILLIAMS, G. D. & SEARL, A. 1993. Structural and sedimentological controls on diagenesis in the Ravenspurn North gas reservoir, UK Southern North Sea. *In*: PARKER, J. R. (ed.) *Petroleum Geology of Northwest Europe: Proceedings of the 4th Conference*. Geological Society, London, 771–785

VAN EYSINGA, F. W. B. 1975. *Geological Time Table* (3rd Edition). Elsevier, Amsterdam.

VAN HOORN, B. 1987. Structural evolution, timing and tectonic style of the Sole Pit inversion. *Tectonophysics*, **137**, 239–284.

VAN WIJHE, D. H. 1987. The structural evolution of the Broad Fourteens Basin. *In*: BROOKS, J. & GLENNIE, K. (eds) *Petroleum Geology of North West Europe*. Graham & Trotman, 315–323.

WINTER, D. A. & KING, B. 1991. The West Sole Field, Block 48/6, UK North Sea. *In*: ABBOTTS, I. L. (ed.) *United Kingdom Oil and Gas Fields, 25 Years Commemorative Volume*. Geological Society, London, Memoirs, **14**, 517–523.

WOODLAND, A. W. (ed.) 1975. *Petroleum and the Continental Shelf of North West Europe: Vol 1. Geology*. Applied Science Publishers, Barking.

ZIEGLER, P. A. 1990. *Geological Atlas of Western and Central Europe* (2nd Edition). Shell International Petroleum Maatschappij.

# Permian (Upper Rotliegend) synsedimentary tectonics, basin development and palaeogeography of the southern North Sea

GARETH T. GEORGE[1] & JEREMY K. BERRY[2]

[1] *School of Earth Sciences, University of Greenwich, Pembroke, Chatham Maritime, Chatham, Kent ME4 4AW, UK*
[2] *Scott Pickford Group, 256 High Street, Croydon, Surrey CR0 1NF, UK*

**Abstract:** Upper Rotliegend sedimentation in the southern North Sea Basin (SNSB) was a response to complex synsedimentary tectonics and cyclical climatic changes. During the early Upper Rotliegend, faulting (?dextral strike-slip) resulted in the formation of a number of isolated sub-basins e.g. the Sole Pit, Silverpit and the Broad Fourteens, which were gradually united as a consequence of the progressive collapse of the Inde-Cleaver Bank High. Basin-margin tectonics controlled the evolution of the basin and the nature and distribution of the various types of alluvial fan and fan delta systems that evolved during pluvial climatic phases. Many of the earlier Upper Rotliegend fans display a systematic lateral displacement along synsedimentary faults. Along the Dowsing Fault Zone, 'dry-type' alluvial fans display a progressive northwestward migration with time. Also, fan deltas, sourced from the Texel–Ijsselmeer High, young progressively to the SE along the trend of the main faults. These systematic lateral displacements have been tentatively explained by invoking dextral oblique transfer of the source area relative to the depositional basin, within the context of a strike-slip tectono-sedimentary model. Alternatively, in an extensional model, the distribution pattern could be explained by sourcing the fans from both the footwall scarps and the transfer zones of dip-slip faults. A rapid decline in synsedimentary faulting, and a gradual change to more pluvial climatic conditions, resulted in the development of larger 'wet type' fans during the later part of the Upper Rotliegend time interval.

In the two recent publications (George & Berry 1993, 1994), the lithostratigraphies, depositional environments and palaeogeographies of the Upper Rotliegend strata in the UK and Dutch sectors of the southern North Sea Basin (SNSB) were reappraised. These studies, which involved the analysis of the wireline logs and core from 126 UK wells (5127 m core) and 80 Dutch wells (2743 m core), indicated that the tectono-sedimentary evolution of the basin was far more complex than previously thought. The work was based on the recognition of five lithostratigraphic units (Units 1–5 from oldest to youngest) which could be recognized across the basin. Since the publications of the above papers, further research has been undertaken on the logs and available core (2880 m) from 52, previously unreleased, offshore UK wells. The locations of the wells comprising the complete data base are shown on Fig. 1.

These new data, and the results of detailed correlations between sequences in the UK and Dutch sectors, have suggested that some modifications are necessary to the lithostratigraphic units designated by George & Berry (1993, 1994). Thus, the aim of this publication is to present an integrated summary of the tectonic and sedimentary evolution of this important gas basin, illustrated with a series of new palaeogeographic maps, based on the revised lithostratigraphic units. The most significant changes pertain to the Units 1, 2 and 3, and the palaeogeographic maps for these units supersede those published by George & Berry (1993, 1994), previously. The revised lithostratigraphic subdivision is discussed in the context of other Upper Rotliegend lithological subdivisions, and reservoir zonation schemes, which have been adopted in individual gas fields, groups of gas fields or lager areas within the Southern Permian Basin.

## Synopsis of previous papers

In this section (i) the tectonic setting, (ii) the lithofacies and depositional models, and (iii) the lithostratigraphic subdivision of the Upper Rotliegend of the SNSB are reviewed by reference to the earlier papers of the authors (George & Berry 1993, 1994) and relevant new research data.

### Tectonic setting

The main structural and palaeogeographical elements of the SNSB during the Upper Rotliegend are indicated on Fig. 2. The southern

*From* Ziegler, K., Turner, P. & Daines, S. R. (eds), 1997, *Petroleum Geology of the Southern North Sea: Future Potential,* Geological Society Special Publication No. 123, pp. 31–61.

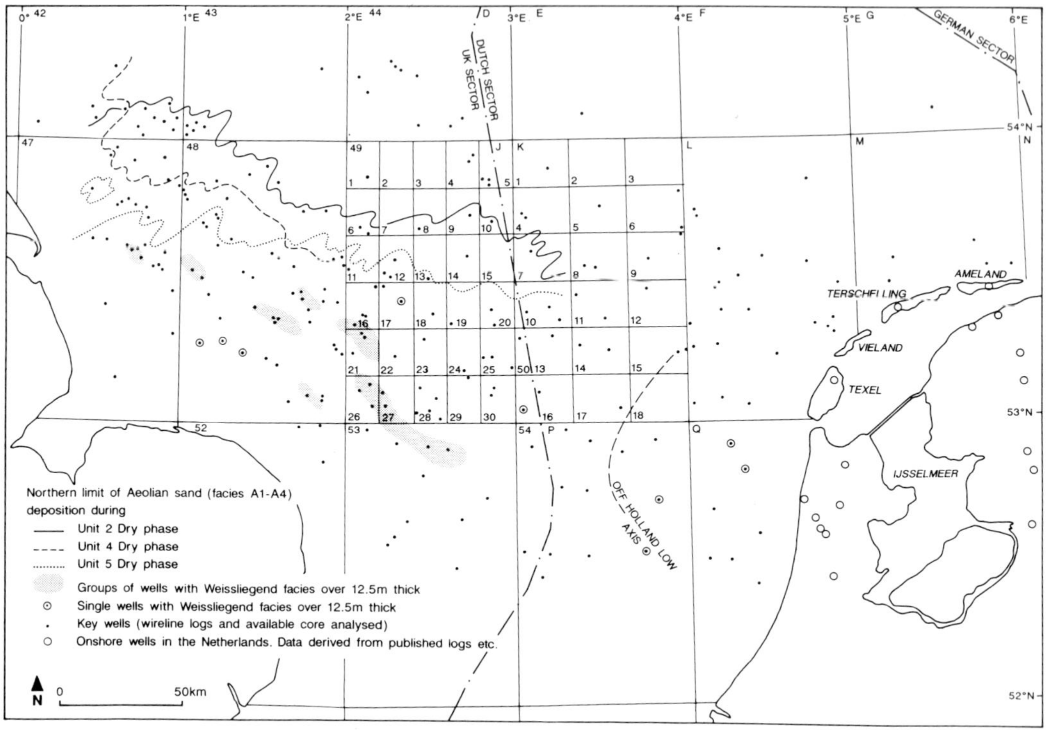


**Fig. 1.** Locality map showing the offshore and onshore well data base, and supplementary sedimentological data referred to in the text. Note that the northern limit of aeolian sand deposition for Unit 4 is approximately coincident with that for Unit 5 in Quadrant 49.

margin of the basin was defined by the Anglo-Brabant Massif which was the host of the East Anglia Granite (Allsop 1987; Chroston *et al.* 1987) and important marginal faults. The South Hewett Fault system was believed to have formed as a response to the buoyant uplift of the East Anglia Granite during the evolution of the Upper Rotliegend basin. In Quadrants P and Q of the Dutch sector the northern margin of the massif was dissected by a series of NW–SE-trending faults which, in part, represent the southeastward continuation of the South Hewett, Dowsing and Swarte Bank lineaments. These faults controlled the positions of a number of palaeogeographic salients of the massif e.g. the Winterton, Ijmuiden and Zandvoort Ridges (Fig. 2) (George & Berry 1993).

During its early evolution, the western margin of the basin was defined by the East Midlands High. This High appears to have been activated as a result of dextral strike-slip movements along the Dowsing Fault Zone and, later, by dip-slip faulting induced by the buoyancy of subsurface Caledonian granites (Fig. 2) e.g. the Newark Granite (Rollin 1982), the Market Weighton Granite (Bott *et al.* 1978) and the ESE extension of the latter into offshore Quadrant 42 (Donato & Megson 1990), named the Amethyst Granite by Leeder & Hardman (1990). More recently, a further Caledonian granite body has been postulated to be present to the east of the Sole Pit Basin (Donato 1993). This proposed granite body has a NW–SE trend with two, approximately circular, culminations (Fig 2). The most southerly cupola is largely confined to the north-central part of Quadrant 49 and is thus located below the northwestern margin of the Cleaver Bank High. However, the most northerly cupola, which straddles the southern contact of Quadrants 43 and 44, is located below the southwestern margin of the Silverpit Basin.

The Texel–Ijsselmeer High, which defines the eastern margin of the basin, was a complex, anticlinal structure which was formed originally during the Variscan Orogeny (van Wijhe 1987*a,b*). Evidence, discussed later, suggests that this structure was reactivated by dextral oblique-slip and dip-slip faulting during the early Upper Rotliegend.

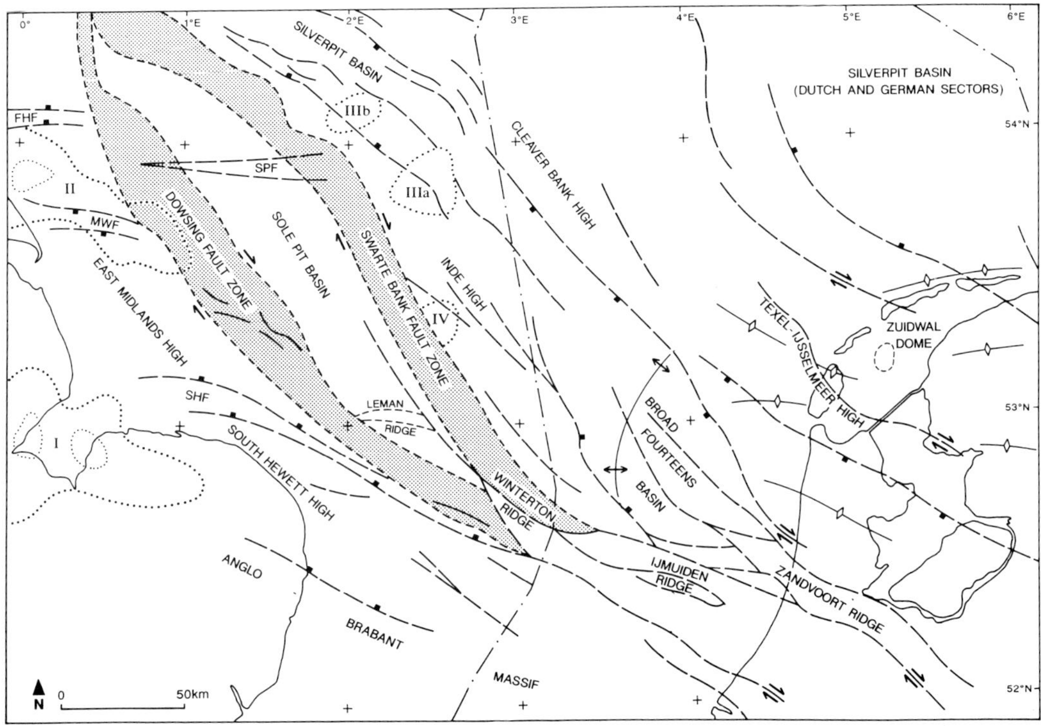


**Fig. 2.** Tectonic framework and depositional sub-basins for the Lower Permian of the Southern North Sea Basin. The faults are based on the maps published by the BGS (1985, Map 2) and Geco (1989), and on published data cited in the text. (See Fig. 8a for key.)

**Table 1.** *List of Rotliegend facies*

| Classes | Depositional regime | Association | Code | Facies |
|---|---|---|---|---|
| Lithogenic facies | Aeolian | Dune | A1 | Dune top |
|  |  |  | A2 | Dune core |
|  |  |  | A3 | Dune base |
|  |  | Interdune | A4 | Dry interdune–interdune sheet |
|  |  |  | A5 | Damp interdune |
|  |  |  | A6 | Wet interdune |
|  | Fluvial | Proximal | F1 | Fluvial fan |
|  |  |  | F2 | Fluvial channel |
|  |  | Distal | F3 | Structured sheetflood |
|  |  |  | F4 | Dewatered sheetflood |
|  |  |  | F5 | Subaqueous sheetflood |
|  | Sabkha | Sabkha | S1 | Lake margin sabkha |
|  |  |  | S2 | Inland sabkha |
|  | Lacustrine | Lacustrine | L1 | Playa lake |
|  |  |  | L2 | Desert lake (clastics) |
|  |  |  | L3 | Desert lake (evaporites) |
| Reworked lithogenic facies |  |  | A● F▲ | Deflation lag/serir |
|  |  |  | A → F | Fluvially redeposited aeolian |
|  |  |  | A* (or Wa*) | Biogenically reworked aeolian |
|  |  |  | F* (or Wf*) | Biogenetically reworked fluvial |
| Modified lithogenic facies (Weissliegend) |  |  | Wa | Homogenized aeolian |
|  |  |  | Wf | Homogenized fluvial |

## Lithofacies and depositional models

The lithofacies scheme presented in Table 1 is based on the analysis of a total of 10 750 m of core. In general, proximal fluvial facies (F1 and F2) and fluvially redeposited aeolian facies (A → F) are more abundant in the Dutch sector than they are in the UK sector. This is especially true of sequences deposited within and around the margins of the Off Holland Low (Fig. 1).

The aeolian depositional model of George & Berry (1993, figs 5 & 6) can be utilized to predict reservoir thickness and integrity. Aeolian sandstone facies (A1–A4), which form the principal reservoirs in the gas province, increase in thickness toward the erg interior from (a) the feather-edge of the desert lake, (b) the basin margin and (c) internal palaeo-highs (Fig. 3). In the UK sector, the central portion of the Sole Pit erg is clearly defined by the 180 m (600 ft) aeolian sandstone (facies A1–A4) isopach (Fig. 3). Around the margins of the erg, sand sheets and small dunes, with sets up to 3 m thick, occur interbedded with (a) sabkha and desert lake facies or (b) proximal to distal fluvial facies. The sabkha, desert lake and fluvial facies pinch-out toward the erg interior, which is characterized by

aeolian-dominated sequences, often containing aeolian dunes with sets up to 15 m thick. The distribution and morphology of these dunes was controlled by the wind regime and the sand budget, as is the case with dunes described from modern deserts (Fryberger et al. 1983; Lancaster 1983; Wasson & Hyde 1983; Brookfield 1984). Diplog data indicate that the aeolian sands were deposited within a (?seasonally) variable NE Trade Wind regime. Strong unimodal winds blowing to the SW, off the desert lake, initially reworked sabkha sediments and redeposited the sand fraction as thin sheets and small barchan dunes. Increasing sand supplies, considered to have been derived from the Dutch sector, and decreasing wind energy, resulted in the formation of progressively larger barchanoid and transverse dunes. A relative decrease in the sand budget and increased seasonal wind variability towards the centre of the erg, led to the occasional deposition of linear and star dunes. These latter, vertically building dunes appear to be confined to areas susceptible to accelerated synsedimentary subsidence (e.g. Sole Pit Basin) where they would be expected to attain a 'minimum survival size' (see Nielson & Kocurek 1987) and thus have greater preservation potentials.

The fluvial facies of the SNSB consist of a mixture of proximal mass flow and channel deposits (facies F1 and F2 respectively) and more distal structured and dewatered sheetflood deposits (facies F3 and F4 respectively). The areal distribution of fluvial facies within the UK sector (Fig. 4) reflects the predominantly north-westerly directed palaeoflow off the Dowsing Fault Zone/East Midlands High and the Anglo-Brabant Massif. In many areas, particularly in the Off Holland Low, the fluvial facies are quite thickly developed and areally extensive. These sequences were deposited on large, elongate, 'wet-type' alluvial fans supplied mainly by braided streams, sourced from the Anglo-Brabant Massif (George & Berry 1994, fig. 7). In other areas the fluvial facies are confined to much smaller 'dry-type' alluvial fans which were supplied mainly by mass-flows initiated on intra-basinal tectonic features (e.g. Dowsing Fault Zone, Texel–Ijsselmeer High). The abundance of dish and pillar structures within the above sequences indicates that they were frequently deposited from subaqueous flows. In many cases the fans were only temporarily inundated with water as a result of flash-flooding and flow expansion on the distal margins of the fans. However, some alluvial fans and terminal fan lobes appear to have been deposited in standing bodies of desert lake water and should thus be classified as fan deltas and braid deltas

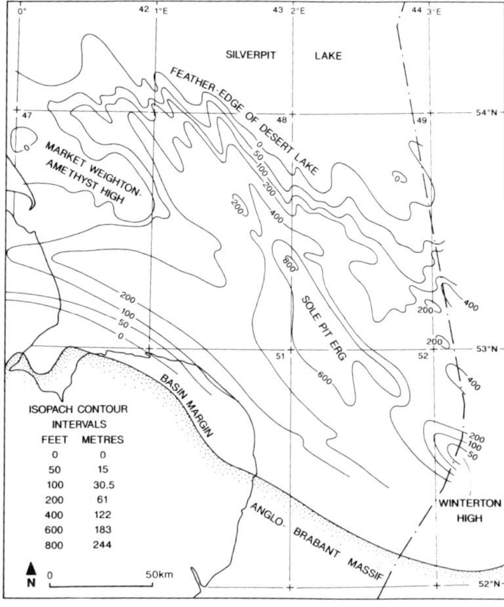


**Fig. 3.** Aeolian sandstone (facies A1–A4) isopach map for the total Upper Rotliegend (Units 1–5) in the UK Sector of the Southern North Sea Basin. Data based on core and wire-line log analyses. (Contour intervals in feet.)

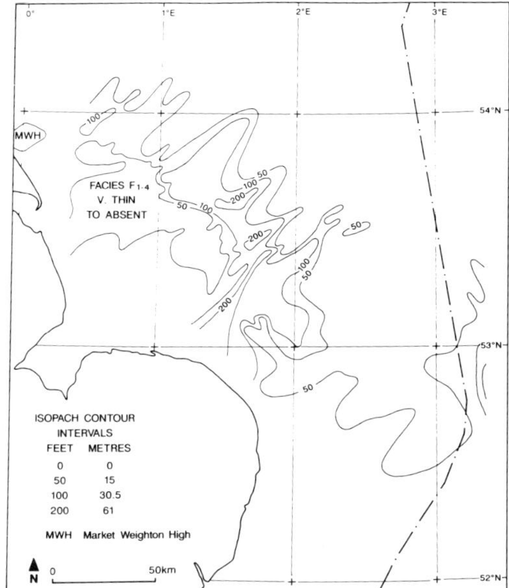



**Fig. 4.** Fluvial sandstone (facies F1–F5) isopach map for the total Upper Rotliegend (Units 1–5) in the UK Sector of the Southern North Sea Basin. Data based on core and wire-line log analyses. (Contour intervals in feet.)

respectively. This distinction between 'wet-type' fan, 'dry-type' fan, fan delta and braid delta has very important implications for facies modelling, palaeogeographic interpretations and hydrocarbon prospect evaluation in the Upper Rotliegend of the SNSB.

## Lithostratigraphic subdivision

Previous studies have shown that it is possible to subdivide the Upper Rotliegend of the SNSB into a series of drying upward cycles (George & Berry 1993, fig. 8, 1994, fig. 5). Good examples of drying upward cycles, from the present study, are illustrated in Figs 5, 7 and 11. Each cycle can be further subdivided into a wet phase (a), a dry phase (b), and sometimes a transitional phase (c) (Figs 5 & 7). From the detailed analysis of wireline logs and cores it is generally possible to recognize five drying upward sequences (designated lithostratigraphic Units 1, 2, 3, 4 and 5 in stratigraphic order) within the most complete Upper Rotliegend successions (George & Berry 1994, figs 3 & 9). In those areas where sedimentation was delayed, due to the incremental tectonic evolution of the basin, one or more of the earlier units may be completely absent or incomplete.

Since the publication of the George & Berry (1993, 1994) papers, the Upper Rotliegend succession of offshore The Netherlands has been reappraised using sequence stratigraphy concepts and spectral analysis (Yang & Nio 1994). These authors have divided the Upper Rotliegend into five supersequences (RO1–RO5, oldest to youngest); supersequence RO1 is confined to the central part of the basin while

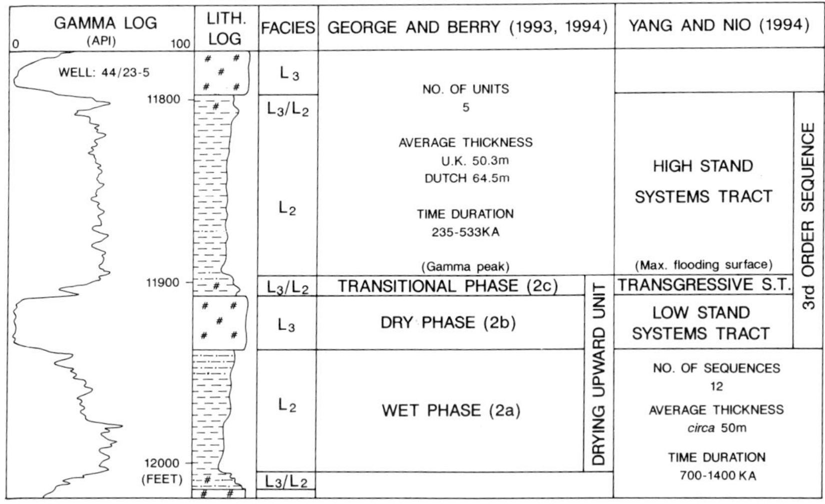



**Fig. 5.** Comparison of the terminologies relating to the drying upward units of George & Berry (1993, 1994) and the third order sequences of Yang & Nio (1994). The lithological interpretation of well 44/23-5 is based on wire-line log analysis.

**Table 2.** *Summary of unit thicknesses (metres)*

|  | UK Sector | | Dutch Sector | |
|---|---|---|---|---|
|  | Range | Average | Range | Average |
| Unit 5 | 19.2–112.7 | 58.5 | 36.6–101.5 | 63.7 |
| Unit 4 | 24.4–106.0 | 46.9 | 36.9–87.2 | 59.7 |
| Unit 3 | 18.3–79.9 | 49.4 | 32.0–128.9 | 73.2 |
| Unit 2 | 22.9–81.4 | 46.6 | 29.9–107.6 | 61.3 |
| Unit 1 | — 54.3 | — | — 109.1 | — |
| Grand average (Units 2–5) | 50.3 | | | 64.5 |

the younger supersequences (RO2, RO3, RO4 and RO5) successively onlap the basin flank. They have subdivided the supersequences into 12 sequences (RO1, RO2.1, RO2.2, RO3.1, RO3.2, RO3.3, RO3.4, RO4.1, RO4.2, RO4.3, RO4.4 and RO5), which they consider to be comparable to the third-order sequences of Vail *et al.* (1977) and Haq *et al.* (1987). The 12 sequences of Yang & Nio (1994) have average thicknesses of *c.* 50 m and, in terms of their sedimentology, they can be considered as 'wetting upward' cycles (lowstand–transgressive–highstand systems tracts). Similar thicknesses have been calculated for the five drying upward cycle, *sensu* George & Berry (1993, 1994), present in the UK sector (50.3 m), and in the Dutch sector (64.5 m) of the SNSB (Table 2). A summary of the main points of difference between the two schemes is presented on Fig. 5. The greater number of sequences recognized by Yang & Nio (1994) can be attributed to two main factors.

(1) Their study included well G/17-1 which is abnormally thick (538 m) for the Dutch sector. The present authors consider that this well, which has eight discrete halite beds, is genetically related to the German basin where up to 13 individual halites are present in the Upper Rotliegend (see later discussion).

(2) Yang & Nio (1994) considered that climatic change was the main factor controlling the deposition of their sequences and they ignored the possible effects of synsedimentary tectonics. In the opinion of the present authors, many of these climatically controlled depositional sequences have been overprinted by intra-basinal, synsedimentary tectonic events which may have increased sediment accumulation rates and the number of sequences deposited in specific areas e.g. the NW flank of the Texel–Ijsselmeer High.

### Revised lithostratigraphic subdivision

The main modification to the lithostratigraphic scheme previously proposed by George & Berry (1993, 1994) requires the adjustment of the base of Unit 5 (youngest unit), particularly in those wells located along the feather-edge of the desert lake. Detailed correlations between the UK and Dutch sectors, in this sedimentologicaly sensitive and complex zone, suggest that Unit 5 is thicker than previously documented. In some wells, the new base of Unit 5 corresponds to the previous base of Unit 4 (e.g. wells 48/25-1 and 49/12-3 in George & Berry 1993, figs 12 and 13). The old 'pick' is considered to represent a fairly localized event, related to the amelioration of the climate that occurred during the late Upper Rotliegend time interval. The new 'pick' facilitates correlations between offshore sequences in the UK and the generally thicker sequences in the Dutch sector where transitional phases are frequently developed at the tops of Units 3, 4 and 5 (i.e. 3c, 4c and 5c) (George & Berry 1993, figs 9 and 10). Increasing the thickness of Unit 5 obviously has a 'knock-on' effect for the underlying lithostratigraphic units. One of the most important consequences of the revision is that Unit 1 (oldest unit), is now considered to be absent from a large part of the north-central area of the basin, thus increasing the size of the Inde–Cleaver Bank High during the early Upper Rotliegend.

A summary of the thickness data for the five units in the UK and Dutch sectors is presented in Table 2. Average thicknesses have been calculated using data from wells where the unit in question is deemed to be complete (viz. it is underlain by an older unit and there is no evidence of faulting). Thus, it is only possible to give maximum recorded thickness values for Unit 1. The thick Upper Rotliegend sequences developed in G/17-1 (538 m) and N/4-1 (482 m) have been excluded from the calculations for the Dutch sector since both of these wells are considered to be genetically linked to the German basin (see below). The data presented in Table 2 confirm the fact that the lithostratigraphic units are thicker in the Dutch sector (average thickness of 64.5 m) than they are in the UK sector (average thickness 50.3 m). Since the thickest Upper Rotliegend sequences, are invariably recorded from wells located within the central zone of the permanent desert lake environment, it would appear that these areas were the most rapidly subsiding. However, it is frequently very difficult to apply the subdivision scheme objectively to these argillaceous/evaporite sequences, and thus it is possible that they may, in some cases, represent more than five lithostratigraphic units. This must be true in the German sector where the total desert lake deposits are often over 1000 m thick and contain up to 13 discrete halite horizons (Davis 1993, encl. 1). Likewise, the argillaceous/evaporite sequences in wells N/4-1 and G/17-1, referred to above, probably represent eight or nine units. Such increases in both the thicknesses of the successions, and the numbers of lithostratigraphic units developed, reflect the fact that Upper Rotliegend deposition was initiated earlier in these more easterly parts of the Southern Permian Basin (see later discussion).

The data summarized on Fig. 6 illustrates how various Upper Rotliegend lithological subdivisions and reservoir zonation schemes, that have been adopted in UK sector, relate to the five lithostratigraphic units discussed in this paper. In this compilation, any transitional phases, developed between the units, have been omitted. Also, the subunits and their correlative zones are represented as uniformly thick intervals, while in reality they display very dramatic thickness changes across the basin. In the sandstone dominated sequences, for example, the wet phase subunits/zones are most thickly developed on the basin margin but they are progressively replaced by dry phase facies towards the Sole Pit erg (compare Figs 3 & 4).

Outside of the Sole Pit erg, there is a very good correlation between the zones composed predominantly of aeolian dune and sheet sandstones (best reservoir) and the dry phases (1b, 2b, 3b, 4b and 5b). Similarly, the zones composed mainly of interdune, fluvial and sabkha sandstones (secondary reservoir) correlate with wet phases (1a, 2a, 3a, 4a and 5a). An apparent discrepancy in the correlation between the zones of Heinrich (1991*a*) and the units can be evaluated by reference to a well from the Cleeton

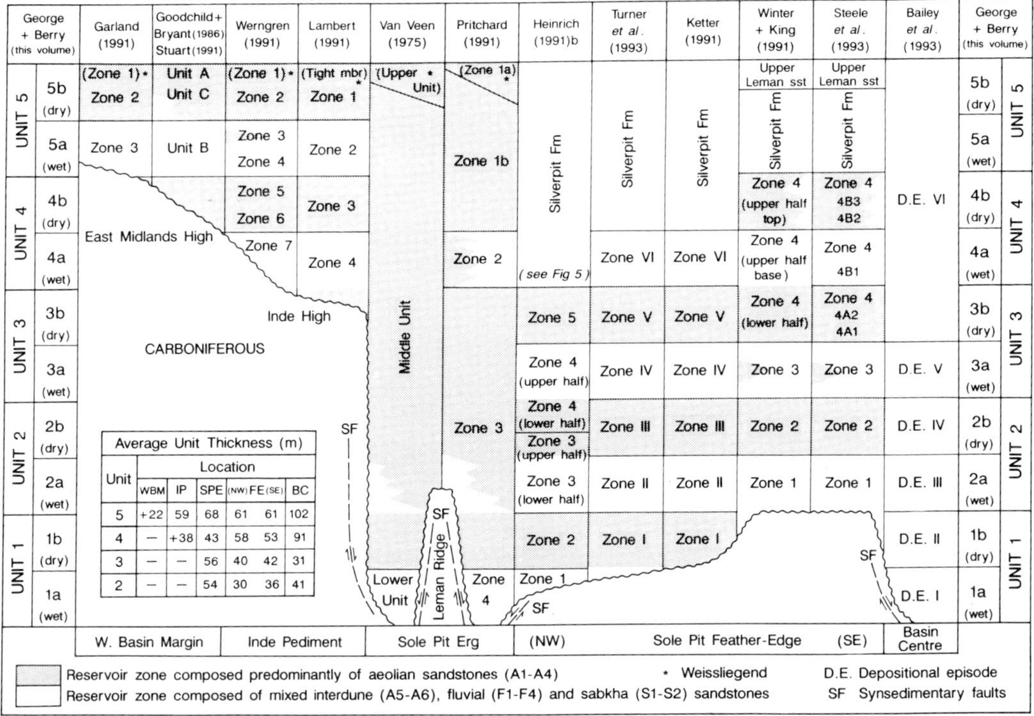


Average Unit Thickness (m)

| Unit | Location | | | | | |
|---|---|---|---|---|---|---|
| | WBM | IP | SPE | (NW) FE | (SE) | BC |
| 5 | +22 | 59 | 68 | 61 | 61 | 102 |
| 4 | − | +38 | 43 | 58 | 53 | 91 |
| 3 | − | − | 56 | 40 | 42 | 31 |
| 2 | − | − | 54 | 30 | 36 | 41 |

**Fig. 6.** Comparison of the various reservoir zonation and lithological subdivision schemes for Upper Rotliegend sequences in the UK Sector of the Southern North Sea Basin. Inset table shows the average thicknesses of Units 2–5 in different parts of the basin.

Field (Fig. 7). In this well, the Units 1, 2 and 3 are clearly defined by increasing upward, sonic transit-time motifs, that generally correspond to upward increasing poroperm trends. Unfortunately, Heinrich (1991*a*) does not discuss the petrophysical details of his zones, therefore it is difficult to explain the reasons why his Zones 3 and 4 do not reflect the sonic motifs.

Within the Sole Pit erg, the correlation of the lithostratigraphic units with the reservoir zones is very subjective due to the fact that the thick sequences are composed predominantly of aeolian sandstones. However a number of valid correlations can be made.

(1) The Lower (wadi) Unit of van Veen (1975) and Zone 4 of Pritchard (1991) correspond to the wet phase of Unit 1 (subunit 1a).

(2) Zone 2 of Pritchard (1991), which is composed of around 18 m of damp interdune (adhesion rippled) and aeolian sheet sandstones, corresponds to the wet phase of Unit 4 (subunit 4a). Further, but more subjective, correlations of the

aeolian dominated sequences can be made by picking the bases of the wet phases at positions where there is a pronounced decrease in the height of the dunes, or where dunes are rapidly replaced by aeolian sheets, and/or thin interdune to distal fluvial facies. These horizons are frequently associated with increases in cementation that produce kicks on the neutron/density and sonic logs (see George & Berry 1993). Preferential cementation along such bounding surfaces is a major cause of vertical heterogeneity within the aeolian dominated reservoirs of the Sole Pit erg, and also in many other aeolian reservoir sequences (Chandler *et al.* 1989; Fryberger 1993).

Although the Weissliegend frequently forms a distinct reservoir zone within the dry phase of Unit 5 (Fig. 6), no direct correlation exists between this diagenetic zone and the lithogenic subunit 5a.

In the basin centre, a good correlation can be made between the major depositional episodes I,

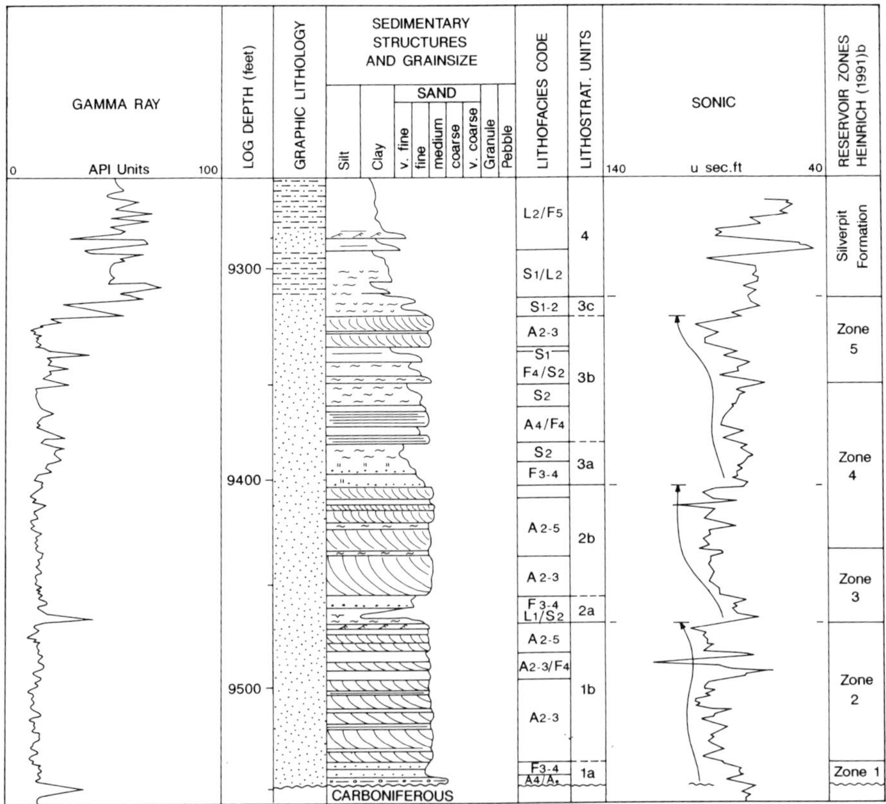

**Fig. 7.** A comparison of the reservoir zonation scheme of Heinrich (1991*a*, fig. 6) and the drying upward units discussed in this paper. The facies interpretation is based on published data and wire-line log analysis.

II, III, IV and V, recognised by Bailey *et al.* (1993), and the subunits 1a, 1b, 2a, 2b and 3a (Fig. 6). However, a number of important points of difference are apparent when comparing the stratigraphic relationships of the depositional episodes (Bailey *et al.* 1993, fig. 6) and subunits within the basin. In the depositional model presented in this paper, (a) the units display a progressive southwesterly onlap onto the margin of the basin, and (b) the dry phase units prograde basinward during each depositional episode (see Fig. 14). The latter point results in the halites being precipitated during a period of progressive lake contraction (cf. Bailey *et al.* 1993, fig. 6).

## Climatic cyclicity

Recent work on the dating and correlation of the Upper Rotliegend of Germany sheds light on the possible time duration of the lithostratigraphic units described above. Data discussed by Glennie (this volume) indicate that the Upper Rotliegend of North Germany was deposited during the lower half of the Tatarian stage, which (depending on the numerical data accepted) represents a time span of between 4 and 10 Ma. However, the maximum time duration of 10 Ma, attributed to the Upper Rotliegend by Gast (1993), may, in the near future, be reduced to 9 or even 8 Ma (Glennie this volume). If one accepts a time span of between 4 and 8 Ma for the Upper Rotliegend, each of the 17 cycles that have been recognized from the very thick (*c.* 2000 m) succession of NE Germany (Helmuth & Sussmuth 1993) would have a duration of 235–470 ka. Estimating the time duration of the five lithostratigraphic units recorded from the much thinner Upper Rotliegend succession of the UK and The Netherlands is even more speculative. Glennie (this volume) has suggested that the Upper Rotliegend of the UK was deposited during the last third of the time interval recorded in North Germany. If this assumption is valid, the Upper Rotliegend of the UK would represent a relatively short time interval of between 1.3 and 2.7 Ma, and each of the five lithostratigraphic units or cycles would have a time duration of 260–533 ka. These values are very much lower than those derived by Yang & Nio (1994) from their sequence stratigraphy and cyclicity analyses. These authors have estimated that the total time duration of the Upper Rotliegend of the Dutch sector was 10.7 Ma, with most of their 12, third-order sequences having time spans of between 500 and 1400 ka and net sediment accumulation rates of 6–11 cm ka$^{-1}$. Considering the recent age data from Germany, a time duration of 10.7 Ma appears to be a gross over-estimate for the deposition of this relatively thin Upper Rotliegend succession. This would also explain, what are considered to be, the unusually slow sedimentation rates for a 'red-bed' depositional environment. In this context it would be particularly informative to compare the respective sedimentation and precipitation rates of the claystones and pure halites present in the desert lake cycles. Thickness data suggest that the cycles from the basin centre that contain relatively thick halite horizons (Fig. 6, Units 2 & 3, 41 m & 31 m respectively) are thinner than those composed entirely of claystones (Fig. 6, Units 4 & 5, 91 m & 102 m respectively). This would imply that the accumulation rate of clay was greater (far greater if one considers differential compaction) than the precipitation rate of halite. Published estimates for the rates of precipitation of halite vary from 5 to 140 mm a$^{-1}$ (500–14 000 cm ka$^{-1}$) (Sonnenfeld 1984) but periodic dissolution could result in much lower halite preservation rates of 0.1–4.0 mm a$^{-1}$ (10–400 cm ka$^{-1}$) (Barnett & Straw 1983).

A cycle periodicity in the order of 260–533 ka, inferred for the five lithostratigraphic units, could possibly be explained by relating climatic fluctuations within the Upper Rotliegend 'northern hemisphere trade-wind desert' to south polar glaciations over Gondwanaland (Glennie 1983). This type of allogenic relationship between high latitude glaciations and the growth of low latitude sand seas ('ice age aridity'), has been discussed by numerous authors including Sarnthein (1978), Lancaster (1990) and Glennie (1994). However, the recurrence interval range for the Upper Rotliegend drying upward cycles is of greater magnitude than (i) the values (*c.* 180–405 ka) postulated for Carboniferous glacio-eustatic cycles (which were also controlled by the growth and decay of Gondwana ice caps), and (ii) the 100 ka and 400 ka Milankovitch (eccentricity) climatic fluctuation bands deduced for Pleistocene glaciations (see discussions in Leeder 1988 and Heckel 1990). Also, the minimum periodicity value of 235 ka is well outside the 100 ka eccentricity cycle estimated for the Early Permian by Berger *et al.* (1989). Conversely, Yang & Nio (1994) have recognized a 100 ka eccentricity Milankovitch cycle and a range of smaller (obliquity and precession) cycles within their third order sequences.

Further progress in our understanding of the importance of Milankovitch cyclicity on the deposition the of the Upper Rotliegend sequences will only be achieved within the constraints of a more precise radiometric age framework. Only then will it be possible to accurately determine climatic periodicities and subsidence/sedimentation rates, and to correlate 'climatically forced' sedimentary cycles across the whole of the Southern Permian Basin.

## Palaeogeography

Palaeogeographic maps are presented for each of the five lithostratigraphic units recognised (Fig. 8a–e). These maps highlight the lithofacies distribution during the wet phases (1a, 2a, 3a, 4a and 5a) because the fluvial processes responsible for their deposition were extremely sensitive to synsedimentary tectonics and resultant base-level changes, and also the fluvial systems had a dramatic modifying effect on the predominantly arid depositional basin. During the succeeding dry phases (1b, 2b, 3b, 4b and 5b), most of the inland basin, excluding a number of discrete perpetual fluvial tracts, was subjected to deflation and/or extensive aeolian sedimentation. A decrease in the amounts of precipitation and run-off, and a corresponding increase in the rate of evaporation, resulted in the precipitation of halite in discrete brine pools (salterns), located within the contracting desert lake. The main areas of halite precipitation are shown on the 'wet phase' palaeogeographic maps (Fig. 8a–d) for convenience. A separate dry phase palaeogeographic map is produced for Unit 5 (Fig. 9) to illustrate how the depositional environment reacted to the climate amelioration that occurred during the late Upper Rotliegend time interval.

### Unit 1 (Fig. 8a)

Early Upper Rotliegend sedimentation in the SNSB was restricted to a number of partially connected to completely detached sub-basins the margins of which were defined by NW–SE-trending faults. The desert lake, which was already well established in the Dutch sector, probably developed from a much earlier feature initiated in offshore and onshore areas of Germany. In the UK sector the Silverpit Basin may have developed as a relatively small pull-apart, isolated from the Sole Pit and Broad Fourteens Basins by a compressional platform (the Inde–Cleaver Bank High and Silverpit Ridge).

Coarse-grained sedimentation commenced during a wet or pluvial phase (1a) and resulted in marginal fans prograding off three principal areas: (i) the Texel–Ijsselmeer High, (ii) the Anglo-Brabant Massif and (iii) the Dowsing Fault Zone. The fan, sourced from the northern margin of the Texel–Ijsselmeer High, extended to the NW to reach the feather-edge of the desert lake in block K/3 and L/1. An abundance of dewatering dish and pillar structures present within the sequences indicates that deposition frequently occurred within the desert lake itself and thus this fan should be classified as a fan delta. This relatively large fan delta appears to have been confined to a NW-trending zone by synsedimentary faults. Progressive erosion of the source area resulted in the fan delta complex receding progressively to the SE.

The Anglo-Brabant Massif was also activated at this time, probably as a result of movements along WNW–ESE-trending strike-slip faults. Prominent highs, such as the Ijmuiden and Zandvoort Ridges, acted as the source areas for northwesterly prograding fans which were confined to a small, developing, sub-basin. Although no released wells have penetrated the proximal facies of these fans, their more distal facies have been recorded in the P/1-1 and P/2-2 wells (George & Berry 1994).

In the UK sector, coarse-grained sedimentation was initiated in the incipient, NW–SE-orientated, Sole Pit Basin. This basin was bounded by the Dowsing Fault Zone to the west and the Swarte Bank Fault Zone to the east. The uplifted shoulder of the former fault provided the source for a series of easterly and northeasterly extending, mass-flow fans, the largest of which occur to the SE in the region of the Camelot (53/1), Leman (49/26 & 49/27) and Vulcan (48/25 & 49/21) Fields. Consequently, many wells in these fields are characterised by a basal development of proximal fluvial facies (see respectively, Holmes 1991, fig. 5; van Veen 1975, fig. 6; Pritchard 1991, fig. 4). In the Leman Field, the basal fluvial deposits assigned to Unit 1 are absent along a WNW–ESE-trending high (Leman Ridge) which appears to have formed a barrier to fan progradation (Fig. 6). This high

**Fig. 8.** Palaeogeographic maps illustrating the tectonic and sedimentary evolution of the Upper Rotliegend basin in the UK and Dutch sectors of the Southern North Sea. Each map represents, with the exception of the areas of halite precipitation, the facies distribution during the wet phase of the unit. During each succeeding dry phase, fluvial activity declined or ceased, aeolian reworking and deposition increased, and the desert lake contracted, frequently resulting in the precipitation of halite. (**a**) Unit 1; (**b**) Unit 2; (**c**) Unit 3; (**d**) Unit 4; (**e**) Unit 5.

**(a)**

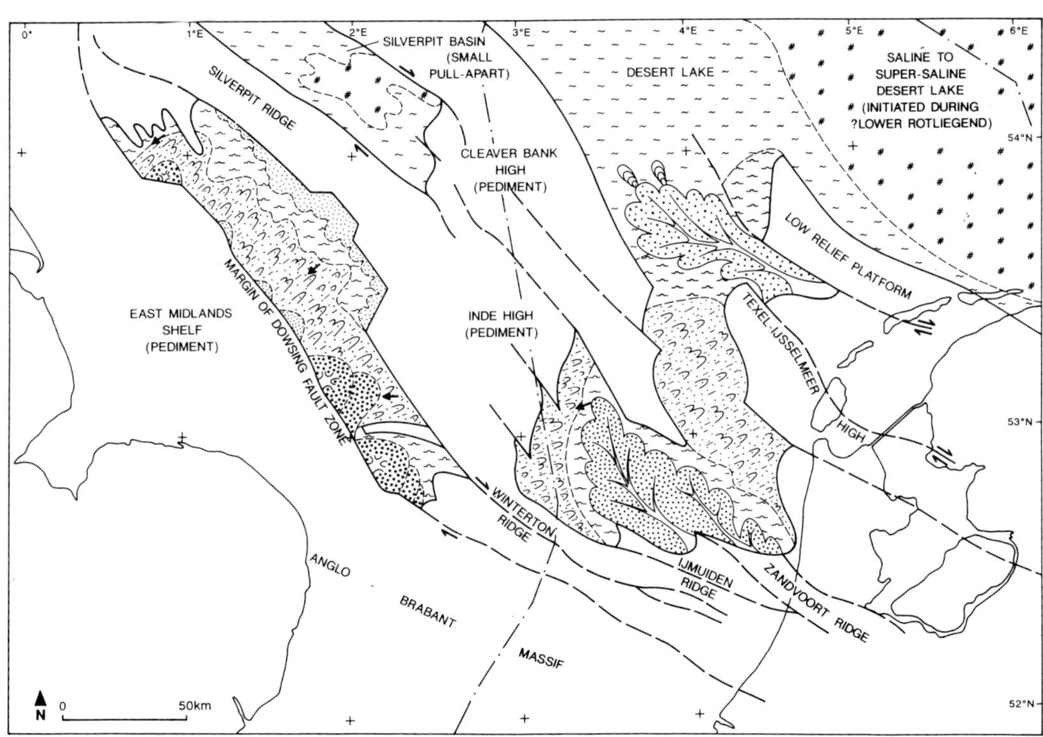


DEPOSITIONAL ENVIRONMENTS

 Aeolian dunes and sheets

 Dry-type alluvial fan (mass-flow dominated)

 Minor sheetflood/sheetwash marginal fans

Wet-type alluvial fan (stream-flow dominated) with terminal lobe or fluvio-lacustrine braid delta complex

Fan delta (mixed mass-flow and stream-flow)

Sabkha and floodplain

Desert lake, "normal" salinity

Saline to super saline lake

Residual deposits

 Fan head entrenchment along linear fault scarp

 Unit 3 Wet phase fans

Unit 2 Wet phase fans

Unit 1 Wet phase fans

OTHER FEATURES

IH    Isolated high

Anticlinal axis in Silesian stata ( after Leeder and Hardman 1990 )

Anticlinal axis

Pull-apart axis

Strike-slip fault

Normal fault

FHF    Flamborough Head faults

SHF    South Hewett faults

SPF    Silver Pit faults

MWF    Market-Weighton faults

I    East Anglia granite

II    Market Weighton - Amethyst granite

III    Cleaver Bank granite

IV    Indefatigable granite ( Leeder and Hardiman, 1990, Fig. 7 )

PALAEOCURRENTS

Palaeowind directions based on dip log analysis

← Mean vector for combined wet and dry phases

← Mean vector for dry phase only

↞ Fluvially reworked aeolian sands (A-F), direction inferred

**(b)**

0°
1°E
2°E
EXTENSION OF
MID NORTH
SEA HIGH
3°E
DESERT LAKE
4°E
5°E
6°E
SALINE TO
SUPER-SALINE
DESERT LAKE
54°N
DESERT LAKE
IH
BRINE POOLS
LOCATED IN
?FAULT-DEFINED
LOWS
FAN ABANDONED
DURING UNIT 2
DOWSING FAULT ZONE
EAST MIDLANDS
SHELF (PEDIMENT)
IH
CLEAVER BANK
/ INDE HIGH
TEXEL-IJSSELMEER HIGH
RECEDING TO S.E.
53°N
SOUTH HEWETT
SUB-BASIN DEVELOPED
PROGRESSIVELY DURING
UNIT 2
WINTERTON HIGH
ANGLO
BRABANT
MASSIF
N
0
50km
52°N

**(c)**

0°
1°E
2°E
3°E
4°E
5°E
6°E
SALINE TO
SUPER-SALINE
DESERT LAKE
54°N
BRINE POOLS
IN FAULT-
DEFINED LOWS
II
MARKET WEIGHTON
DESERT LAKE
EAST MIDLANDS
SHELF (PEDIMENT)
AMETHYST HIGH
FLUVIO-LACUSTRINE
BRAID DELTA
COMPLEX
INDE HIGH
ZUIDAL DOME
I
REMAINS OF
TEXEL-IJSSELMEER
HIGH
53°N
ANGLO
BRABANT
MASSIF
IJMUIDEN
RIDGE
ZANDVOORT RIDGE
REMAINS OF
(COLLAPSED)
WINTERTON HIGH
N
0
50km
52°N

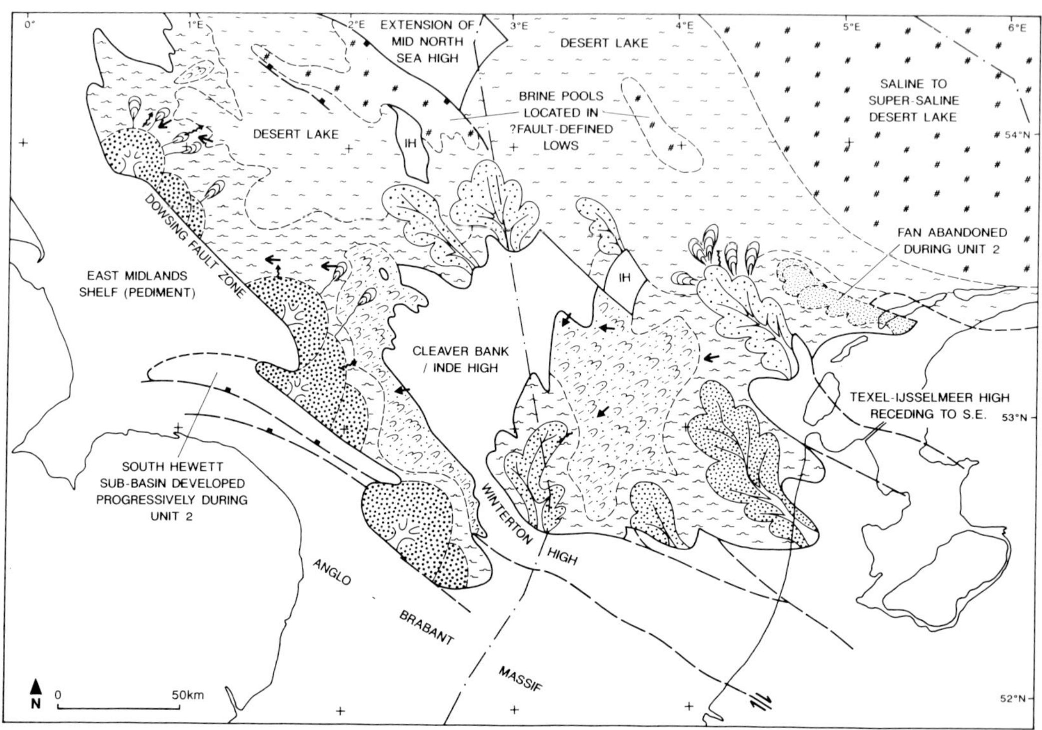

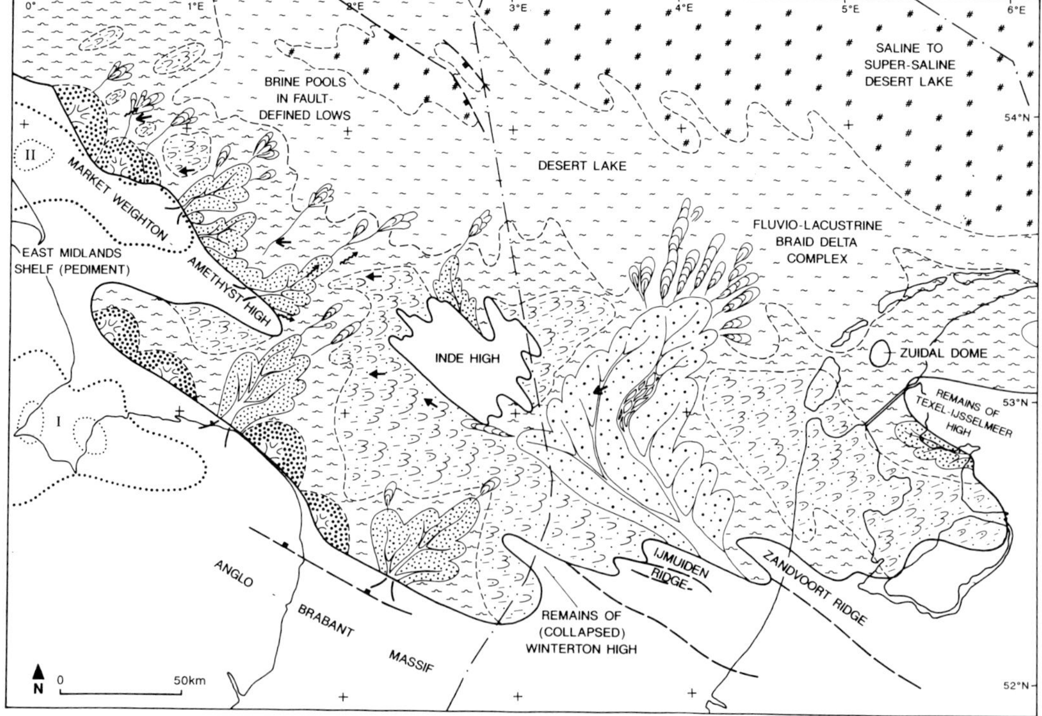

**(d)**

0°
1°E
2°E
3°E
4°E
5°E
6°E
FLUVIO-LACUSTRINE BRAID DELTA COMPLEX
ISOLATED BRINE IN ?FAULT DEFINED LOWS
SALINE TO SUPER-SALINE DESERT LAKE
54°N
DESERT LAKE
UPPER SLOTCHEREN SANDSTONE FLUVIO-LACUSTRINE BRAID DELTA COMPLEX
REMAINS OF MARKET WEIGHTON / AMETHYST HIGH
IH
53°N
ANGLO BRABANT MASSIF
N
0
50km
52°N

**(e)**

0°
1°E
2°E
3°E
4°E
5°E
6°E
IH
IH
54°N
DESERT LAKE
UPPER SLOCHTEREN SANDSTONE DEVELOPED AS A FLUVIO-LACUSTRINE BRAID DELTA COMPLEX
53°N
ANGLO BRABANT MASSIF
N
0
50km
52°N

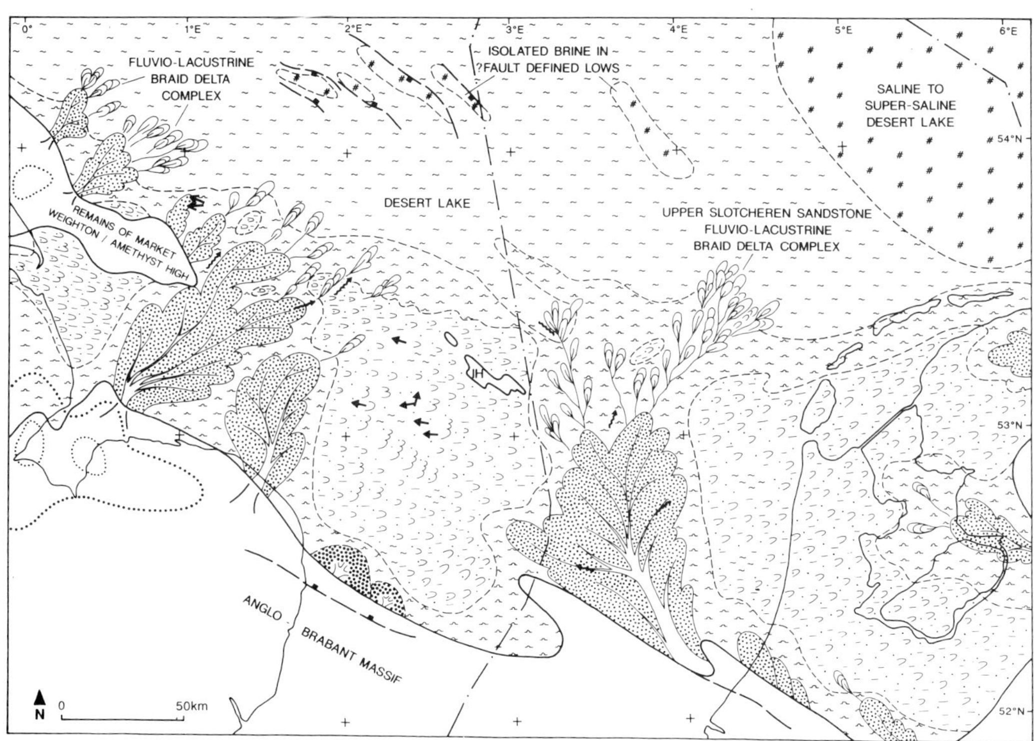

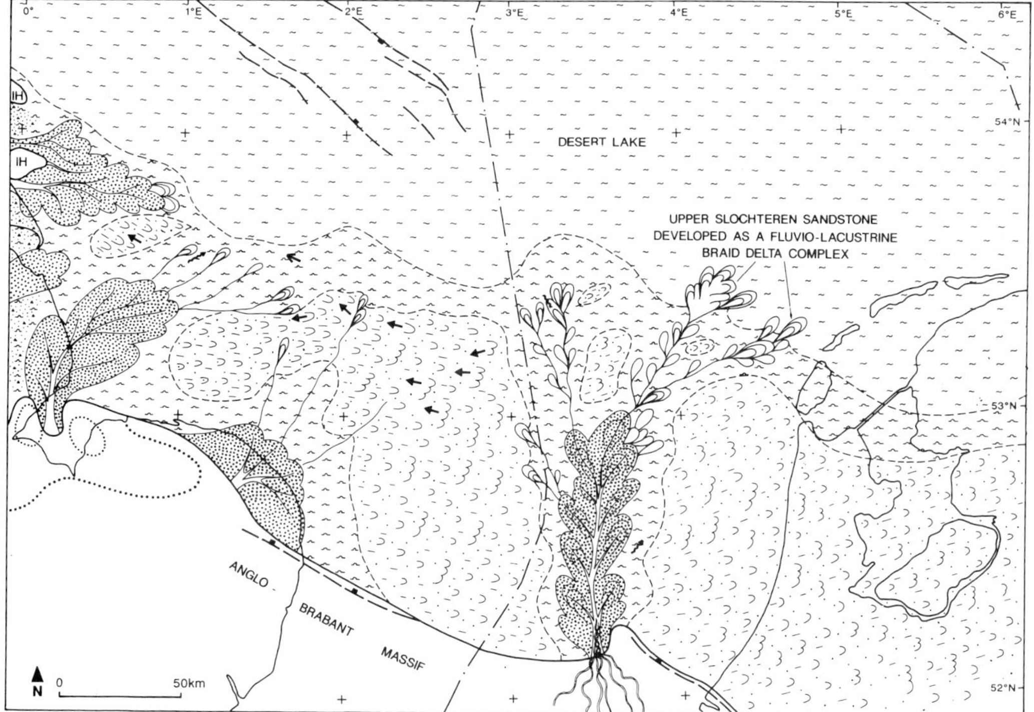

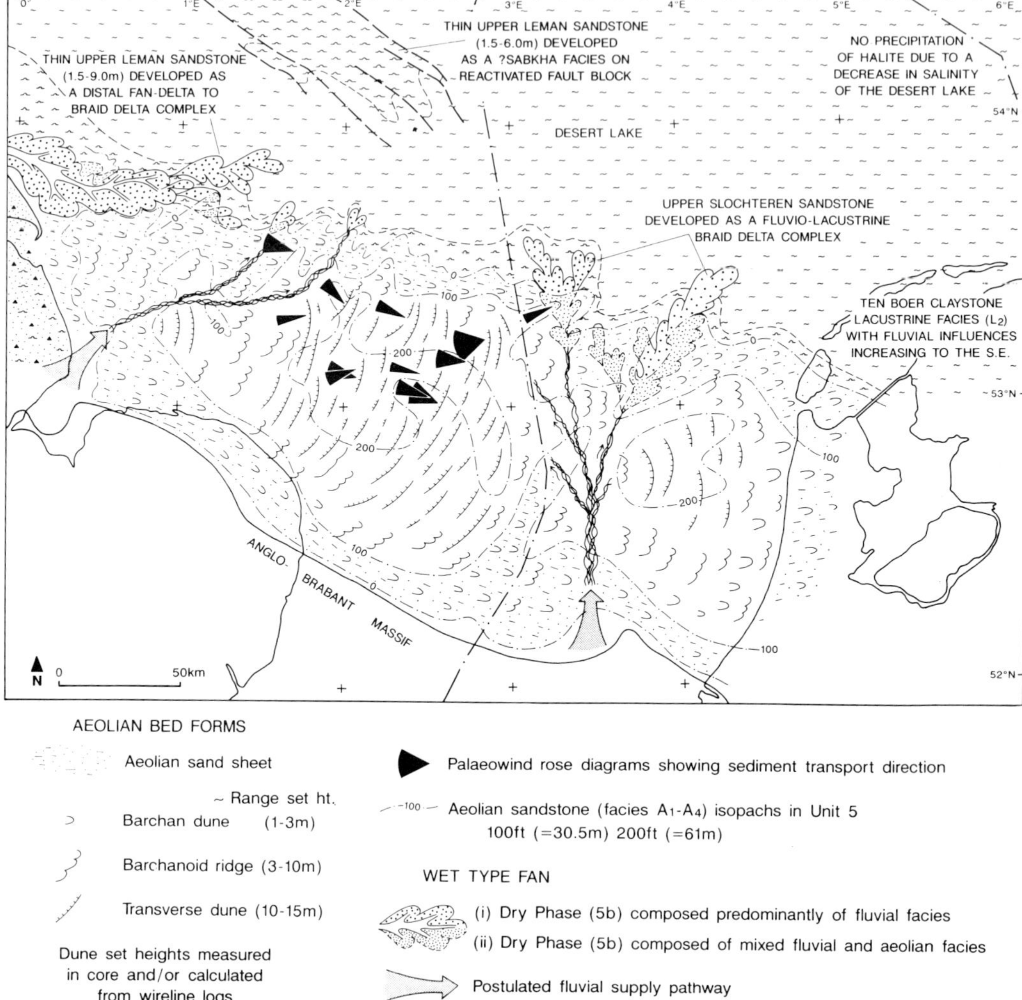


**Fig. 9.** Palaeogeographic map for the 5b dry phase showing the areas where fluvial activity continued until the termination of Upper Rotliegend deposition. (Aeolian isopachs in feet.)

was most likely related to movements along WNW–ESE faults that are known to be present in the southern part of the Leman Field (van Veen 1975, fig. 3; Hillier & Williams 1991, fig. 3b).

Further to the NW along the line of the Dowsing Fault Zone, in the region of the Clipper (48/19) and Barque (48/13 & 48/14) Fields, the fluvial facies are generally thinner and lithologically variable. In this area they form the lower part of the 'C' zone of Farmer & Hillier (1991a, b). At the northwestern termination of the Sole Pit Basin, in the region of the Cleeton (42/29) and Ravenspurn (42/30 & 43/26) Fields, fluvial deposits are thin to absent and comprise distal sheetfloods and

sabkha facies. They have been recorded in Zone 1 of Heinrich (1991a) (Fig. 7) and from the basal part of Zone 1 of Turner *et al.* (1993) (Fig. 6).

During the succeeding dry phase (1b), the previously discussed alluvial fans were abandoned, sometimes deflated, and buried by aeolian sands; while the fan deltas were similarly abandoned and covered by prograding lake margin sabkhas. Northeasterly and easterly winds deposited dune sands (15–45 m thick) in many areas including the Camelot, Leman and Frigate (48/20) Fields, and as far north as the Cleeton Field (Fig. 7). Similar thicknesses of dune sand were deposited in parts of the Broad Fourteens Basin.

Extreme desiccation resulted in progressive lake contraction and simultaneous sabkha progradation, and ultimately the precipitation of halites in super-saline pools (salterns) within the Silverpit desert lake. The thick, pure halite deposits that occur in wells 43/25-1 (25 m), 44/21-1 (24 m) and 44/22-2 (38 m), correspond to those recorded by Bailey *et al.* (1993) from their 'depositional episode' II (Fig. 6). Bailey *et al* (1993) have also suggested that there was some input of fluvial sands, from the northern margin of the basin at this time.

## Unit 2 (Fig. 8b)

Rapid basin expansion took place during Unit 2 as a result of further synsedimentary faulting, back-stepping along earlier faults and by the erosional retreat of existing fault scarps. In the Dutch sector, detritus which had accumulated on marginal highs during the preceding dry phase (1b), was flushed into the basin via northwestward prograding fans. The detritus derived from the Anglo-Brabant Massif contained a very large proportion of polycrystalline quartz and thus it can be readily distinguished from the less compositionally mature detrius derived from the Texel–Ijsselmeer High. Proximal polymict conglomerates (facies F1) derived from the former provenance (i.e. Zandvoort Ridge) have been recorded in well Q/4-2. Further to the NE, fan delta complexes continued to be sourced from the flanks of the Texel–Ijsselmeer High which was being subjected to active synsedimentary faulting. Two northwestward extending fan deltas have been recognized, their characteristic sequences forming part of the Lower Slochteren Sandstone in the K/10, L/8, L/11 and L/14 blocks. In core, these sequences display abundant dewatering dish and pillar structures indicating that they were deposited from subaqueous mass flows. Some of the larger mass flows extended beyond the distal margins of the fans to deposit subaqueous terminal lobes, particularly in Block L/4. This paleogeographic model is very different from that proposed by Atherton & Gibbs (1992, fig. 8) who have postulated the presence of three fluvial channel complexes prograding northeastward, northward and northwestward through Quadrants L, M and N, respectively, during the Lower Slochteren Sandstone time interval.

In the UK sector, the South Hewett Fault developed as a northwestward trending splay of the reactivated Dowsing Fault Zone. Detritus was shed northeastward into the basin via mass-flow dominated, marginal fans, that were themselves further extended by ephemeral streams with terminal lobes. Thick, multistorey, proximal fluvial deposits have been recorded in cores from the Clipper Field (Fig. 10a). In this area the 2a wet phase was protracted and fluvial influxes frequently interrupted aeolian sedimentation during the succeeding 2b dry phase (Fig. 10a). Subunits 2a and 2b, together with the overlying 3a wet phase, correspond approximately to the upper two-thirds of the poor quality 'C' reservoir zone of Farmer & Hillier (1991b).

In the West Sole Field and Hyde Fields the 2a wet phase is represented by mixed fluvial facies with thin aeolian interbeds. These deposits equate with Zone 1 of Winter & King (1991) and Steele *et al.* (1993) (Fig. 6). The Ravenspurn Fields were again receiving distal fluvial deposits derived from an alluvial fan advancing to the E and NE, off the Dowsing Fault Zone. These deposits consist of lithologically variable sequences of channellised and thickening upward sheetfloods with interbedded sabkha and lacustrine facies. They equate with the Zone II sequences of Ketter (1991) and Turner *et al.* (1993) (Fig. 6), although these authors have derived their fluvial deposits from the south. In the Cleeton Field, the 2a wet phase deposits consist of mixed sheetflood and sabkha facies (Fig. 7), which were derived from the Dowsing Fault Zone to the SW (Heinrich 1991a).

The eastern margin of the Sole Pit Basin was defined by NW–SE-trending faults, forming the most westerly elements of the Swarte Bank Fault Zone. The faults in question transect the southwestern part of Block 49/28 where they have been mapped by Arthur *et al.* (1986, fig. 3). Arthur *et al.* (1986) have suggested that the thickness variation observed in the Leman Sandstone Formation of this area was the result of differential subsidence that occurred across these synsedimentary faults.

Log and core data indicate that fan deltas were also being sourced from the northwestern margin of the Inde–Cleaver Bank High during this wet phase. Their deposits comprise a large proportion of the 'Basal Leman Sandstone', illustrated by Alberts & Underhill (1991, fig. 12.4a, b), that extends northward, from the northern flank of the high, into the desert lake.

During the dry phase of Unit 2, abandoned fans and ephemeral water courses were again partially deflated and covered by aeolian sand which was being transported by, predominantly, westerly blowing winds. Thick aeolian sands, often in excess of 30 m, were deposited in many parts of the Sole Pit Basin. Wireline log and core

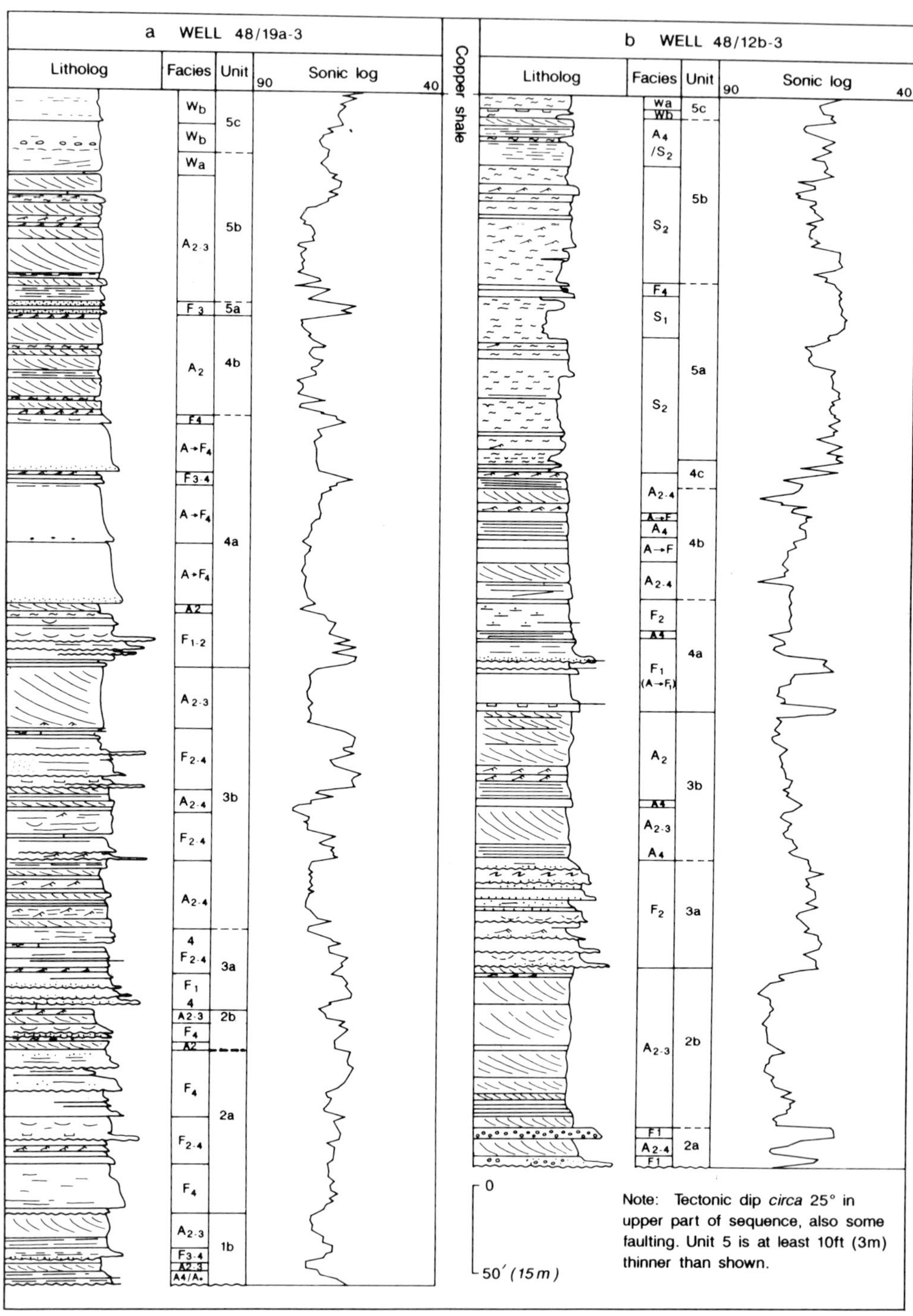



**Fig. 10.** Interpretation and subdivision of (**a**) well 48/19a-3 and (**b**) well 48/12b-3. Both wells were almost completely cored: 48/19a-3 (95%); 48/12b-3 (80%). (See Table 1 for the list of facies).

data indicate that dune set thicknesses frequently increase upward in individual wells, suggesting a vertical, time-related, change from small (barchan) to larger (transverse) dunes (e.g. well 48/12b-3, Fig. 10b) (see also Arthur *et al.* 1986). Abundant sand supplied from the Dutch sector, slow basin subsidence and fairly constant westerly blowing winds would have created the ideal conditions for the development of transverse dune fields (see also Winter & King 1991).

Desert lake contraction during the 2b dry phase resulted in the northward expansion of sabkha flats. Salt precipitation occurred over a large part of the Silverpit Basin, with up to 15 m of halite being recorded in wells 43/25-1, 44/21-2, 44/23-4, 44/23-5 (Fig. 5, 9 m of halite in dry phase 2b) and 44/29-1. The complete absence of Unit 2 in well 44/28-1, in which there is no evidence of post-depositional faulting, indicates that it was located on a small, isolated high.

## Unit 3 (Fig. 8c)

Basin expansion continued during Unit 3 as a result of (i) back-stepping along the main basin margin faults, (ii) continued erosion of the Texel–Ijsselmeer High and (iii) the break-up of the Inde-Cleaver Bank High into a small block, isolated from the Winterton High.

Extensive tectonic uplift occurred along the northern flank of the Anglo-Brabant Massif in the Dutch sector. This high sourced a very large 'wet-type' alluvial fan which was supplied by braided streams, that terminated within the desert lake as a braid delta complex (George & Berry 1994).

In the UK sector, alluvial fans continued to be sourced from the shoulders of fault scarps developed along both the Dowsing and South Hewett Faults. Again, in parts of the Clipper Field, fluvial processes were particularly active, and frequently interrupted aeolian sand deposition in the succeeding dry phase. In well 48/19a-3 for example, three distinct fluvial packages are developed (Fig. 10a). Core data indicate that the lowermost fluvial package is coarser grained and contains a much higher proportion of mass-flow (F1) to stream-flow (F2) facies than those developed higher in Unit 3. This suggests that the fans were becoming more distal (relative to the source area), and more stream-flow dominated, with time. In some parts of the Clipper Field (e.g. 48/19a-4), heavy rainfall, associated with the deposition of the earlier mass-flow fan, resulted in the slumping and homogenisation of previously deposited sands. Further to the NW in well 48/12b-3, the thick, multistorey channel-

fill sequence recorded from subunit 3a (Fig. 10b), provides additional evidence that the larger fans had developed axial feeder systems and were becoming stream-flow dominated.

A major expansion of the sand seas occurred during this dry phase (3b), which resulted in the deposition of thick (30–60 m) aeolian dune sands over many parts of the basin (e.g. well 48/12b-3, Fig. 10b). The thickest dune sands (*c.* 60 m) were deposited in the erg located in the Sole Pit Basin which appears to have been steadily subsiding. Basin subsidence and variable winds, blowing to the W and NW, may have provided the optimum conditions necessary for the deposition and preservation of vertically building linear or star dunes (George & Berry 1993).

Halites continued to be deposited in the deeper parts of the Silverpit desert lake. In the UK sector between 8 and 12 m of halite is present in the wells 43/21-1, 43/21-2, 44/23-4 and 44/23-5 (Fig. 5).

## Unit 4 (Fig. 8d)

A detailed palaeogeographic account of the basin during Unit 4 has been given by the authors previously (George & Berry 1993, 1994) and only an updated summary is presented here.

During this time interval, intra-basinal faulting rapidly declined and the Anglo-Brabant Massif became the principal source terrain for alluvial and fluvial systems. The activation of the northern margin of this massif in the UK sector is considered to be the result of extensional faulting related to the buoyant uplift of the East Anglia Granite. Exploitation of this source area resulted in a much larger drainage basin becoming available and consequently the alluvial fan systems were frequently larger, more elongate and stream-flow, rather than mass-flow, dominated. The latter point is illustrated by the occurrence of multistorey, channel-fill sequences in subunit 4a of wells 48/18a-4 (Fig. 11), 48/19a-3 and 48/12b-3 (Fig. 10a & b). Above the channel-fill sequence in well 48/19a-3, thick, homogenised, redeposited sands (Facies A → F) have been logged. The thickness of these sands suggest that they could have been redeposited from grain flows initiated by synsedimentary movements along the Dowsing Fault Zone. If this interpretation is correct, these deposits may be classified as 'seismites' (Seilacher 1984). Similar redeposited sands have also been recorded from subunit 4a in well 48/12b-3 (Fig. 10b), and from a number of wells in block 49/11 where they occur associated with sabkha facies. In the relatively distal depositional setting

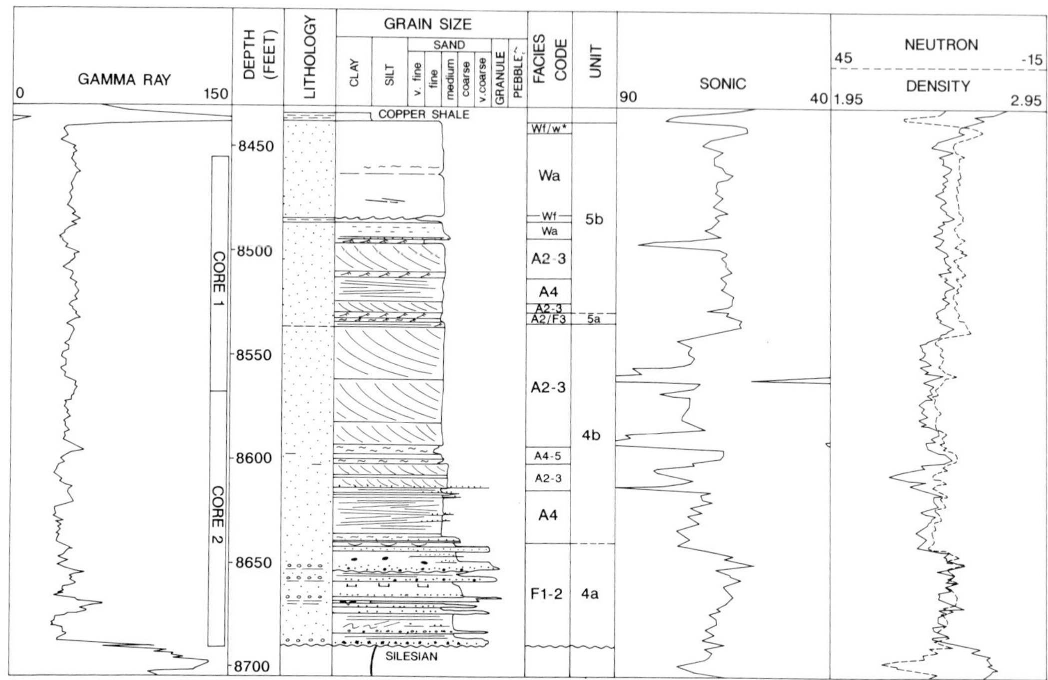

**Fig. 11.** Core and wireline logs from well 48/18a-4 illustrating the development of an excellent example of a drying upward cycle in Unit 4. Note the multistorey channel-fill in the wet phase (4a) and the upward increase in the scale of the aeolian dunes in the dry phase (4b). The 5a wet phase is represented by a very thin fluvial sheetflood (facies F3) which is overlain by aeolian sandstones that again display an upward increase in dune set thicknesses which continues into the Weissliegend (facies Wa).

of block 49/11, one must assume that the 'failed' sands were retransported some distance down palaeoslope.

The East Midlands Shelf, which was progressively eroded during the time of deposition of Unit 4, provided a more local source for alluvial fans and fan deltas prograding to the NE and E across the northern margin of Quadrant 47 into Quadrant 42 (the Frobisher and Ravenspurn Fields). The distal deposits of these fans have been recorded from Zone VI of the Ravenspurn Field (Ketter 1991; Turner *et al.* 1993) (Fig. 6), where they give rise to the 'ratty' gamma ray response that characterises the transitional passage of the Leman Sandstone Formation into the Silverpit Claystone Formation.

By this time, the Inde High consisted of a very small, elongate ridge defined by a series of en echelon NW–SE-trending faults with an outlier centred on well 49/19-2a. The latter feature is considered to represent the central part of the 'main horst block' of the Indefatigable Field described by Pearson *et al.* (1991, fig. 2), while the ridge itself appears to coincide with the southeastward continuation of this structure

through the Sean Field (see Hobson & Hillier 1991, fig. 4) with an offset through well 49/25-3.

In the Dutch sector, the Texel–Ijsselmeer High had ceased to be a major source for fluvial sediments. However, further faulting occurred along the margin of the Anglo-Brabant Massif and it continued to source large 'wet-type' alluvial fans, which again occupied the Off Holland Low during Unit 4. The axial braided streams of these fans extended into the desert lake where they deposited complex systems of overlapping, fluvio-lacustrine lobes, similar to those developed in Unit 3. These fluvio-lacustrine lobes, or braid deltas, represent the transitional phases 3c and 4c in wells drilled along the feather-edge of the desert lake.

During the penultimate dry phase (4b), the fans, described above, were largely abandoned, partly deflated and buried under advancing dune fields. However, dunes deposited within the axis of the Off Holland Low were commonly reworked by periodically active braided streams flowing northward to the desert lake. Thus, the margins of this fluvial tract contain a high proportion of reworked aeolian sands

(Facies A → F). Palaeowinds were directed predominantly to the W and WNW. Similar palaeowind trends have been confirmed by Winter & King (1991, p. 520) and illustrated by Werngren (1991, fig. 6) from well 49/28-8. Note that Zones 5–6 and Zones 1–2 of Werngren (1991) represent, respectively, the 4b and 5b dry phases of the present study (Fig. 6). These winds deposited extensive barchan and transverse dune fields with sand accumulations of between 30 and 45 m thick and frequently thicker in the Sole Pit erg. In the latter area, individual dunes were often in excess of 15 m high. Further north, diplog data from the type well (49/22-4a) of the Victor Field, give consistent westerly (4b dry phase) and northwesterly (5b dry phase) directed palaeowind directions (see Lambert 1991, fig. 4). In the above well, a basal dune set with westerly dips of c. 24–28° represents the top 5.75 m of the 3b dry phase (Fig. 6). In well 48/18a-4 (Fig. 11), the 4b dry phase is composed of an aeolian sandstone sequence (33 m thick) that begins with a sand sheet (facies A4) that passes up into thin dunes and interdunes (facies A2-5) that are finally replaced by stacked dunes (facies A2-3). The stacked dunes display an upward increase in set height, suggesting a vertical transition from small barchan dunes to larger transverse dunes.

In the UK sector, a number of thin halites horizons were precipitated during the 4a dry phase. Eastward into the Dutch and German sectors these halites horizons appear to unite and increase in thickness.

## Unit 5 (Figs 8e & 9)

*Wet phase (5a).* During the final pluvial phase (Fig. 8e), many of the large, stream-flow dominated alluvial fans, that were initiated in subunit 4a, were re-established (George & Berry 1993). In the UK sector, two alluvial fan systems, sourced from the Anglo-Brabant Massif, prograded northeastward into the basin. The proximal deposits of the most southerly of these fans have been recorded in the onshore wells of Somerton-1 and East Ruston-1. These systems were extended further to the E and NE by ephemeral streams associated with terminal lobes. Although the offshore extension of the Market Weighton–Amethyst High had been almost completely peneplained by this time, it probably extended onshore as a topographic feature. Thus, it was able to support quite a large alluvial fan that extended eastward into the basin. The proximal deposits of this fan have been described from the Amethyst Field

(Garland 1991, Zone 2) and from the Rough Field (Goodchild & Bryant 1986, Unit B; Stuart 1991, Unit B; Ellis 1993, Unit B) (Fig. 6).

The remnant of the Inde High was finally covered by sediments which are generally less than 60 m thick. These sediments represent Unit 5, sometimes with a thin development of the underlying 4b dry phase. They frequently begin with very thin, residual, breccio-conglomerates (facies A● and/or F▲) which are overlain by sabkha (facies S1), aeolian sheet (facies A4) and aeolian dune (facies A2-3) sandstones. Towards the north, the aeolian sandstones are progressively replaced by sabkha and desert lake (facies L2) deposits which commonly contain, well developed, nodular anhydrite (Alberts & Underhill 1991).

In the Dutch sector, the fluvial system that re-occupied the Off Holland Low, was more confined than its predecessor and therefore it could be classified as a braid plain rather than a true alluvial fan. This system was presumably derived from a lower relief, but more extensive, catchment area, developed within, rather than along the margins of, the Anglo-Brabant Massif. This braid plain was supplied by more or less perpetual streams that flowed northward into the desert lake where they deposited a braid delta complex (George & Berry 1994).

*Dry phase (5b).* During the 5b dry phase, the depositional basin reached its maximum size. Extensive aeolian sands, often over 60 m thick, were deposited in large areas of the Sole Pit and Broad Fourteens Basins (Fig. 9). Maximum aeolian sandstone thicknesses of c. 88 m have been recorded in the Sole Pit erg where individual dune sets are frequently over 15 m high. In the Amethyst Field, Garland (1991, fig. 7, Zone 2) has illustrated preserved duneforms, up to 20 m high, from the 5b dry phase. Diplog data indicate that the winds that deposited these dunes were blowing predominantly to the WNW (Fig. 9). Dune foreset dips, measured from orientated core, provide further evidence of northwesterly directed (vector mean 301°) palaeowinds (Carter & Seele 1993, fig. 7).

In the UK and Dutch sectors of the SNSB, perennial fluvial activity continued throughout the dry phase of Unit 5 in a number of well-defined tracks (Fig. 9). A 'wet type' alluvial fan, that had developed earlier during the 5a wet phase, advanced eastward from the Market Weighton High, to reach the desert lake where it terminated as a distal braid delta to fan delta complex. These deposits represent the Upper Leman Sandstone (1.5–15.0 m thick) developed in blocks 1, 2, 3, 6 and 7 of Quadrant 48 (see also, Winter & King 1991, fig. 6). The predicted source area of this fan coincides with a westward extending re-entrant of the Late Carboniferous (Barren Red Beds) subcrop, shown on the Pre-Permian geology maps of Besly (in press) and Cameron (1993). Further north, particularly in blocks 21, 22 and 23 of Quadrant 44, a genetically unrelated development of the Upper Leman Sandstone (1.5–6.0 m thick) appears to represent sabkha progradation across a high developed within the desert lake. It is also possible that this development of the Upper Leman Sandstone represents distal fluvial deposits derived from a northerly source terrain (Mid North Sea High).

Increased rainfall during this dry phase also resulted in an expansion of the desert lake and a corresponding decrease in its salinity. Consequently, halite precipitation did not occur in the central zone of the desert lake in the UK or Dutch sectors of the SNSB (Fig. 9). The progressive southward retreat of aeolian deposition, recorded in he successive dry phases from Units 2, 3 and 5 (see Fig. 1), also suggests that the Upper Rotliegend climate was becoming less arid with time.

*Weissliegend*

Although the Weissliegend is classified as a modified lithogenic facies association (Table 1) (George & Berry 1993), a number of points concerning its occurrence and mode of origin are pertinent to the palaeogeograhic interpretation of the dry phase of Unit 5 (Fig. 9). The Weissliegend is developed within the uppermost *c.* 50 m of Unit 5, directly below the Copper Shale (Kupferschiefer). It is characterized by discoloured (grey-green), aeolian sandstones (facies Wa) and, generally less common, fluvial sandstones (facies Wf). Both facies usually display, one or more types of soft-sediment deformation structures (e.g. facies Wa – homogenized bedding, oversteepened and contorted foresets, soft-sediment microfaults; facies Wf – slumps, flame structures, dish and pillar structures, soft-sediment microfaults). Bioturbation, initiated during the Zechstein transgression, is responsible for only very minor amounts of sediment reworking and this is confined to the top metre, or so, of some sequences.

The occurrences of fluvial facies (Wf) within the predominantly modified aeolian facies (Wa) of the Weissliegend, indicate that fluvial processes continued to be active, in some areas (e.g. blocks 48/15 and 48/19), during the dry phase of Unit 5. In well 48/19a-3 (Fig. 10a), for example, the Weissliegend, which has a total thickness of 14 m, can be subdivided into two parts. The lowermost 5 m of the sequence consists of aeolian dune sandstones (facies Wa) in which the foreset laminae have only been slightly contorted and oversteepened. The remaining 9 m of the sequence is composed of fluvial sandstones (facies Wf) that display erosional scours, intraformational sandstone clasts, vague dish and pillar structures and also, abundant inclusions of carbonaceous and coaly debris. These features also indicate that the climate at this time was becoming less arid and, as a consequence of increased precipitation, it was frequently able to support perennial 'wet type' alluvial fans with vegetated interfluves.

The soft-sediment deformation features present in the aeolian facies (Wa) from the Weissliegend are believed to have formed as a result of post-depositional degassing of dune sands during the rapid Zechstein transgression (Glennie & Buller 1983). In those Weissliegend sequences composed of both aeolian (Wa) and fluvial (Wf) facies, some, or all, of the deformation could have resulted from either (a) the slumping and reworking of dune crests, during the Zechstein transgression (Glennie 1972; Blaszczyk 1981) or (b) the degassing and resedimentation of aeolian sands triggered by of heavy rainfall and fluvial inundation, prior to the Zechstein transgression (Heward 1991). In the UK sector, wells with relatively thick Weissliegend sequences (over 12.5 m) are concentrated within NW–SE-trending zones (i.e. parallel to the structural trend) that cut across the insopachs for aeolian sandstones present in Unit 5 (Figs 1 & 9). Such a relationship could be explained if the soft-sediment deformation, characteristic of the Weissliegend, was induced as a result of the preferential venting of trapped air through synsedimentary, structural highs, present at the time of the Zechstein transgression. In this situation sediment reworking during, or prior to, the Zechstein transgression could also have caused, localized, penecontemporaneous deformation.

## Synsedimentary faulting and basin evolution

In this section the interplay between faulting and sedimentation is analysed in an attempt to elucidate the early evolution of the SNSB. Emphasis is placed on the study of post-Variscan faults and fault zones, many of which were instrumental in controlling basin configuration during the Upper Rotliegend time interval.

## Regional tectonic framework

The late Palaeozoic tectonic evolution of NW Europe and adjacent areas has been analysed in detail by Arthaud & Matte (1977), and has been further discussed by numerous researchers including Dewey (1982), Glennie (1986, 1990) and Ziegler (1982, 1990). The above authors are in general agreement that the European Variscides evolved as a result of right-lateral megashear caused by the relative movement of Gondwana and Laurussia. This stress field produced a complex set of NW–SE-trending dextral strike-slip zones and related pull-apart structures that transected the Variscan fold belt and extended onto the foreland (Ziegler 1982, encl. 12, 1990, encl. 4). It is believed that post-Variscan faulting, which was associated with magmatic activity in many areas (see Glennie 1990), had more or less ceased by late Early Permian (late Autunian) times. However, data from the present study indicates that faulting (dip-slip and/or strike-slip) continued well into Late Permian (late Saxonian) times (see also Glennie this volume). The strike-slip component of these Upper Rotliegend faults could have been generated by the reactivation of the NW–SE Variscan faults during E–W extension, in a similar manner to that discussed by Bartholomew *et al.* (1993, fig. 3) to account for the Mesozoic evolution of parts of the North Sea rift system. Oudmayer & Jager (1993) have also discussed the relationship between selective fault reactivation and oblique-slip during the structural evolution of the SNSB.

## Fault systems

No seismic data were directly analysed for this study, the main faults being derived from published 1 : 1 000 000 scale maps (BGS 1985, map 2; Geco 1989), research publications (e.g. Glennie & Boegner 1981; van Wijhe 1987a; van Hoorn 1987), and from detailed analysis of Upper Rotliegend facies distributions.

In the UK sector of the SNSB, the main faults that defined the margins of the Sole Pit Basin, namely, the Dowsing Fault Zone, the Swarte Bank Fault Zone and the South Hewett Fault, have been mapped in detail at a number of stratigraphic levels (e.g. Glennie & Boegner 1981; BGS 1985; van Hoorn 1987). The complex en echelon to anastomosing pattern of faults comprising the two former lineaments have been discussed by Glennie & Boegner (1981) and van Hoorn (1987) and they have also been illustrated on the 1:1 000 000 scale top pre-Permian map published by the BGS (1985, map 2). Those authors, referred to previously, concur that the basin originated during the Early Permian as a pull-apart induced by dextral strike-slip faulting. Detailed seismic mapping has indicated that many of these faults appear to be upward branching flower structures (Glennie & Boegner 1981; Gibbs 1986; van Hoorn 1987). However, some of these strike-slip flowers were induced during much later episodes of faulting which were associated with the inversion of the Sole Pit Basin (Glennie & Boegner 1981). The actual amount of lateral displacement along the Dowsing Fault Zone may well be greater than that originally thought, when one compares the pre-Permian subcrop map of Glennie & Boegner (1981, fig. 2b) with the more recent one of Leeder & Hardman (1990, fig. 2). In the latter version, the subcrop pattern, particularly that of the Barren Red Measures developed above the 'Symon Unconformity', offers more potential for dextral translation along the line of the Dowsing Fault.

Regional seismic data have indicated that the South Hewett lineament is composed of a series of normal faults that pass onto the margin of the Anglo-Brabant Massif, down-throwing progressively to the NE (see van Hoorn 1987, fig. 11). These faults are considered to have formed as result of crustal extension related to he buoyancy of the subsurface East Anglia Granite. Similar, predominantly E–W-trending, extensional faults also define the margins of the subsurface Market Weighton–Amethyst Granite located on the East Midlands High.

Although many of the NW–SE-trending Rotliegend basins of Central Europe are associated with dextral strike-slip fault zones (Glennie this volume), much of this faulting occurred during the late phases (Stephanian–early Permian) of the Variscan orogeny and/or as a result of fault reactivation during the Late Jurassic to Early Tertiary (Betz *et al.* 1987; Schwab 1987; Ziegler 1990). In the Dutch sector of the SNSB the principal fault systems have been indicated on pre-Cretaceous and pre-Permian geological maps published by van Wijhe (1987b, figs 2 & 4 respectively). Along the northern margin of the Anglo-Brabant Massif the principal faults converge to form important palaeogeographic highs. The faults that define the margins of the Sole Pit Basin (South Hewett, Dowsing and Swarte Bank lineaments) converge to the SE to form the Winterton Ridge and the Ijmuiden Ridge while the faults that define the Broad Fourteens Basin converge to form the Zandvoort Ridge (Fig. 2). The simultaneous formation of basins at zones of fault divergence (transtension) and highs at zones of fault convergence (transpression) is very characteristic of strike-slip structural regimes. The complex fan-shaped morphology of the Broad Fourteens Basin (Fig. 8a) is similar to the grabens, formed as a response to Autunian dextral shear, in other parts of NW Europe (Ziegler 1990, p. 61).

Further north in the Dutch sector, the location of the Texel–Ijsselmeer High was also controlled by a series of essentially NW–SE-orientated faults that run sub-parallel to fold axes in the Westphalian A subcrop (van Wijhe 1987*b*, fig. 4). Three of the most important faults cross the islands of Texel, Terschelling and Ameland (Fig. 2), the latter two faults extending further to the NW to unite with those forming the Central North Sea Graben. Relative displacements shown on the pre-Permian subcrop map of van Wijhe (1987*b*, fig. 4) imply that dextral strike-slip has occurred along the Terschelling lineament. Later (Cimmerian) dextral strike-slip, along this complex fault zone, was also instrumental in forming the Mesozoic Vlieland Basin (Herngreen *et al.* 1991). These authors also provide evidence that the Rotliegend sandstones thin dramatically over the Zuidwal volcanic dome and consequently this structure must have formed an outlier of the Texel–Ijsselmeer High until late Upper Rotliegend (?Unit 4) times.

A further fault believed to be important in the structural evolution of this area runs between the islands of Ameland and Schiermonnikoog and extends southeastward onshore to the west of Groningen (van Wijhe 1987*b*, fig. 4). The pre-Permian subcrop map of van Wijhe (1987*b*) indicates a possible, significant, dextral displacement of the contact between the Westphalian B and Westphalian C plus younger strata. This fault could possibly be the principal strike-slip displacement zone in the NE part of The Netherlands. If this is the case, the latter fault may represent a stronger link between the Central North Sea Graben and the Lower Saxony Basin than the Terschelling lineament, as suggested by Herngreen *et al.* (1991).

## Synsedimentary tectonics and basin-fill geometry

In this section the importance of synsedimentary tectonics in controlling the architecture and sedimentologicl character of the basin-fill is assessed. Criteria considered to be diagnostic of, but not exclusive to, strike-slip sedimentary basins (see Crowell 1975; Reading 1980; Steel & Gloppen 1980; Gloppen & Steel 1981; Miall 1984; Nilsen & McLaughlin 1985; Steel 1988) are assessed within the constraints of the new sedimentological data and the existing structural data.

Equivocal evidence for strike-slip faulting, or, more correctly, oblique-slip faulting, is provided by the spatial distribution of the marginal alluvial fans deposited during Units 1, 2 and 3 along (a) the active shoulders of the Dowsing Fault Zone and (b) the flanks of the Texel–Ijsselmeer High. In both these areas a systematic lateral displacement of the alluvial fans and fan deltas can be demonstrated (Fig. 12).

Along the major part of the Dowsing Fault Zone the marginal fans display a progressive northwestward migration with time (Fig. 12). This lateral migration is associated with westward backstepping of the Unit 2 and 3 fans from a splay point along the fault zone (Fig. 12). Such a distribution can be explained by invoking dextral oblique transfer of the source area relative to the depositional basin (Fig. 13a), a concept which has been discussed by Crowell (1982, fig. 9) and Steel (1988, figs 6 & 7) to account for the distribution of fans in the Ridge and Hornelen strike-slip basins, respectively. Alternatively, the fans could have been sourced from the erosion of the footwall scarps of a series of en enchelon, extensional faults. In this situation, footwall uplift would have to have been transferred, progressively, to successive, left stepping, strands of the fault set (Fig. 13b).

Further to the NW, in a region where the Dowsing Fault veers to the west, another group of fans young to the west (Fig. 12). In terms of strike-slip tectonics, the curvature of the fault could represent a restraining bend (Crowell 1982). Such restraining bends of strike-slip faults are frequently associated with contemporaneous reverse faulting and 'push up' folds or 'pop up' blocks which could provide potential source areas for the fans referred to previously (Fig. 13a). In an extensional regime it would be equally possible to source the fans from a relay ramp developed between the overlapping segments of two en echelon normal faults (Fig. 13b). The geometries and kinematics of relay ramps, and associated relay structures, have been discussed recently by Larsen (1988), Peacock & Sanderson (1994) and Trudgill & Cartwright (1994). In this extensional model the curvature of the Dowsing lineament could represent a transfer fault (*sensu* Gibbs 1984) connecting the original normal faults. A model illustrating the progressive breakdown of a relay ramp to produce a composite fault with 'along strike bend' has been discussed by Peacock & Sanderson (1994) and the model has been

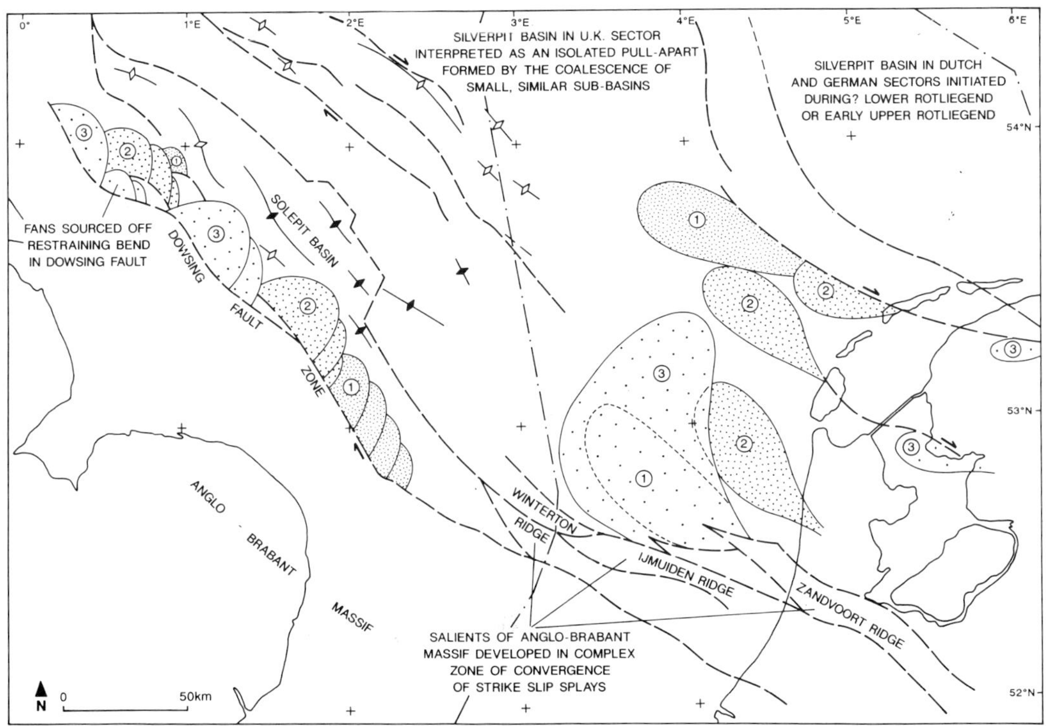



**Fig. 12.** Schematic map showing the systematic lateral succession of alluvial fans and fan deltas along synsedimentary (?oblique-slip) faults during the wet phases of Units 1, 2 and 3. Along the Dowsing Fault Zone, marginal alluvial fans display a progressive northwestward migration with time. Note that many of the fans, including those derived from the South Hewett Fault Zone have been omitted for clarity.

further amplified by Trudgill & Cartwright (1994). Relay structures and transfer faults have been shown to exert a significant control over contemporaneous drainage and sedimentation patterns in many extensional tectonic settings (Leeder & Gawthorpe 1987; Morley *et al.* 1990; Gawthorpe & Hurst 1993).

In offshore and onshore areas of The Netherlands the temporal and spatial relationships of the main alluvial fans and fan deltas to synsedimentary faults are more speculative due to the limited released well and seismic data. Nevertheless, the fans sourced from the Texel–Ijsselmeer High clearly young progressively to the SE along the trend of the main synsedimentary faults (Fig. 12). Also note that the transport paths of the fans are oblique to the controlling faults, which is in contrast to the Dowsing fans which were oriented approximately normal to the active fault shoulders. The fans themselves invariably display thinning and fining upward signatures (George & Berry 1994, fig. 6). The, more or less, continuous source area retreat required to sustain such

retrogradational fans could quite easily be generated by oblique strike-slip faulting. Since these Lower Slochteren fans were deposited during a predominantly drying upward phase (George & Berry 1994, fig. 3), their retrogradational (thinning and fining upward) character cannot be related to base-level rise resulting from lake expansion. It would also to be possible to derive the fans from transfer zones developed between overlapping, synthetic dip-slip faults (Morley *et al.* 1990). However, the structural evolution of these zones, as discussed by Peacock & Sanderson (1994) and Trudgill & Cartwright (1994), should result in a continuously developing and enlarging catchment area. Consequently any alluvial fans or fan deltas derived from relatively large transfer zones should display evidence of progradation (i.e. coarsening upward trends).

Although he fans derived from the Anglo-Brabant Massif display no apparent systematic spatial variation (Fig. 12), the orientation of the largest fan indicates that axial basin transport was initiated during the wet phase of Unit 3.

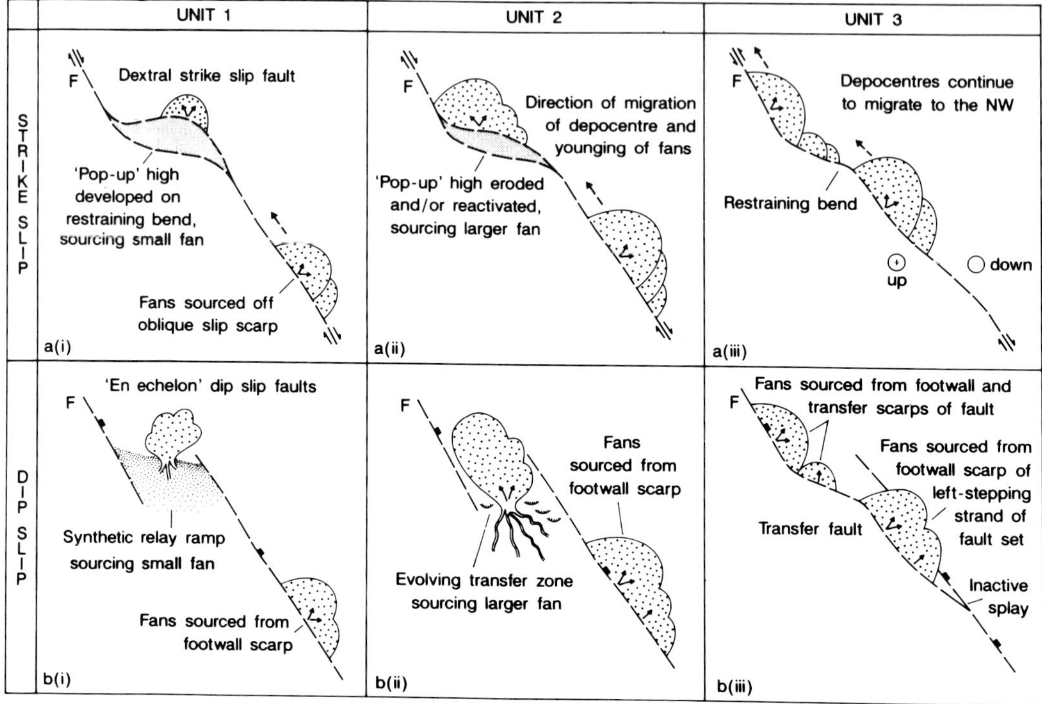



**Fig. 13.** Schematic maps illustrating the possible sediment supply pathways for alluvial fans derived from the Dowsing Fault Zone: (**a** i–iii) strike-slip model, (**b** i–iii) dip-slip model. Previously abandoned fans are not shown.

## Discussion

The data presented and discussed in this paper illustrate how both synsedimentary tectonics and cyclical climatic changes controlled the deposition of the Upper Rotliegend sequences in the UK and Dutch sectors of the SNSB. Although the evidence is equivocal, it seems likely that dextral strike-slip faulting was an important component of the synsedimentary tectonic regime during the development and filling of the basin. Glennie (this volume) has also came to a similar conclusion, and he has suggested that both thermal subsidence and strike-slip faulting were responsible for the development of many of the Upper Rotliegend sub-basins, both in the Southern and Northern Permian Basins.

The main features of the sub-basins/basin which, when considered collectively, are believed to be diagnostic of strike-slip tectonics can be summarized as follows.

(a) Elongate, fault-defined basins associated with highs developed at zones of convergence, and along the restraining bends of the marginal faults.

(b) Rapid lateral facies changes and pronounced asymmetry of the basin-fill geometry in both lateral and longitudinal profiles (Fig. 14). The above features can be related to both, differential subsidence/deposition along synsedimentary faults and the complexity of the dispersal systems (see below).

(c) Complex dispersal systems related to continuously evolving and geographically variable provenances (Fig. 14). Mass flow fans were propagated at right angles and oblique to NW–SE-trending synsedimentary faults which were themselves reactivated by back-stepping and/or splaying. Also, during the later development of the basin, a very prominent axial drainage system was established.

(d) Systematic lateral displacement of marginal alluvial fans and fan deltas along synsedimentary faults (Fig. 12). Individual fans young in the direction of lateral shift (dextral) of the source areas. Those developed along the Dowsing Fault Zone young predominantly to the NW while those developed on the opposite side of the basin, along the faults defining the Texel–Ijsselmeer High, young to the SE.

Many strike-slip basins, besides being elongate and narrow, are also very deep. In the present example, this characteristic is only evident when one considers the whole of the Southern Permian Basin. In the West Schleswig

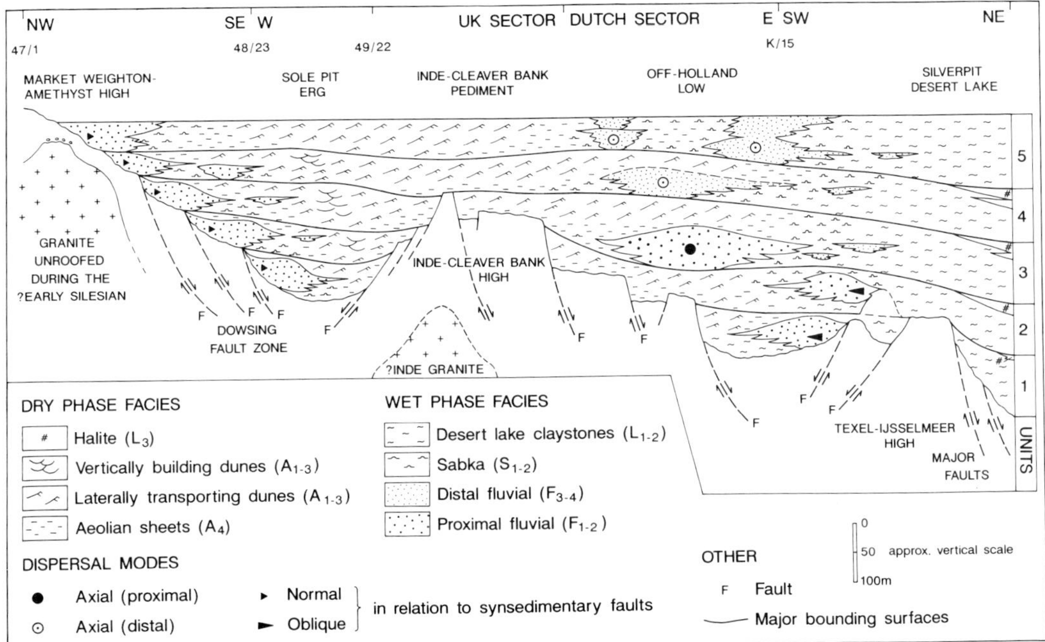


**Fig. 14.** Schematic section across the UK and Dutch sectors of the Southern Permian Basin illustrating the vertical and lateral facies architecture, the asymmetry of the basin fill, and the complex dispersal systems. (Modified from George & Berry 1994.)

Platform of offshore Germany, the Upper Rotliegend expands rapidly to reach a maximum recorded thickness of 1440 m in the J/11-1 well where the sequence consists of a basal sandstone (Hauptsandstein), 60 m thick, which is overlain by 1380 m of claystones and evaporites with 13 halite horizons (Davis 1993, encl. 1). This well, located 29 km south of J/5-1, displays a similar sequence that is approximately 1200 m thick (Glennie this volume, fig. 4). It is interesting to note that well J/11-1 lies very close to the postulated offshore extension of the Elbe Line which could represent the principal displacement zone of the Southern Permian Basin in the framework of a strike-slip basinal model. The well also lies very close to the axis of the basin predicted by Glennie (this volume, fig. 4), which extends eastward into onshore North Germany where the Upper Roliegend has a maximum thickness of around 2000 m. The rapid differential basin subsidence required to accommodate around 2000 m of sediment in a relatively short geological time would also seem to favour a strike-slip (oblique-slip) depositional setting.

The features discussed above are characteristic of, but not unique to, strike-slip sedimentary basins (see discussions in Nilsen & McLaughlin 1985; Mitchell & Reading 1989).

In many strike-slip basins, for example, multidirectional dispersal systems, involving lateral transport from the opposite sides of basins and longitudinal transport from the apexes of basins, are well documented (Steel & Gloppen 1980; Link 1982; McLaughlin & Nilsen 1982; Link *et al.* 1985; Nilsen & McLaughlin 1985). Conversely, equally complex, sediment dispersal patterns have been illustrated from extensional depositional settings (Frostick & Reid 1987; Leeder & Gawthorpe 1987; Morley *et al.* 1990; Leeder 1993; Leeder & Jackson 1993; Gawthorpe & Hurst 1993; Gabrielsen *et al.* 1995). Considering the evidence available at the present time, a compromise sedimentary model involving both dip-slip and strike-slip faulting appears to be the best option for the Upper Rotliegend of the SNSB. Such a model has been proposed by Bartholomew *et al.* (1993) to account for architecture and geometry of many Mesozoic sub-basins of the North Sea. In this model, the reactivation of NW–SE- and NE–SW-trending pre-Permian basement lineaments, within an E–W extension tectonic regime, has produced basins bounded by dextral and sinistral oblique-slip faults. In this type of tectonic setting although the actual amount of horizontal (oblique-slip) displacement may be limited, it is enough to create complex sediment dispersal paths and basin architectures which can have a significant effect on hydrocarbon trapping mechanisms (Bartholomew *et al.* 1993).

## Implications for future hydrocarbon exploration and reservoir modelling

The tectono-sedimentary depositional models discussed in this paper have a number of important implications for (i) sedimentological and reservoir modelling of existing gas fields, and also for (ii) further exploration in those areas of the Southern Permian Basin which are considered to have greater play fairway risks e.g. the feather-edge of the productive basin and areas on, or adjacent to, palaeo-highs. Many of these sites have not been fully investigated.

Within the feather-edge zone of the basin, play fairway risks are high due to the fact that the potential aeolian reservoirs display extreme vertical and lateral heterogeneity. In this, and other, sedimentologically complex zones, reservoir drainage and deliverability can be enhanced by drilling horizontal wells to target high permeability aeolian sandstones (see Steele *et al.* 1993; Maskall *et al.* 1994). The success of field developments involving horizontal wells is dependent on a detailed understanding of both the structure of the field and the geometry and architecture of the lithofacies present. This data enables development teams to determine the best possible heading to intersect, and steer through, high permeability lithofacies. In this context it is extremely important to establish the sediment supply pathways of any non-reservoir fluvial facies, especially if these facies truncate or replace the reservoir. The important role that synsedimentary palaeo-highs played in providing sediment flux has been stressed in this paper.

Concerning the reservoir potential of the fluvial facies discussed in this paper, an important distinction has been made between 'wet-type' fans, 'dry-type' fans, fan deltas and braid deltas. Such a distinction is important when developing facies models for prospect evaluations. 'Wet-type' alluvial fans offer the most attractive prospects due to their relatively large size and the fact that their streamflow deposited sandstones have the best reservoir characteristics of the fluvial facies. In the Dutch sector, fan deltas, derived from the Texel–Ijsselmeer High, have extended the northern depositional limit of potential reservoir into the desert lake. The mass-flow deposited conglomerates and sandstones, associated with these Lower Slochteren fan deltas, generally have low permeabilities.

However, since they lie directly on the Westphalian source rocks, and are overlain by thick desert lake argillaceous facies, they exhibit good stratigraphic trap configuration. Coarsening and thickening upward braid deltas sequences have also extended the northern limit of sandstone deposition into the desert lake at the northern termination of the Off Holland Low. In distal locations these sandstone sequences generally have too many vertical permeability barriers to be viable reservoirs but both vertical and lateral continuity can improve updip due to the amalgamation of fluvial sandstones and the development of thin aeolian sandstones. Such facies variations can result in braid delta reservoirs displaying both depositional and diagenetic heterogeneity (Gawthorpe *et al.* 1990). However, it is interesting to note that Gaupp *et al.* (1993) have discussed how lake margin sandstones, from the Upper Rotliegend of Northwest Germany, can display enhanced reservoir properties when radial chlorite cement has inhibited later quartz diagenesis. Consequently, in this gas province, the best reservoirs frequently occur in shoreline sandstones located in structural lows. In these lows, relatively dense, saline, Rotliegend formation waters have caused the precipitation of chlorite grain coatings, rather than the pore-plugging illite, developed in sandstones located on structural highs invaded by less saline Carboniferous formation waters (Gaupp *et al.* 1993; Burri *et al.* 1993).

A complete understanding of the tectonic regime is essential to the interpretation of the evolution of drainage systems and for predicting the spatial and temporal distribution of fluvial facies which can, in some situations, represent viable secondary reservoir targets. This is extremely important in the Upper Rotligend fluvial sequences where it is very difficult to acquire reliable palaeocurrent data from diplogs. In this context it is of paramount importance to distinguish between strike-slip and dip-slip depositional models. The complexity of the fluvial sediment dispersal systems discussed from the Upper Rotliegend of the SNSB appear to match those described from recognized strike-slip basins. However, in extensional tectonic regimes, complex drainage systems can also develop as a result of the progressive evolution of a series of footwall scarps and associated transfer zones.

Although the southern North Sea Basin is a mature gas province, it is still considered to have undiscovered potential, particularly within the feather-edge zone of the desert lake and in a variety of locations adjacent to synsedimentary faults. It is hoped that the data discussed in this paper will stimulate further regional structural studies in order to reconcile the case for a strike-slip, a dip-slip or even a mixed mode origin for both the individual synsedimentary tectonic elements and the entire Southern Permian Basin.

The authors would like to thank G. O. Simonian and J. Turner (GAPS Geological Consultants) for permission to publish much of the data presented in this paper, and H. Foxwell and H. Sutton for drawing the figures. The text was greatly improved by the helpful comments of the referees, the editors and G. Manby.

# References

ALBERTS, M. A. & UNDERHILL, J. R. 1991. The effect of Tertiary structuration on Permian gas prospectivity, Cleaver Bank area, southern North Sea, UK. *In:* SPENCER, A. M. (ed.) *Generation, accumulation, and production of Europe's hydrocarbons.* Special Publications of the European Association of Petroleum Geoscientists, **1**, 161–173.

ALLSOP, J. M. 1987. Patterns of late Caledonian intrusive activity in eastern and northern England from geophysics, radiometric dating and basement geology. *Proceedings of the Yorkshire Geological Society*, **46**, 335–353.

ARTHAUD, F. & MATTE, P. 1977. Late Paleozoic strike-slip faulting in southern Europe and northern Africa: Result of a right-lateral shear zone between the Appalachians and the Urals. *Geological Society of America Bulletin*, **88**, 1305–1320.

ARTHUR, T. J., PILLING, D., BUSH, D. & MACCHI, L. 1986. The Leman Sandstone Formation in Block 49/28 sedimentation, diagenesis and burial history. *In:* BROOKS, J., GOFF, J. & VAN HOORN, B. (eds) *Habitat of Palaezoic Gas in NW Europe.* Geological Society Special Publications, **23**, 251–266.

ATHERTON, A. F. & GIBBS, P. B. 19892. The structure, stratigraphy, and hydrocarbon potential north of the Frisian Islands Offshore Netherlands. *In:* SPENCER, A. M. (ed.) *Generation, Accumulation and Production of Europe's Hydrocarbons II.* Special Publications of the European Association of Petroleum Geoscientists, **2**, 85–101.

BAILEY, J. B., ARBIN, P., DAFFINOTI, O., GIBSON, P. & RITCHIE, J. S. 1993. Permo-Carboniferous plays of the Silver Pit Basin. *In:* PARKER, R. J. (ed.) *Petroleum Geology of Northwest Europe: Proceedings of the 4th Conference.* The Geological Society, London, 707–715.

BARNETT, J. M. & STRAW, W. T. 1983. Sedimentation rate of salt determined by micrometric analysis. *Geological Society of America Abstract with Programs*, **15**, 521.

BARTHOLOMEW, I. D., PETERS, J. M. & POWELL, C. M. 1993. Regional structural evolution of the North Sea: oblique slip and reactivation of basement lineaments. *In:* PARKER, R. J. (ed.) *Petroleum Geology of Northwest Europe: Proceedings of the 4th Conference.* The Geological Society, London, 1109–1122.

BERGER, A., LOUTRE, M. F. & DEHANT, V. 1989. Astronomical frequencies for pre-Quaternary palaeoclimate studies. *Terra Nova*, **1**, 474–479.

BESLY, B. M. in press. Carboniferous. *In:* GLENNIE, K. W. (ed.) *Petroleum Geology of the North Sea* (4th edition). Blackwell Scientific Publications.

BETZ, D., FUHRER, F., GREINER, G. & PLEIN, E. 1987. Evolution of the Lower Saxony Basin. *Tectonophysics*, **137**, 127–170.

BGS 1985. *Map 1: Pre-Permian Geology of the United Kingdom (south). Map 2: Contours on the top of the pre-Permian surface of the United Kingdom (south).* Scale 1:1 000 000. British Geological Survey.

BLASZCYZK, J. K. 1981. Palaeomorphology of Weiss-liegendes top as the control on facies variability on ore-bearing series of Lubin copper field, south-western Poland. *Geologia Sudetica*, **16**, 195–217.

BOTT, M. H. P, ROBINSON, J. & KOHNSTAMM, M. A. 1978. Granite beneath Market Weighton, East Yorkshire. *Journal of the Geological Society, London*, **135**, 535–543.

BROOKFIELD, M. E. 1984. Eolian facies. *In:* WALKER, R. G. (ed.) *Facies Models.* Geological Society of Canada, 91–118.

BURRI, P., FAUPEL, J. & KOOPMANN, B. 1993. The Rotliegend in northwest Germany, from frontier to fairway. *In:* PARKER, J. R. (ed.) *Petroleum Geology of Northwest Europe: Proceedings of the 4th Conference.* The Geological Society, London, 741–748.

CAMERON, T. D. J. 1993. Triassic, Permian and pre-Permian of the Central and Northern North Sea. *In:* KNOX, R. W. O'B & CORDEY, W. G. (eds) *Lithostratigraphic Nomenclature of the UK North Sea.* British Geological Survey.

CARTER, D. C. & STEELE, R. P. 1993. A practical method for estimating the orientation of eolian dune sands from deviated wells: An example from the Rotliegendes Group of the Southern North Sea. *Journal of Sedimentary Petrology*, **63**, 746–750.

CHANDLER, M. A., KOCUREK, G., GOGGIN, D. J. & LAKE, L. W. 1989. Effects of stratigraphic heterogeneity on permeability in eolian sandstone sequence, Page Sandstone, northern Arizona. *American Association of Petroleum Geologists Bulletin*, **73**, 658–668.

CHROSTON, P. N., ALLSOP, J. M. & CORNWELL, J. D. 1987. New seismic refraction evidence on the origin of the Bouger anomaly low near Hunstanton, Norfolk. *Proceedings of the Yorkshire Geological Society*, **46**, 311–319.

CROWELL, J. C. 1975. The San Gabriel Fault and Ridge Basin. *In:* CROWELL, J. C. (ed.) *San Andreas Fault in Southern California,* California Division of Mines and Geology Special Reports, **118**, 208–233.

——1982. The tectonics of Ridge Basin, Southern California. *In:* CROWELL, J. C. & LINK, M. H. (eds) *Geological History of Ridge Basin, Southern California,* Society of Economic Paleontologists and Mineralogists, Pacific Section, 25–42.

DAVIS, N. J. 1993. *The Lower Permian Rotliegendes of the German North Sea: A revision of the existing nomenclature and the introduction of a new lectostratotype well*. MSc thesis, University of Greenwich.

DEWEY J. F. 1982. Plate tectonics and the evolution of the British Isles. *Journal of the Geological Society, London*, **139**, 371–412.

DONATO, J. A. 1993. A buried granite batholith and the origin of the Sole Pit Basin, UK Southern North Sea. *Journal of the Geological Society, London*, **150**, 255–258.

—— & MEGSON, J. B., 1990. A buried batholith beneath the East Midland Shelf of the Southern North Sea Basin. *Journal of the Geological Society, London*, **147**, 133–140.

ELLIS, D. 1993. The Rough Gas Field: distribution of Permian aeolian and non-aeolian reservoir facies and their impact on field development. *In*: NORTH, C. P. & PROSSER, D. J. (eds) *Characterization of Fluvial and Aeolian Reservoirs*. Geological Society, London, Special Publications, **73**, 265–277.

FARMER, R. T. & HILLIER, A. P. 1991*a*. The Barque Field, Blocks 48/13a, 48/14, UK North Sea. *In*: ABBOTS, I. L. (ed.) *United Kingdom Oil and Gas Fields, 25 Years Commemorative Volume*. Geological Society, London, Memoirs, **14**, 395–400.

——1991*b*. The Clipper Field, Blocks 48/19a, 48/19c, UK North Sea. *In*: ABBOTS, I. L. (ed.) *United Kingdom Oil and Gas Fields, 25 Years Commemorative Volume*. Geological Society, London, Memoirs, **14**, 417–423.

FROSTICK, L. E. & REID, I. 1987. Tectonic control of desert sediments in rift basins ancient and modern. *In*: FROSTICK, L. & REID, I. (eds) *Desert Sediments: Ancient and Modern*. Geological Society, London Special Publications, **35**, 53–68.

FRYBERGER, S. G. 1993. A review of aeolian bounding surfaces, with examples from the Permian Minnelusa Formation, USA. *In*: NORTH, C. P. & PROSSER, D. J. (eds) *Characterization of Fluvial and Aeolian Reservoirs*. Geological Society, London, Special Publications, **73**, 167–197.

——, AL-SARI, A. M. & CLISHAM, T. J. 1983. Eolian dune, interdune, sheet sand and siliciclastic sabkha sediments of an offshore prograding sand sea, Dhahran area, Saudi Arabia. *American Association of Petroleum Geologists Bulletin*, **63**, 280–312.

GABRIELSEN, R. H., STEEL, R. J. & NOTTVEDT, A. 1995. Subtle traps in extensional terranes: a model with reference to the North Sea. *Petroleum Geoscience*, **1**, 223–235.

GARLAND, C. R. 1991. The Amethyst Field, Blocks 48/8a, 47/9a, 47/13a, 47/14a, 47/15a, UK North Sea. *In*: ABBOTS, I. L. (ed.) *United Kingdom Oil and Gas Fields, 25 Years Commemorative Volume*. Geological Society, London, Memoirs, **14**, 387–393.

GAST, R. E. 1993. Sequenzanalyse von aolischen Abfolgen im Rotliegenden und deren Verzahnung mit Kustensedimenten *Geologisches Jahrbuch*, **A131**, 117–139.

GAUPP, R., MATTER, A., PLATT, J., RAMSEYER, K. & WAZEBUCK, J. 1993. Diagenesis and fluid evolution of deeply buried Permian (Rotliegende) gas reservoirs, Northwest Germany. *American Association of Petroleum Geologists*, **77**, 1111–1128.

GAWTHORPE, R. L. & HURST, J. M. 1993. Transfer zones in extensional basins: their structural style and influence on drainage development and stratigraphy. *Journal of the Geological Society, London*, **150**, 1137–1152.

——, —— & SLADEN, C. P. 1980. Evolution of Miocene footwall-derived coarse-grained deltas, Gulf of Suez, Egypt: implications for exploration. *American Association of Petroleum Geologists Bulletin*, **74**, 1077–1086.

GECO 1989. *Tectonic Map of the North Sea and adjacent onshore areas*. Scale 1 : 1 000 000.

GEORGE, G. T. & BERRY, J. K. 1993. A new lithostratigraphy and depositional model for the Upper Rotliegend of the UK Sector of the Southern North Sea. *In*: NORTH, C. P. & PROSSER, D. J. (eds) *Characterization of Fluvial and Aeolian Reservoirs*. Geological Society, London, Special Publications, **73**, 295–323.

——1994. A new palaeogeographic and depositional model for the Upper Rotliegend, offshore The Netherlands. *First Break*, **12**, 147–158.

GIBBS, A. D. 1984. Structural evolution of extensional basin margins. *Journal of the Geological Society, London*, **141**, 609–620.

——1986. Strike-slip basins and inversion: a possible model for the Southern North Sea gas area. *In*: BROOKS, J., GOFF, J. C. & VAN HOORN, B. (eds) *Habitat of Palaeozoic Gas in NW Europe*. Geological Society, London, Special Publications, **23**, 23–35.

GLENNIE, K. W. 1972. Permian Rotliegendes of northwest Europe interpreted in light of modern desert sedimentation studies. *American Association of Petroleum Geologists Bulletin*, **56**, 1048–1071.

——1983. Early Permian (Rotliegendes) palaeowinds of the North Sea. *Sedimentary Geology*, **34**, 245–265.

——1986. Development of NW Europe's Southern gas basin. *In*: BROOKS, J., GOFF, J. & VAN HOORN, B. (eds) *Habitat of Palaeozoic Gas in NW Europe*. Geological Society, London, Special Publications, **23**, 3–22.

——1990. Rotliegend sediment distribution: a result of late Carboniferous movements. *In*: HARDMAN, R. F. P. & BROOKS, J. (eds) *Tectonic Events Responsible for Britain's Oil and Gas Reserves*. Geological Society, London, Special Publications, **55**, 127–138.

——1994. Quaternary dunes of SE Arabia and Permian (Rotliegend) dunes of NW Europe: Some comparisons. *Zentralblatt fur Geologie und Palantologie*, **1**, 1199–1215.

——1997. Recent advances in understanding the southern North Sea Basin: a summary. *This volume*.

—— & BOEGNER, P. 1981. Sole Pit inversion tectonics. *In*: ILLING, L. V. & HOBSON, G. D. (eds) *The Petroleum Geology of the Continental Shelf of NW Europe*. Heyden, 110–210.

—— & BULLER, A. T. 1983. The Permian Weissliegend of NW Europe: the partial deformation of aeolian dune sands caused by the Zechstein transgression. *Sedimentary Geology*, **35**, 43–81.

GLOPPEN, T. G. & STEEL, R. J. 1981. The deposits, internal structure and geometry of six alluvial fan–fan delta bodies (Devonian, Norway) – a study in the significance of bedding sequence in conglomerates. *In*: ETHERIDGE, F. G. (ed.) *Nonmarine deposits – Models for exploration*. Society of Economic Paleontologists and Mineralogists, Special Publications, **31**, 49–69.

GOODCHILD, M. W. & BRYANT, P. 1986. The geology of the Rough field. *In*: BROOKS, J., GOFF, J. & VAN HOORN, B. (eds) *Habitat of Palaeozoic Gas in NW Europe*. Geological Society, London, Special Publications, **23**, 223–235.

HAQ, B. U., HARDENBOL, J. & VAIL, P. R. 1987. Chronology of fluctuating sea levels since the Triassic (250 million years ago to present). *Science*, **235**, 1156–1167.

HARDING, T. P., VIERBUCHEN, R. C. & CHRISTIE-BLICK, N. 1985. Structural styles, plate-tectonic settings, and hydrocarbon traps of divergent (transtensional) wrench faults. *In*: BIDDLE, K. T. & CHRISTIE-BLICK, N. (eds) *Strike-slip Deformation, Basin Formation and Sedimentation*. Society of Economic Palaeontologists and Mineralogists, Special Publications, **37**, 51–77.

HECKEL, P. H. 1990. Evidence for global (glacial-eustatic) control over Upper Carboniferous (Pennsylvanian) cyclothems in midcontinent North America. *In*: HARDMAN, R. F. P. & BROOKS, J. (eds) *Tectonic Events Responsible for Britain's Oil and Gas Reserves*. Geological Society, London, Special Publications, **55**, 35–47.

HEINRICH, R. D. 1991*a*. The Cleeton Field, Block 42/29, UK North Sea. *In*: ABBOTS, I. L. (ed.) *United Kingdom Oil and Gas Fields, 25 Years Commemorative Volume*. Geological Society, London, Memoirs, **14**, 409–415.

HEINRICH, R. D. 1991*b*. Ravenspurn South Field, Blocks 42/49, 42/30, 43/26, UK North Sea. *In*: ABBOTS, I. L. (ed.) *United Kingdom Oil and Gas Fields, 25 Years Commemorative Volume*. Geological Society, London, Memoirs, **14**, 469–475.

HELMUTH, H.-J. & SUSSMUTH, S. 1993. Die lithostratigraphische Gliederung des jungeren Oberrotliegenden (Oberrotliegendes 11) in Nordostdeutschland. *Geologisches Jahrbuch*, **A131**, 31–55.

HERNGREEN, G. F. W., SMITH, R. & WONG, TH. E. 1991. Stratigraphy and tectonics of the Vlieland basin, The Netherlands. *In*: SPENCER, A. M. (ed.) *Generation, accumulation and production of Europe's hydrocarbons*. Special Publications of the European Association of petroleum Geoscientists, **1**, 175–192.

HEWARD, A. P. 1991. Inside Auk – the anatomy of an aeolian oil reservoir. *In*: MIALL, A. D. & TYLER, N. (eds) *The three-dimensional facies architecture of terrigenous clastic sediments and its implications for hydrocarbon discovery and recovery*. SEPM Concepts in Sedimentology and Palaeontology, **3**, 44–56.

HILLIER, A. P. & WILLIAMS, B. P. J. 1991. The Leman Field, Blocks 49/26, 49/27, 49/28, 53/1, 53/2, UK North Sea. *In*: ABBOTS, I. L. (ed.) *United Kingdom Oil and Gas Fields, 25 Years Commemorative Volume*. Geological Society, London, Memoirs, **14**, 451–458.

HOBSON, G. D. & HILLIER, A. P. 1991. The Sean North and South Fields, Block 49/25a, UK North Sea. *In*: ABBOTS, I. L. (ed.) *United Kingdom Oil and Gas Fields, 25 Years Commemorative Volume*. Geological Society, London, Memoirs, **14**, 485–490.

HOLMES, A. J. 1991. The Camelot Fields, Blocks 53/1a, 53/2, UK North Sea. *In*: ABBOTS, I. L. (ed.) *United Kingdom Oil and Gas Fields, 25 Years Commemorative Volume*. Geological Society, London, Memoirs, **14**, 401–408.

KETTER, F. J. 1991. The Ravenspurn North Field, Blocks 42/30, 33/26a, UK North Sea. *In*: ABBOTS, I. L. (ed.) *United Kingdom Oil and Gas Fields, 25 Years Commemorative Volume*. Geological Society, London, Memoirs, **14**, 495–467.

LAMBERT, R. A. 1991. The Victor Field, Blocks 49/17, 49/22, UK North Sea. *In*: ABBOTS, I. L. (ed.) *United Kingdom Oil and Gas Fields, 25 Years Commemorative Volume*. Geological Society, London, Memoirs, **14**, 503–508.

LANCASTER, N. 1983. Controls of dune morphology in Namib sand sea. *In*: BROOKFIELD M. E. & AHLBRANDT, T. S. *Eolian Sediments and Processes*. SEPM Developments in Sedimentology, **38**, 261–289.

——1990. Palaeoclimatic evidence from sand seas. *Palaeogeography, Palaeoclimatology, Palaeoecology*, **76**, 279–290.

LARSEN, P.-H. 1988. Relay structures in a Lower Permian basement-involved extension system, East Greelland. *Journal of Structural Geology*, **10**, 3–8.

LEEDER, M. R. 1988. Recent developments in Carboniferous geology: a critical review with implications for the British Isles and N.W. Europe. *Proceedings of the Geologists' Association*, **99**, 73–100.

——1993. Tectonic controls upon basin development, river channel migration and alluvial architecture: implications for hydrocarbon reservoir development and characterization. *In*: NORTH, C. P. & PROSSER, D. J. (eds) *Characterization of Fluvial and Aeolian Reservoirs*. Geological Society, London, Special Publications, **73**, 7–22.

—— & GAWTHORPE, R. L. 1987. Sedimentary models for extensional tilt-block/half-graben basins. *In*: COWARD, M. P., DEWEY, J. F. & HANCOCK, P. L. (eds) *Continental Extensional Tectonics*. Geological Society, London, Special Publications, **28**, 139–152.

—— & HARDMAN, M. 1990. Carboniferous geology of the Southern North Sea Basin and controls on hydrocarbon prospectivity. *In*: HARDMAN, R. F. P. & BROOKS, J. (eds) *Tectonic Events Responsible for Britain's Oil and Gas Reserves*. Geological Society, London, Special Publications, **55**, 87–105.

—— & JACKSON, J. A. 1993. The interaction between normal faulting and drainage in active extensional basins, with examples from the western United States and central Greece. *Basin Research*, **5**, 79–102.

LINK, M. H. 1982. Provenance, paleocurrents and paleogeography of Ridge Basin, southern California. *In*: CROWELL, J. C. & LINK, M. H. (eds) *Geologic History of Ridge Basin, Southern California*. Pacific Section, Society of Economic Paleontologists and Mineralogists, 265–276.

——, ROBERTS, M. T. & NEWTON, M. S. 1985. Walker Lake Basin, Nevada: An example of late Tertiary (?) to Recent sedimentation in a basin adjacent to an active strike-slip fault. *In*: BIDDLE, K. T. & CHRISTIE-BLICK, N. (eds) *Strike-slip deformation, basin formation, and sedimentation*. Society of Economic Paleontologists and Mineralogists Special Publications, **37**, 105–125.

MCLAUGHLIN, R. J. & NILSEN, T. H. 1982. Neogene non-marine sedimentation and tectonics in small pull-apart basins of the San Andreas fault system, Sonoma County, California. *Sedimentology*, **29**, 865–876.

MASKALL, R., REID, I. & URQUHART, M. 1994. Core handling, sedimentology, and petrophysical log response of a horizontal core, Barque Field, United Kingdom Sector, Southern North Sea. *American Association of Petroleum Geologists Bulletin*. **78**, 432–444.

MIALL, A. D. 1984. *Principles of Sedimentary Basin Analysis*. Springer-Verlag, New York.

MITCHELL, A. H. G. & READING, H. G. 1989. Sedimentation and tectonics. *In*: READING, H. G. (ed.) *Sedimentary Environments and Facies*. Blackwell Scientific, Oxford, 471–591.

MORLEY, C. K., NELSON, R. A., PATTON, T. L. & MUNN, S. G. 1990. Transfer zones in the East African rift system and their relevance to hydrocarbon exploration in rifts. *American Association of Petroleum Geologists Bulletin*, **74**, 1234–1253.

NIELSON, J. & KOCUREK, G. 1987. Surface processes, deposits, and development of star dunes: Dumont field, California. *Geological Society of America Bulletin*, **99**, 177–186.

NILSEN, T. H. & MCLAUGHLIN, R. J. 1985. Comparison of tectonic framework and depositional patterns of the Hornelen strike-slip basin of Norway and the Ridge and Little Sulphur Creek strike-slip basins of California. *In*: BIDDLE, K. T. & CHRISTIE-BLICK, N. *Strike-slip Deformation, Basin Formation and Sedimentation*. Society of Economic Paleontologists and Mineralogists, Special Publications, **37**, 79–103.

OUDMAYER, B. C. & JAGER, DE J. 1993. Fault reactivation and oblique-slip in the Southern North Sea. *In*: PARKER, J. R. (ed.) *Petroleum Geology of Northwest Europe: Proceedings of the 4th Conference*. Geological Society, London, 1281–1290.

PEACOCK, D. C. P. & SANDERSON, D. J. 1994. Geometry and development of relay ramps in normal fault systems. *American Association of Petroleum Geologists*, **78**, 147–165.

PEARSON, J. F. S., YOUNGS, R. A. & SMITH, A. 1991. The Indefatigable Field, Blocks 49/18, 49/19, 49/23, 49/24, UK North Sea. *In*: ABBOTS, I. L. (ed.) *United Kingdom Oil and Gas Fields, 25 Years Commemorative Volume*. Geological Society, London, Memoirs, **14**, 443–450.

PRITCHARD, M. J. 1991. The V-Fields, Blocks 49/16, 49/21, 48/40a, 48/25b, UK North Sea. *In*: ABBOTS, I. L. (ed.) *United Kingdom Oil and Gas Fields, 25 Years Commemorative Volume*. Geological Society, London, Memoirs, **14**, 497–502.

READING, H. G. 1980. Characteristics and recognition of strike-slip fault systems. *In*: BALLACE, P. F. & READING, H. G. (eds) *Sedimentation in Oblique-slip Mobile Zones*. International Association of Sedimentologists, Special Publications, **4**, 7–26.

ROLLIN, K. E. 1982. *Interpretation of the Main Features of the Humber–Trent 1:250 000 Bouguer Anomaly Map*. Institute of Geological Sciences, Applied Geophysics Unit, Report No. **74**.

SARNTHEIN, M. 1978. Sand deserts during glacial maximum and climatic optimum. *Nature*, **272**, 43–46.

SCHWAB, K. 1987. Compression and right-lateral strike-slip movement at the Southern Hunsruck Borderfault (Southwest Germany). *Tectonophysics*, **137**, 115–126.

SEILACHER, A. 1984. Sedimentary structures tentatively attributed to seismic events. *Marine Geology*, **55**, 1–12.

SONNENFELD, P. 1984. *Brines and Evaporites*. Academic Press Inc.

STEEL, R. J. 1988. Coarsening-upward and skewed strike-slip transfer fault movement in sedimentary basins. *In*: NEMEC, W. & STEEL, R. J. (eds) *Fan Deltas: sedimentology and tectonic settings*. Blackie, 75–83.

—— & GLOPPEN, T. G. 1980. Late Caledonian (Devonian) basin formation, western Norway: signs of strike-slip tectonics during infilling. *In*: BALLANCE, P. F. & READING, H. G. (eds) *Sedimentation in Oblique-slip Mobile Zones*. International Association of Sedimentologists, Special Publications, **4**, 79–103.

STEELE, R. P. ALLAN, R. M. ALLINSON, G. J. & BOOTH, A. J. 1993. Hyde: a proposed field development in the Southern North Sea using horizontal wells. *In*: PARKER, J. R. (ed.) *Petroleum Geology of Northwest Europe: Proceedings of the 4th Conference*. The Geological Society, London, 1465–1472.

STUART, I. A. 1991. The Rough Gas Storage Field, Blocks 47/3d, 47/8b, UK North Sea. *In*: ABBOTS, I. L. (ed.) *United Kingdom Oil and Gas Fields, 25 Years Commemorative Volume*. Geological Society, London, Memoirs, **14**, 477–484.

TRUDGILL, B. & CARTWRIGHT, J. 1994. Relay-ramp forms and normal-fault linkages, Canyonlands National Park, Utah. *Geological Society of America Bulletin*, **106**, 1143–1157.

TURNER, P., JONES, M., PROSSER, D. J., WILLIAMS, G. D. & SEARL, A. 1993. Structural and sedimentological controls on diagenesis in the Ravenspurn North gas reservoir, UK Southern North Sea. *In*:

PARKER, J. R. (ed.) *Petroleum Geology of Northwest Europe: Proceedings of the 4th Conference.* Geological Society, London, 771–785.

VAIL, P. R., MITCHUM, R. M. & THOMPSON, S. 1977. Seismic stratigraphy and global changes of sea level. *In*: PAYTON, C. E. (ed.) *Seismic stratigraphy – applications to hydrocarbon exploration.* American Association of Petroleum Geologists Memoirs, **26**, 83–98.

VAN HOORN, B. 1987. Structural evolution, timing and tectonic style of the Sole Pit inversion. *Tectonophysics*, **137**, 239–284.

VAN WIJHE, D. H. 1987*a*. The structural evolution of the Broad Fourteens Basin. *In*: BROOKS, J. & GLENNIE, K. W. (eds) *Petroleum Geology of North West Europe.* Graham & Trotman, 315–335.

——1987*b*. Structural evolution of inverted basins in the Dutch offshore. *Tectonophysics*, **137**, 171–219.

WASSON, R. J. & HYDE, R. 1983. Factors determining desert dune type. *Nature*, **304**, 337–339.

WERNGREN, O. C. 1991. The Thames, Yare and Bure Fields, Block 49/28, UK, North Sea. *In*: ABBOTS, I. L. (ed.) *United Kingdom Oil and Gas Fields, 25 Years Commemorative Volume.* Geological Society, London, Memoirs, **14**, 491–496.

WINTER, D. A. & KING, B. 1991. The West Sole Field, Block 48/6, UK North Sea. *In*: ABBOTS, I. L. (ed.) *United Kingdom Oil and Gas Fields, 25 Years Commemorative Volume.* Geological Society, London, Memoirs, **14**, 517–523.

YANG, C.-S. & NIO, S.-D. 1994. Application of high-resolution sequences stratigraphy to the Upper Rotliegend in the Netherlands offshore. *In*: WEIMER, P. & POSAMENTIER, H. W. (eds) *Siliciclastic Sequence Stratigraphy: recent developments and applications.* American Association of Petroleum Geologists Memoirs, **58**, 285–316.

ZIEGLER, P. A. 1982. *Geological Atlas of Western and Central Europe.* Shell International Petroleum Maatschappij B.V.

——1990. *Geological Atlas of Western and Central Europe.* Shell Internationale Petroleum Maatschappij B.V.

# Climatic cyclicity and accommodation space in arid to semi-arid depositional systems: an example from the Rotliegend Group of the UK southern North Sea

JOHN HOWELL[1,2] & NIGEL MOUNTNEY[1]

[1] STRAT Group, Department of Earth Sciences, University of Liverpool, Brownlow Street, Liverpool L69 3BX, UK
[2] Formerly at: School of Earth Sciences, University of Birmingham, Edgbaston, Birmingham B15 2TT, UK

**Abstract:** The arid to semi-arid, continental deposits of the Permian Rotliegend Group of the UK southern North Sea (SNS) provide an ideal opportunity to test and further the concepts of sequence stratigraphy in fully non-marine depositional systems. Although these depositional systems are detached from the effects of sea-level change, cyclic fluctuations in the climatic regime result in variations in the vertical facies successions laid down. The effects of increasing or decreasing aridity upon depositional processes are considered. Climatic cycles, bounded at the point of lowest aridity are defined. These climatic cycles have been correlated between different, co-existent depositional environments. This enables depositional systems, such as those present in the Rotliegend, to be considered chrono-stratigraphically. The application of these concepts to an extensive data set illustrates that, whilst the internal architectures of sedimentary cycles are controlled by the depositional processes and their positions within the basin, the thickness of the cycles directly reflects the rate of accommodation creation.

Lithostratigraphic correlations, which are time-independent, provide limited information regarding depositional palaeogeographies and facies distributions. This study proposes an alternative time stratigraphic interpretation of the predominantly continental deposits of the Rotliegend Group of the SNS. Such an approach is of great importance to the future petroleum geology potential of the SNS. By understanding how depositional systems evolve over time in response to climate change, and by having the ability to predict facies transitions between individual parts of a basin, valuable insights may be gained into the geometry and nature of subtle, yet important, stratigraphic hydrocarbon plays. As the major gas fields of the SNS approach maturity, it is of vital importance to develop a high resolution understanding of the chronostratigraphic behaviour of the basin in order to develop a predictive model of where potential future reserves are likely to be found. The key to improving our current understanding of the basin is through detailed stratigraphic analysis in order to develop improved palaeo-geographic reconstructions.

Sequence stratigraphy has been successfully applied to marine and marginal marine rocks and is based upon the recognition and correlation of key surfaces caused by changes in relative sea level (Posamentier & Vail 1988; Van Wagoner *et al.* 1990). The occurrence of these surfaces, which include sub-aerial unconformities and marine flooding horizons, is independent of factors such as sediment supply and are consequently correlatable in both strike and dip directions. Away from the marine realm, typically above the first nick point in the fluvial equilibrium profile, other controls upon depositional processes progressively overprint the effects of sea-level change (Shanley & McCabe 1994). Within a dominantly continental, inter-montane basin, such as the Southern Permian Basin of NW Europe, changes in relative sea-level cannot be recognized or used as a basis for correlation. However, such deposits may still be considered within a time stratigraphic framework by taking other aspects of the sequence stratigraphic model. These aspects include; (a) the systematic correlation of changes in depositional processes caused by fluctuations in extra-basinal parameters, (b) the recognition of time stratigraphic surfaces such as supersurfaces (Kocurek 1988; Havholm & Kocurek 1994), and (c) the concept of the rate accommodation space creation versus the rate of sediment supply as a control on preserved stratigraphic architecture.

The data for this study have been collected from core and wireline logs in 55 wells, together with additional wireline log and tops data from a further 188 released wells (Fig. 1). Using the facies scheme of George & Berry (1993), bed-to-bed correlations have been made for closely

*From* Ziegler, K., Turner, P. & Daines, S. R. (eds), 1997, *Petroleum Geology of the Southern North Sea: Future Potential*, Geological Society Special Publication No. 123, pp. 63–86.

J. HOWELL & N. MOUNTNEY

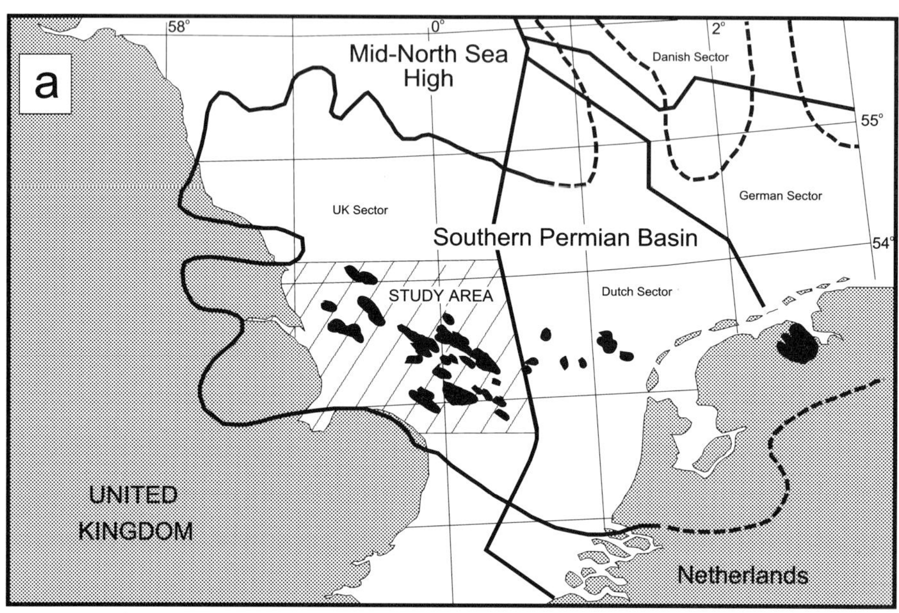


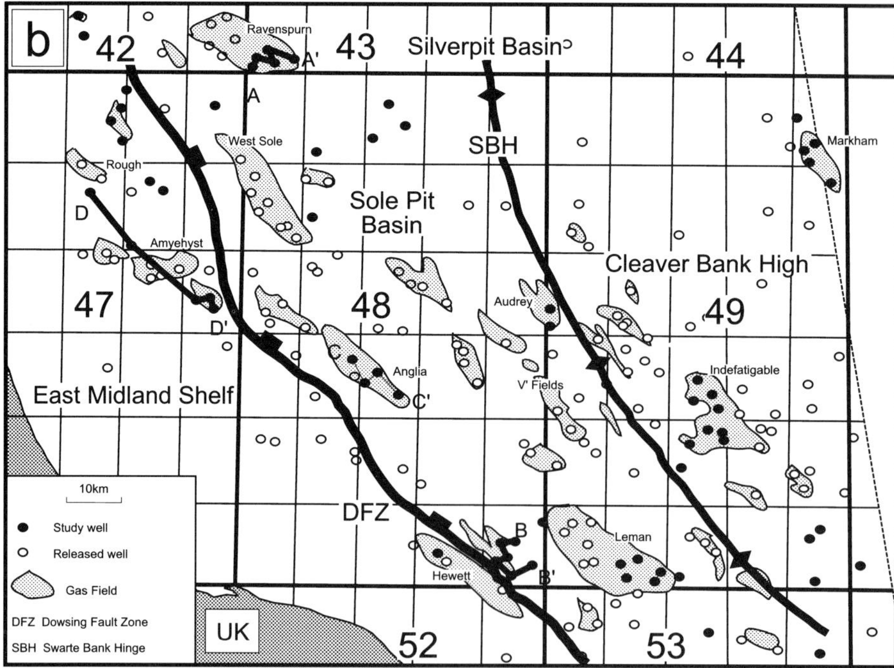


**Fig. 1.** (**a**) Location of study area within the Southern Permian Basin. Principal gas fields indicated. (**b**) Distribution of study wells, released wells, principal gas fields and major structural features within the study area. Location of cross sections in Figs 3–6 indicated: A–A′ Ravenspurn section; B–B′ Hewett section; C–C′ Anglia section and, D–D′ Amethyst section.

spaced wells in nine fields or sub-areas within the basin, four of which are presented in detail within this study (transects A–D in Fig. 1). Using these correlations, changes in depositional processes have been interpreted to result from cyclic fluctuations in the palaeo-climatic conditions within the basin. Up to 12 climatically controlled depositional cycles bounded by the inferred point of minimum aridity have been identified and occur with a variety of drying- and drying/wetting-upward motifs. These cycles represent allostratigraphic depositional units which can be used to trace the history of the Upper Rotliegend within the study area of the Southern Permian Basin. The varied expression of each depositional cycle within different parts of the basin is considered as a balance between subsidence (accommodation creation) and sediment supply within the different depositional settings. The concepts of climatic stratigraphy are discussed within the broader context of sequence stratigraphy.

## Background

The Permian Rotliegend Group provides an ideal succession to test and advance the concepts of time-based stratigraphic correlations within depositional settings de-coupled from changes in sea level (cf. Howell 1992; George & Berry 1993; Kocurek & Havholm 1994; Havholm & Kocurek 1994; Yang & Nio 1994). During the Early Permian the UK SNS was an inter-montane basin, structurally bounded to the south and west by stable platform areas of the East Midland Shelf and to the north by the Mid North Sea High (Fig. 1). The eastern boundary to the study area is the division between the British and Dutch sectors of the North Sea and is political rather than physical. The Permian basin fill unconformably overlies an uplifted Carboniferous sub-crop (Glennie 1986) and comprises predominantly continental deposits of the Upper Rotliegend Group and carbonate/evaporite deposits of the Zechstein Supergroup (Fig. 2). Lithostratigraphically, the Upper Rotliegend comprises the Leman Sandstone Formation and the Silverpit Claystone Formation. The former is predominantly fluvial and aeolian sandstones whilst the latter is comprised of a mixture of playa lake mudstones and siltstones, with discrete halite horizons (Fig. 2) (Glennie 1986). The diachroneity of the two formations (Fig. 2b) means that although descriptively useful, this lithostratigraphic sub-division is of little value for correlation purposes.

In 1965, the first wells were drilled in the SNS, following the discovery of gas within the Rotliegend sandstones in the Groningen field (onshore Netherlands) in 1959. It was rapidly recognized that these deposits were laid down within an inter-montane desert, in a series of concentric facies belts passing from marginal fluvial and alluvial fan systems, through erg and erg margin deposits to a playa lake within the centre of the basin (Glennie 1972; Blanche 1973; Rhys 1975; Marie 1975).

The earliest discoveries within the area (West Sole, Leman, Indefatigable Fields) were confined to the erg centre and its immediate margin. Subsequently, progressively more subtle stratigraphic plays have been discovered on the basin margins and more recently in the zone of interdigitation between the reservoir formation (Leman Sandstone) and the non-reservoir formation (Silverpit). The area is now a mature hydrocarbon province with an extensive well database (Fig. 1).

*Previous work*

The earliest studies on the Rotliegend Group (Glennie 1972) recognized both the basic facies distribution and the diachroneity which exists between the formations (Fig. 2) (Glennie 1986). This facies distribution was refined slightly by Marie (1975). Van Veen (1975) proposed a basic three-fold division of the Leman Sandstone into a lower (predominantly fluvial) unit overlain by middle and upper (predominantly aeolian) units. This division was used subsequently by a number of workers at a field scale (e.g. Hillier & Williams 1991) although it cannot be extended outside the central part of the Sole Pit Basin.

Martin & Evans (1988) identified and correlated 'interdune sabkha' deposits, related to water-table rise, over considerable distances within the 'V' Fields which represented the earliest published autocyclic correlations within the Rotliegend. With the exception of these studies, the majority of work undertaken on the Rotliegend during the late seventies and eighties was concentrated upon diagenesis (e.g. Nagtegaal 1979; Rossel 1982; Lee *et al.* 1989; Purvis 1989) and structure (e.g. Glennie & Boegner 1981; Bulat & Stoker 1987). Glennie (1983) described the aeolian dune morphology as straight-crested to barchanoid with a westerly transport direction, a factor disputed by Steele (1985) and Sneh (1988), neither of whom discussed viable alternatives. Little was done to improve the understanding of the sedimentology and stratigraphy until George & Berry (1993)

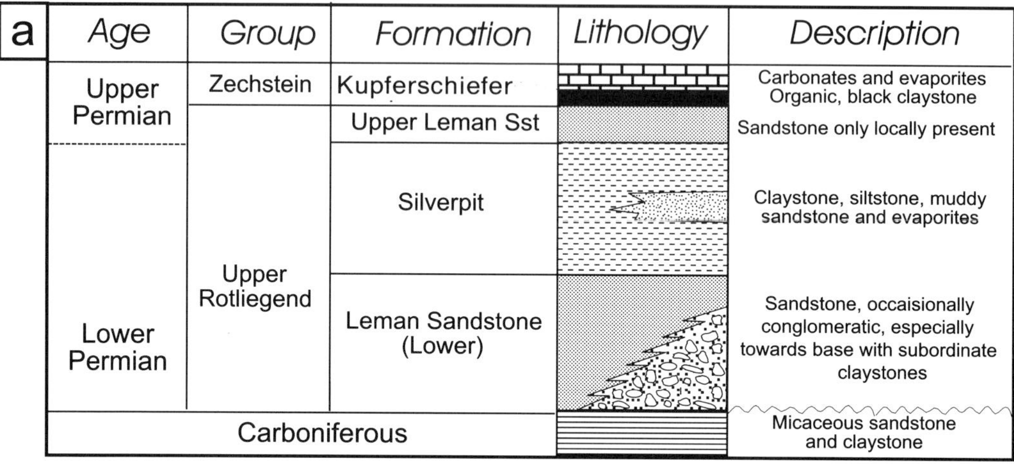



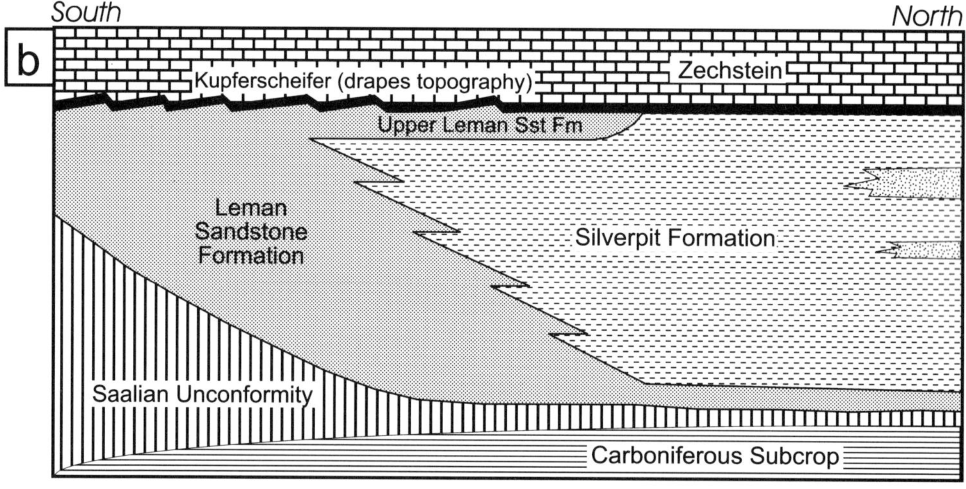



**Fig. 2.** (**a**) Lithostratigraphy of the Upper Rotliegend Group in the UK Southern North Sea. (**b**) Relationship between the formations, note the diachroneity of the Leman and Silverpit Formations and the limited distribution of the Upper Leman Sand Formation (modified after Glennie 1986).

published a new lithostratigraphic subdivision based upon the concepts of correlating drying-upward depositional cycles.

Within other aeolian and semi-arid depositional systems, the importance of time stratigraphic surfaces was recognized with the identification of climatically driven bounding surfaces (Stokes 1968; Talbot 1985). These bounding surfaces are of stratigraphic significance because they formed independently of bedform migration (Hunter 1977; Loope 1984; Kocurek 1988; Fryberger 1993) always separating older from younger strata (Kocurek & Havholm 1994), in a way comparable to

sequence boundaries within marine rocks (Posamentier *et al.* 1988). Across these surfaces, rocks of significantly different ages may be juxtaposed against one another (Kocurek *et al.* 1991). Clemmensen *et al.* (1989) were able to recognize discrete allocyclic packages within an erg margin setting from the Jurassic of Utah. Climatic cyclicity was considered to be the control upon depositional architecture. Most recently, supersurfaces have been used to delineate genetic stratigraphic units within the Page Sandstone in Utah (Havholm & Kocurek 1994). Fluctuations in water table were cited as the driving mechanism responsible for their formation.

George & Berry (1993) applied the concept of climatic cyclicity to an extensive UK Rotliegend data set recognizing a series of five drying-upward cycles. This work represented a major step forward in studies of the Rotliegend, being the first to consider the succession in a time stratigraphic framework, although these authors considered it impossible to correlate these cycles within the erg interior and did not attempt to explain the occurrence of solely drying upward cycles. Work undertaken concurrently (Howell 1992) and discussed in this paper indicates that the Rotliegend Group comprises a maximum of 12 variable drying- and drying/wetting-upward cycles, which can be correlated throughout the basin, including the erg interior where cycle boundaries are expressed as supersurfaces.

The discrepancy in the number of recognized cycles in this study and that of George & Berry (1993) may be partially due to the preservation of several scales of cyclicity. In a spectral analysis study of well logs from the Dutch Rotliegend, Yang & Nio (1994) and Yang & Baumfalk (1994) postulated the presence of twelve 'third-order' sequences contained within five 'super-sequences'.

## Sedimentology of the Rotliegend Group

Fifty-five wells including over 4.5 km of core have been studied from the UK SNS (Fig. 1). The facies from these cores have been described using the scheme developed for the Rotliegend by George & Berry (1993) (Table 1). A full discussion of the identification and classification of these facies can be found in Howell (1992).

The study wells were grouped into fields and informal sub-areas (Fig. 1) within which well spacing was small enough (typically less than 5 km) to correlate the major beds with reasonable confidence. These areas were: the East Midland Shelf (North and South Amethyst Fields); the feather edge of the Silverpit basin (Ravenspurn Field); the Central Sole Pit (Anglia Field); the Southern Sole Pit (Hewett and Leman Fields), the South Cleaver Bank High (Indefatigable Field) and the North Cleaver Bank High (Markham Field). A number of individual wells were also considered from the Northern Sole Pit region (including the Ann and Audrey Fields), but well spacing was too great to accurately correlate at a bed-to-bed scale. Depositional cross sections correlating the principal bedforms were produced, using the criteria documented for semi-arid aeolian bedform

**Table 1.** *Facies associations in the Rotliegend (reprinted from George & Berry 1993 with permission)*

| Class | Association | Facies name | Code |
| --- | --- | --- | --- |
| Lithogenic | Aeolian Dune | Dune top | A1 |
| | | Dune core | A2 |
| | | Dune base | A3 |
| | Aeolian Interdune | Dry interdune/sandsheet | A4 |
| | | Damp interdune | A5 |
| | | Wet interdune | A6 |
| | Proximal Fluvial | Fluvial fan | F1 |
| | | Fluvial channel | F2 |
| | Distal Fluvial | Structured sheetflood | F3 |
| | | Dewatered sheetflood | F4 |
| | | Subaqueous sheetflood | F5 |
| | Sabkha | Lake margin sabkha | S1 |
| | | Inland sabkha | S2 |
| | Lacustrine | Playa lake | L1 |
| | | Desert lake (permanent) | L2 |
| | | Desert lake (evaporative) | L3 |
| Modified lithogenic | Modified (Weissliegend) | Homogenized aeolian | Wa |
| | | Homogenized fluvial | Wf |
| Reworked lithogenic | Reworked | Serir | F· |
| | | Deflation lag | A· |
| | | Fluvially redeposited aeolian | A → F |
| | | Biogenically reworked: | |
| | | Aeolian facies | A*/Wa* |
| | | Fluvial facies | F*/Wf* |

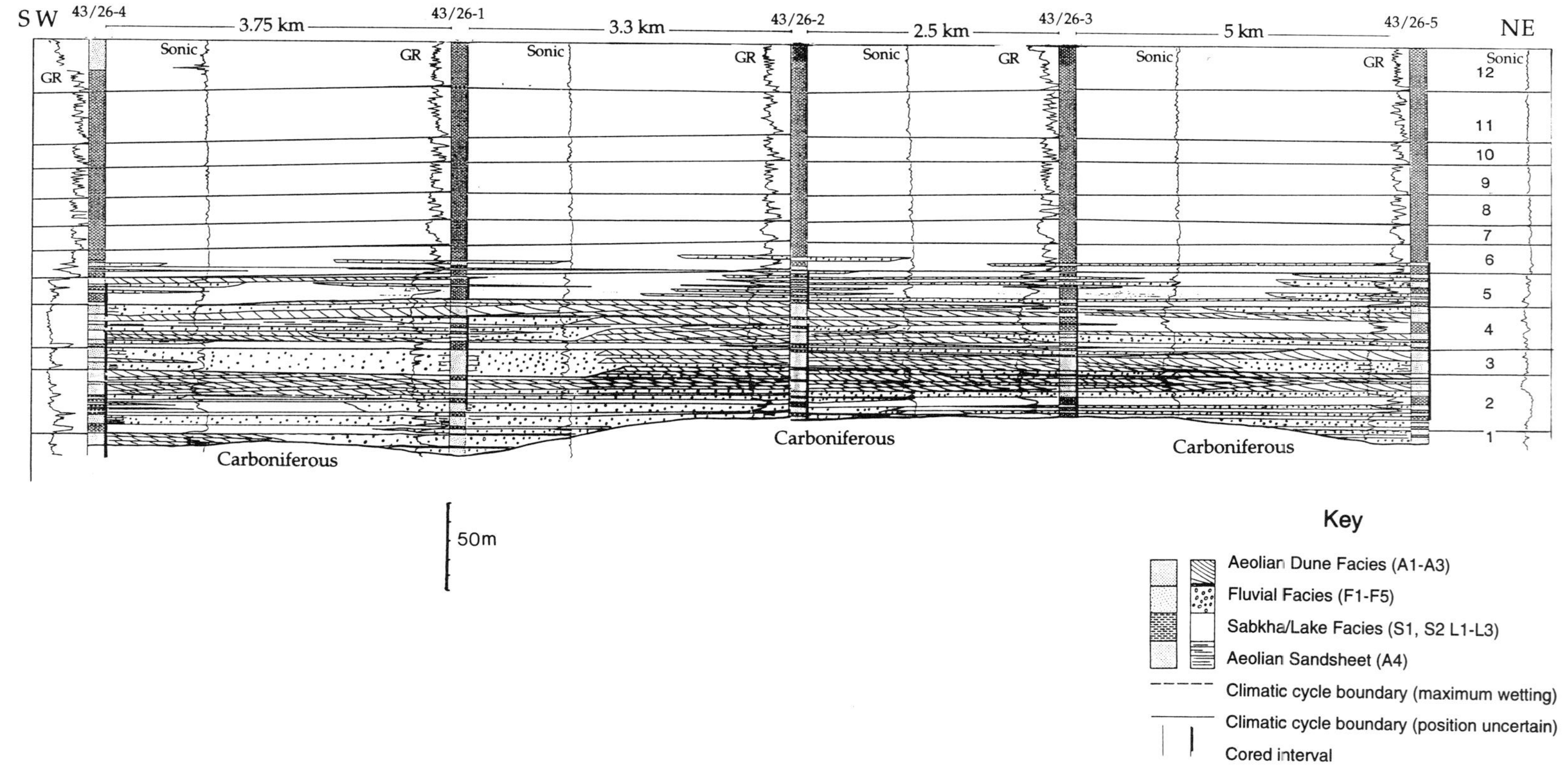


**Fig. 3.** Depositional cross section (A–A′) from the Ravenspurn Field. This section is representative of the desert margin, lake margin, and desert lake. Note the majority of the section is represented by sabkha and lake deposits (Silverpit Formation). Numbers refer to climatic cycles identified from core and wireline studies.

architecture (Hunter 1977; Kocurek 1981, 1988; Rubin & Hunter 1982; Langford & Chan 1989; Herries 1993). Figures 3–6 illustrates a selection of these cross sections from the principal depositional settings. A full discussion of each of these sections is beyond the scope of these paper and the interested reader is referred to Howell (1992). In addition to the basic well-to-well, lithological correlations which have been made, a series of allostratigraphic depositional units have also been identified. These are considered to represent field-wide changes in depositional process and are independent of localised, autocyclic mechanisms. Changes in the Upper Rotliegend climate are considered to be the driving mechanism responsible for the deposition of these units.

## Climatic cyclicity

### Problems associated with correlating arid region continental successions within a time stratigraphic context

The problems of applying sequence stratigraphic concepts to non-marine strata have been reviewed by Walker (1990), Howell (1992) and, more recently, by Shanley & McCabe (1994). They may be summarized as follows:

- the lack of a single overriding control upon base level where systems are completely detached from sea level;
- the interaction of a number of depositional processes (fluvial, aeolian, lacustrine) produces complex facies geometries and juxtaposes facies of different ages;
- continental successions, particularly aeolian ones, contain a far higher frequency of hiatuses and minor disconformities than their marine counterparts;
- both aeolian and ephemeral fluvial depositional processes involve significant erosion as well as deposition, unlike marine systems in which deposition is more or less continuous;
- continental deposits contain a very low frequency of fauna, precluding effective biostratigraphic subdivision;
- many of the surfaces associated with continental bedform migration (particularly aeolian) are highly diachronous;
- continental facies, especially fluvial ones, are frequently laterally very restricted;
- facies tracts within continental depositional systems are less uniform than their marine counterparts in which predictable changes occur within a dip direction and which should also occur in a vertical sense: the absence of such a relationship (i.e. facies tract dislocations) are consequently relatively easy to identify;

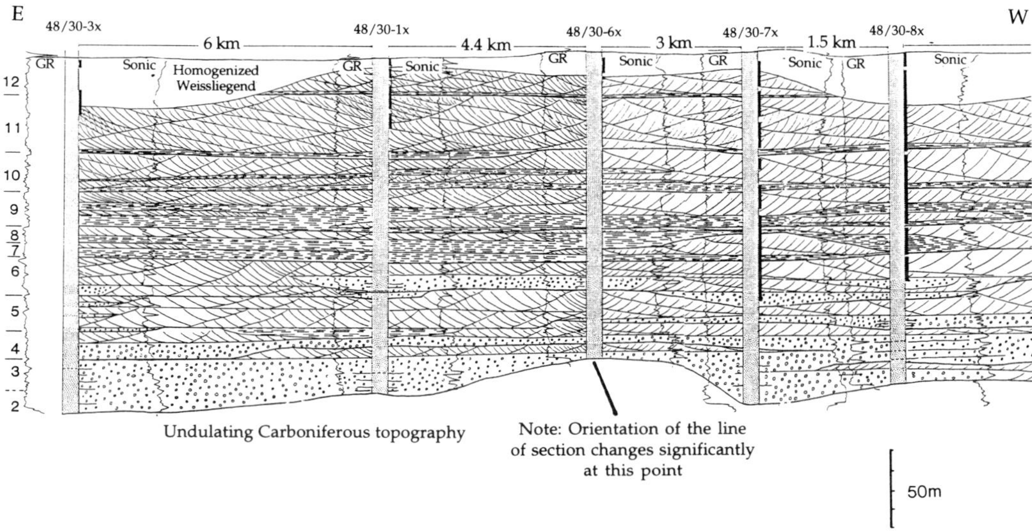



**Fig. 4.** Depositional cross section (B–B′) for the Hewett Field. This section is representative of the erg centre. Note the upward transition from fluvial dominated deposition at the base to aeolian deposition at the top. Note the presence of thick laterally extensive aeolian sandsheets and their inferred relationship to climatic cycles (see Fig. 3 for key).

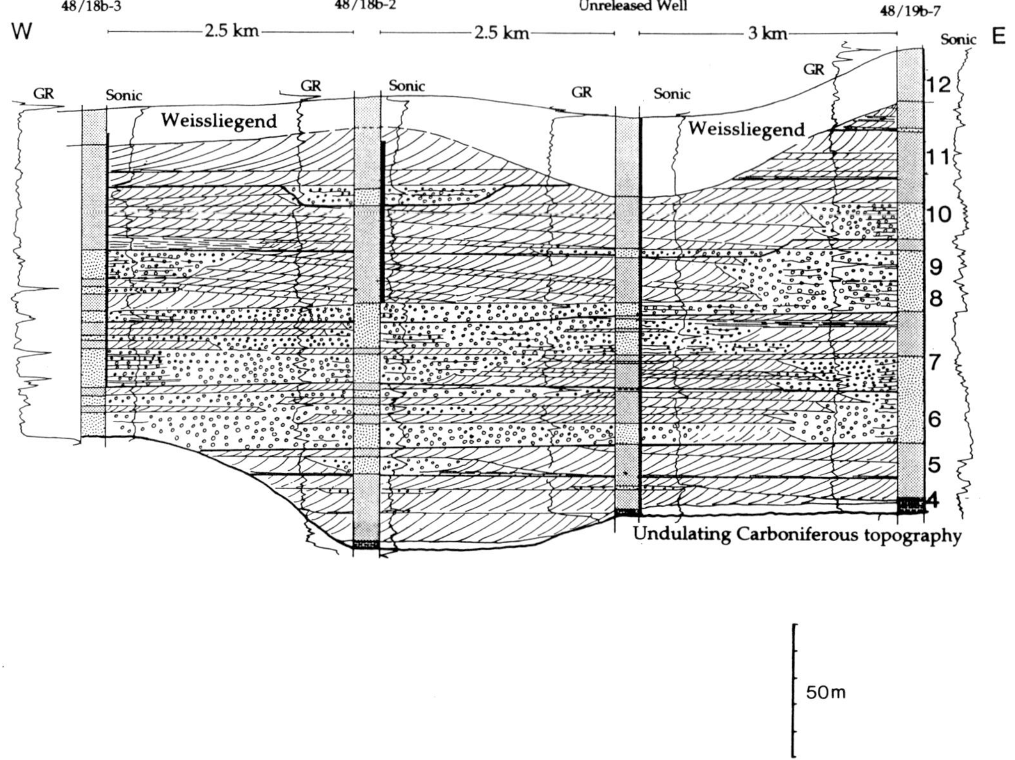



**Fig. 5.** Depositional cross section (C–C′) for the Anglia Field, This section is representative of the margin of the Sole Pit. Note the predominance of thick stacked conglomeratic beds particularly in cycles 6–10 (see Fig. 3 for key).

- arid to semi-arid depositional systems exhibit topography at the time of deposition which will strongly influence concurrent and subsequent facies distribution.

Shanley & McCabe (1994) illustrate that, away from the marine realm, the effects of sea-level change on stratal geometries progressively decreases whilst the effects of other factors increases. It consequently follows that within continental basins, completely detached from the marine realm, the effects of sea-level change should be negligible and climate, sediment supply and accommodation creation will be the principal controls on stratal geometry and architecture.

### Responses of depositional processes to changing climate

Arid and semi-arid depositional systems include a variety of sub-environments, typical of the facies associations identified within this study (Table 1; Figs 3–6). These include, aeolian erg and erg margin; intra-basinal, ephemeral fluvial systems; basin margin alluvial fan systems; desert lake and lake margin sabkha (Glennie 1970, 1986; Ahlbrandt & Fryberger 1982; George & Berry 1993). Each of these depositional environments will respond in a different but predictable manner to a change in the relative aridity of the climate. The first stage in understanding climatic cyclicity is to consider the changes which will occur in each of these depositional settings if the climate becomes more or less arid (Fig. 7).

### Desert lake

Many arid depositional systems, including those active during the deposition of the Rotliegend Group, include an intra-basinal, hydrologically closed lacustrine system (Fig. 3). Such lakes are typically fed by ephemeral streams and ground water (Eugster & Hardie 1975). As much of the water is drawn into the lake by capillary action, the input of coarse (sand grade) clastic material is very low. The bulk of clastic sedimentation is

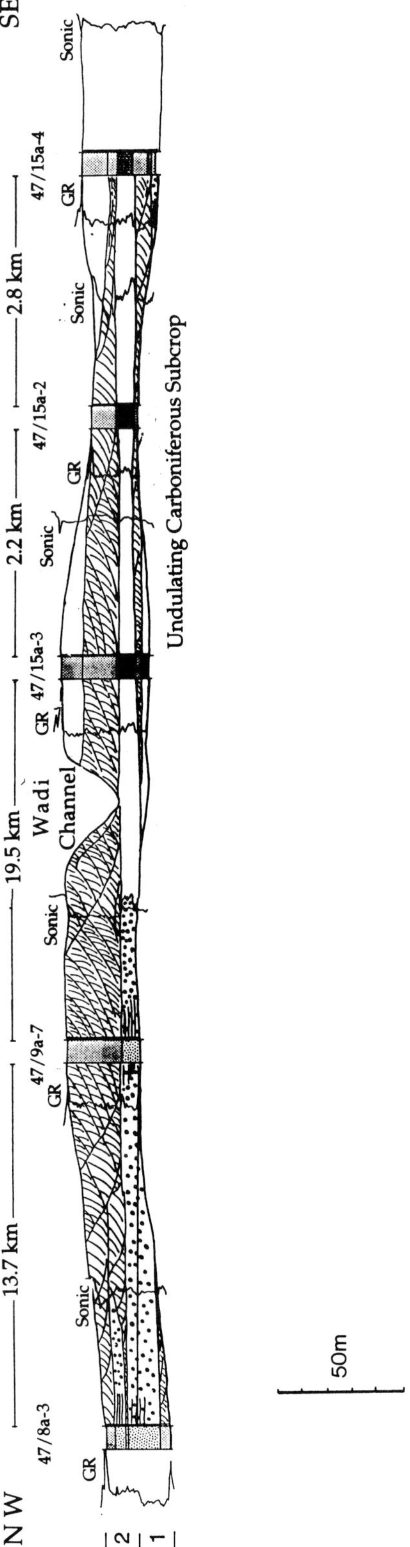


**Fig. 6.** Depositional cross section (D–D′) for the Amethyst Field. This section is representative of deposition within the basin. Note this very thin interval represents only the latest stages of deposition at the basin margin (see Fig. 3 for key).

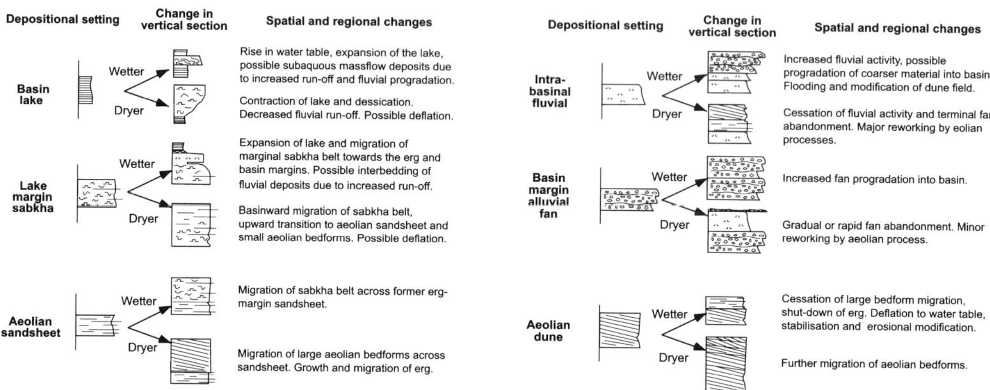


**Fig. 7.** Idealized changes in depositional processes within the principal sub-environments due to increased or decreased aridity.

wind blown loess and clay and silt carried in suspension by ephemeral fluvial systems (Fig. 3). With the exception of coarse grained, poorly sorted deposits within the Rough Field, which may possibly represent lacustrine fan-delta deposits, no evidence was observed for lacustrine delta systems within the UK sector of the Rotliegend. The mudstones may be interbedded with evaporites typically precipitated during more arid periods. The Silverpit Lake (Fig. 3), in common with many similar basin lakes is considered to have been shallow and bounded by low lying sabkha plains (Glennie 1970; George & Berry 1993).

Increased climatic aridity results in a decrease in the size of the lake, decreased fluvial input and a migration of the facies belts. This facies tract migration is represented vertically by a transition from red, planar laminated lake mudstones to chaotically bedded lake-margin sabkha facies with adhesion structures and halokinetic collapse features (Glennie 1970). Evaporite precipitation (particularly halite) occurs locally within the UK sector and becomes regionally extensive within the Dutch and German sectors (George & Berry 1994).

A decrease in the relative aridity will result in a rise in the water table, an expansion of the lake and a flooding of the marginal sabkha belt. Increased run-off from hinterland and intra-basinal fluvial systems may result in increased sediment supply (Fig. 7) and a decrease in desert lake salinity.

## Lake margin sabkha

The principal depositional processes within the lake margin sabkha are the adhesion of wind blown sand and mud to a damp substrate, resulting in a variety of adhesion structures (Nagtegaal 1973; Olsen *et al.* 1989) and periodic sheetflooding. Precipitation and dissolution of salts within the sabkha results in disruption of the original sedimentary fabric (Glennie 1970). An increase in the relative aridity will result in a contraction of the playa lake and a basinward migration of the sabkha belt. In vertical section, this is represented by a transition to sandier facies, such as an extra-dunal sandsheet, or, assuming sufficient sediment supply, aeolian bedform migration across the former sabkha plain (Fig. 7). This will be represented by a transition from adhesion to wind ripple strata and a reduction in the amount of halokinetic disruption. A decrease in the aridity is associated with a water table and lake level rise. Flooding of the sabkha will be extensive and relatively rapid because it is low lying and flat. Prior to flooding, the sabkha plain may experience increased fluvial activity associated with increased run-off from the hinterland and within the basin (Fig. 7). However, it seems likely that a time lag will exist between the water table rise and the increased supply of sediment via fluvial systems.

## Erg margin aeolian sandsheet

Between the lake margin sabkha and the main erg typically lies a transitional, flat-lying erg-margin sandsheet. In this environment wind blown sand is deposited in low-angle, planar sheets with wind rippled bedforms (Breed & Grow 1979; Alhbrandt & Fryberger 1982). The height of the water table, the availability of sand and the frequency of flooding by ephemeral fluvial systems are key factors controlling the distribution and size of erg margin sandsheets (Kocurek & Nielsen 1986). Examples of regionally extensive aeolian sandsheet deposits may be seen in a number of the erg centre cross sections (Figs 4, 5 & 6). Sandsheet development reflects either a zone of reduced sediment supply where material is trapped upwind in the erg centre, or sediment by-passing where the water table lies close to the surface.

An increase in aridity and resultant fall of the water table in association with an increase in mean wind velocity will promote expansion of the adjacent erg. In vertical section an upward transition from wind-rippled aeolian sandsheet deposits to cross bedded aeolian dune and draa deposits will be observed (Figs 4 & 7). A decrease in aridity will result in a rise in the water table and an upward transition to muddier sabkha facies, with a gradual increase in the proportion of adhesion and halokinetic structures and a decrease in wind-ripple laminations (Fig. 7).

## Erg centre

Much has been written on aeolian bedform scale and migration and the formation of regionally extensive, isochronous bounding surfaces (super bounding surfaces or supersurfaces; see Kocurek 1988; Fryberger 1993; Kocurek & Havholm 1994 for reviews). These are created by a change in basin conditions rather than 'normal' bedform migration (Stokes 1968; Talbot 1985; Kocurek 1988). Examples may be seen in a number of sections (Figs 4 & 6). Only recently have these surfaces been considered in the context of relative climatic conditions and the surrounding, marginal depositional environments (Kocurek et al. 1991; Kocurek & Havholm 1994).

An increase in relative aridity will cause a drop in the water table and, providing there is a sufficient upwind sand supply (Havholm & Kocurek 1994), aeolian bedform migration and the passage of large aeolian bedforms into the area will be promoted. Migration and deposition will continue until the available accumulation space is filled, after which time bypass will occur. Interdune deposits related to bedform migration exhibit a transition from predominantly wet to dry as the accumulation builds above the water table. A supersurface may be formed as the sediment budget changes from positive to negative (Kocurek & Havholm 1994).

A decrease in the relative aridity, associated with a decrease in mean wind velocity and a rise in water table level will result in the aeolian system shutting down. Large bedforms either become dormant or are superficially modified by subsequent lesser winds (cf. Kocurek et al. 1991). As the rising water table meets the depositional surface further deflation is halted and a relatively planar supersurface is formed. As less sediment is moved within the system, continued subsidence may accelerate the rate at which the rising water table meets the depositional surface. Before complete deflation occurs, the low velocity winds, associated with pluvial conditions may plane the tops of older large bedforms and in-fill interdune topography with smaller scale dunes. Increased fluvial sheetflooding, associated with a wetter climate may also assist in generating the planar surfaces (Kocurek & Nielsen 1986) observed in the depositional cross sections (Fig. 4).

## Basin margin alluvial fan systems

In this depositional setting sediment is introduced by alluvial fans entering the basin from the hinterland. These fans contain coarse grained, commonly conglomerate material, deposited by ephemeral flash-floods, confined within broad, preferentially etched flow paths (Jolley et al. 1990) (Fig. 5). These systems are the most difficult in which to observe climatically driven cycles, because the autocyclicity associated with the depositional process and the number of potential controlling variables is greatest.

Following an increase in relative aridity, precipitation within the hinterland will be reduced, and the quantity of surface run-off, and consequently the amount of sediment supplied to the alluvial fan systems will also be reduced. In extreme cases these fan systems may be shut down, abandoned and reworked by aeolian processes. This change in fluvial sedimentation style may either occur suddenly or as a progressive back stepping and gradual fan abandonment. The degree of subsequent aeolian reworking will be dependent upon the proportion of conglomeratic material within the original deposits. Concentration of the coarse component will create a lag and armour the surface against subsequent erosion and deflation (Fig. 7).

A decrease in relative aridity will result in increased run-off, possibly accompanied by incision, and alluvial fan progradation. Successive coarse-grained conglomeratic facies will be deposited in vertical succession (Fig. 7). The allocyclic control on alluvial fan facies are difficult to identify and quantify given the lateral discontinuity of alluvial deposits and the highly autocyclic nature of fan processes. The alluvial fan facies are the most difficult to consider within the context of allocyclicity. Studying several closely spaced wells may identify regionally significant changes and reduce potential errors (Fig. 5). It is also necessary to consider vertical trends rather than single beds.

## Intra-basinal fluvial systems

Fluvial activity within arid continental settings occurs either as permanent large fluvial systems, with an extra-basinal origin or as small scale ephemeral systems. Fluvial deposits are almost exclusively ephemeral within the UK Rotliegend depositional system, although more permanent systems have been described from the Dutch Rotliegend (Almon 1981; George & Berry 1994). Consequently, only the latter (ephemeral systems) will be considered here.

Intra-basinal ephemeral fluvial systems may originate as distal equivalents to the alluvial fan systems or they may have their origins within the basin, either due to intra-basinal precipitation or the emergence of a temporarily raised water table. Intra-basinal derived systems will contain a greater proportion of reworked aeolian grains and mudstone rip up clasts whilst extra-basinally sourced systems will contain a greater proportion of exotic material.

An increase in aridity will cause the cessation of deposition, and may result in a complete aeolian reworking of deposits. Coarser material will be concentrated if the sediment includes an extra-basinal component. A decrease in aridity results in increased fluvial activity. Alluvial fan progradation on the basin margins may force distal alluvial fan facies into the dune field, whilst the periodicity of flood events sourced within the basin will increase with a raised water table. The nature of intra-basinal fluvial deposits may also change as the geometry of the dune field changes through the wet period. Flash flood events may contribute to the formation of aeolian sandsheets (Kocurek & Nielsen 1986; Langford & Chan 1989) and a geometric feedback loop may be established. In this situation early stage fluvial deposits will be confined to interdune corridors and will contain abundant reworked aeolian material. Such deposits are more likely to be structured due to tractional currents and will be interbedded with claystones formed by suspension settling from dammed waters (Glennie 1970).

## Cycle boundaries and recognition

Figure 8 illustrates an idealized complete climatic cycle within the Rotliegend from maximum wetness through maximum aridity and back to maximum wetness. The expression of this cycle will vary between different parts of the basin and includes the changes described previously.

It is logical that the boundary between successive climatic cycles should be located at a maxima, either wet or dry. Yang & Nio (1994) proposed that the points picked represent the maximum rate of change, wetting or drying, but then proceed to use maxima. Points based upon rate of change only apply to eustatic sea-level curves (inflection points) in which case they equate to the maximum or minimum water depths (Jervey 1988; Posamentier et al. 1988).

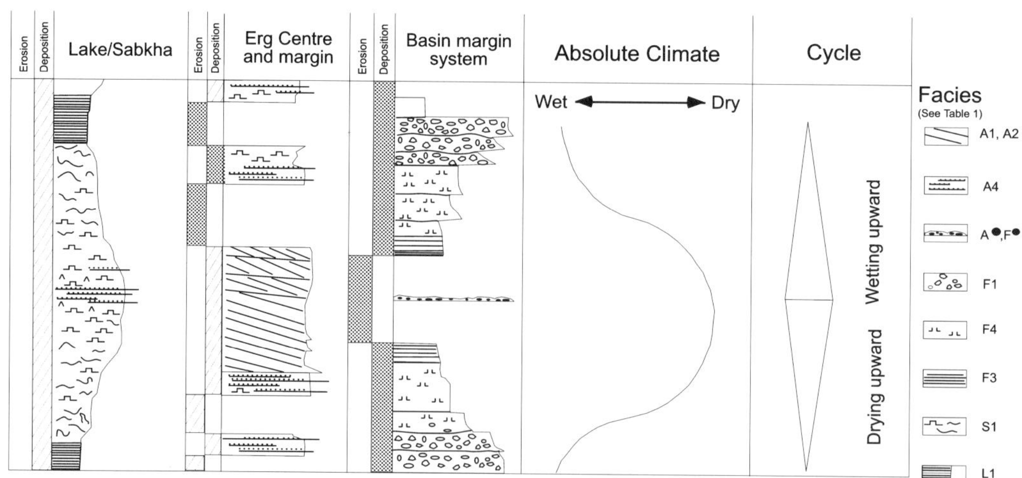


**Fig. 8.** Idealized complete climatic cycle in each of the principal depositional environments. All logs idealized, no vertical scale implied.

If maximum aridity could be accurately correlated with glacio-eustatic lowstands of sea level this would be the most appealing boundary as it would equate to the sequence boundary (Posamentier *et al.* 1988). Such a link is, however, somewhat speculative and as yet unproved. Previous workers (Clemmensen *et al.* 1989; George & Berry 1993) have described drying-upward and drying/wetting-upward cycles bounded at the point of maximum wetness. Within this study, the inferred point of maximum wetness was chosen as a unit boundary because it is easily identified in vertical section. It is marked by the deepest-water lake deposits, the point of maximum alluvial progradation, and the formation of the aeolian supersurfaces.

Climatically defined depositional units were recognized by plotting the relative aridity of each successive bed against the preceding one (Fig. 9), thereby enabling trends of increasing and decreasing aridity, which are independent of depositional setting or process, to be identified. Specific facies may have very different significance, depending upon the position within the basin, and should only be considered in the vertical context and with due consideration for the nature of the contact (erosional or gradational). For example, as will be discussed, the occurrence of sandsheet facies in a lake margin sabkha complex represents the point of maximum aridity whilst the same facies in an erg succession represents deposition during a wet period.

Figure 9 shows a plot of relative climate for a well from the Northern Sole Pit region. This illustrates the minor changes in relative climate, which stack to define the background trends. It is these trends which are correlatable around the basin. It is also interesting to note the variety of preserved cycles, a point which is discussed later. The technique of plotting relative climate and correlating events is even more reliable if a cluster of closely spaced wells are correlated such that climatically driven trends are identified from a series of wells rather than an individual one. This enables localised autocyclic events, such as localized erosion, to be better identified and considered.

## Complete depositional cycles within sub-environments

Complete depositional cycles, bounded by points of maximum wetness, are considered from three major basinal environments (Fig. 8): (a) the lake and its marginal sabkha complex, (b) the erg centre and its immediate margin, and (c) the basin margin (proximal and distal fluvial) system.

### Lake and lake-margin sabkha

Within the lake, maximum wetness equates to maximum lake level, given that any time lag is likely to occur over a geologically short time interval. Increased aridity results in contraction of the lake and migration of the sabkha-facies belt across older lake deposits. The expression of the point of maximum aridity will depend upon the position within the basin and the degree of drying out, but may include the development of a sandsheet and the migration of small-scale aeolian bedforms across the former lake bed. Evaporites may develop at the centre of the basin. Flooding associated with decreasing aridity will be rapid across the flat plain (Fig. 3, base of cycle 10) but may be preserved as a gradual change if the rise in water table only slightly exceeds the sedimentation rate (Fig. 3, top of cycle 6).

### Erg and erg margin

As previously discussed, the point of maximum wetness within the erg environment is characterized by non-deposition, stabilization of the erg centre and aeolian sandsheet formation associated with high water table and frequent fluvial flooding. However, within the Rotliegend of the SNS, no palaeosols have been observed, suggesting that either soil-forming processes do not operate, or else they have been subsequently reworked during the following arid phase. Large quantities of fluvial sands accumulate at the margins of the basin during wet periods and only minor aeolian modification occurs (Kocurek *et al.* 1991).

As aridity increases, the previously deposited fluvial sands are reworked, progressively larger aeolian bedforms develop and migrate across the area. These fill accommodation space created during the previous non-depositional phase and have a very high preservation potential. Once the space is filled, further bedforms will develop above the filled accumulation space, although these have a lower preservation potential. A further increase in the wind velocity may result in erosion and deflation forming an erosional supersurface (Kocurek & Havholm 1994) with little or no deposition occurring prior to the stabilisation of the surface associated with the onset of the subsequent wet period. Subsequent cycles invariably show a drying upward motif (e.g. any of the aeolian dominated cycles in Fig. 4).

As the climate becomes more humid and the water table rises, and a laterally extensive sandsheet may be developed over the regionally

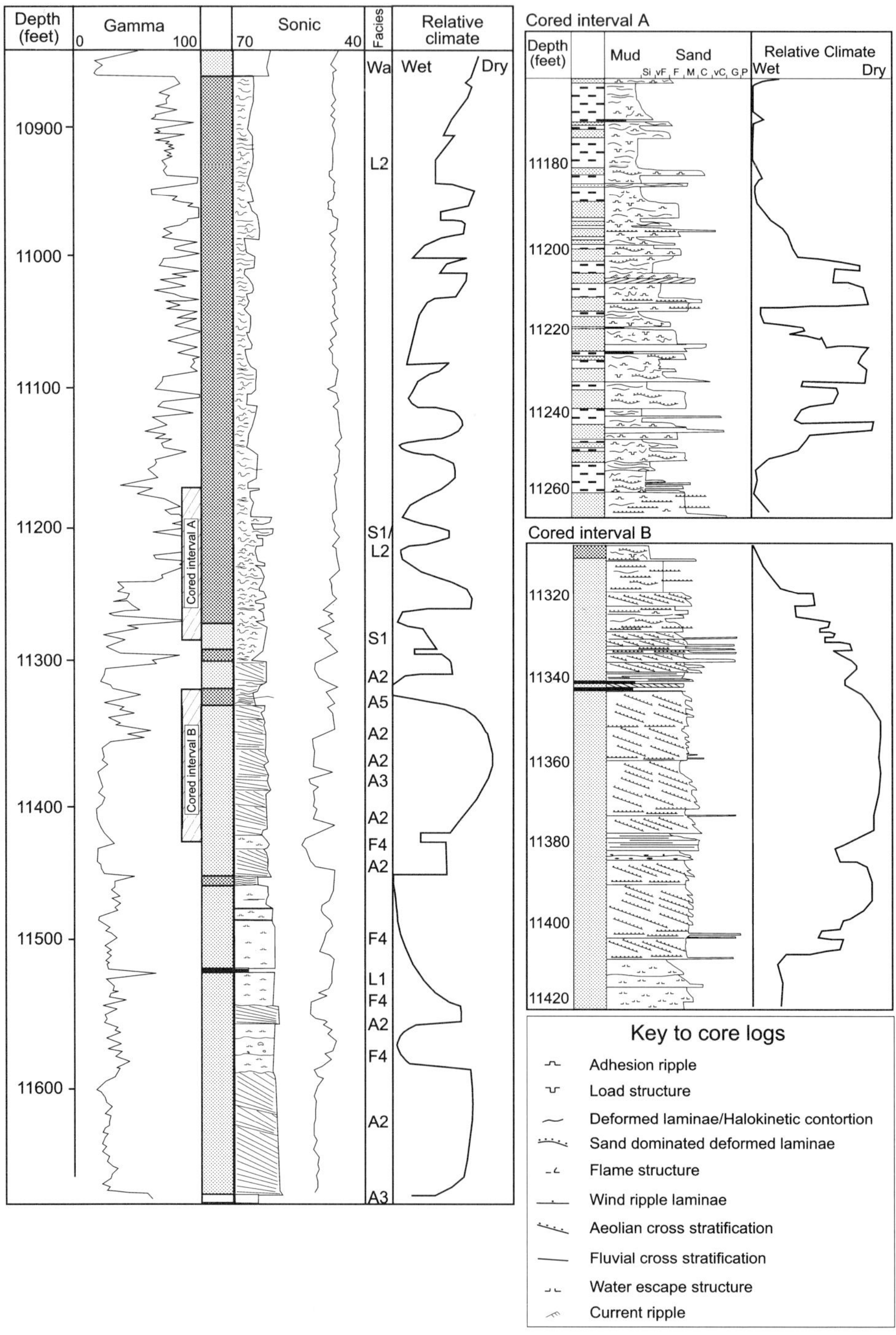



**Fig. 9.** Representative well illustrating the plotting of a relative climate curve from both core and wireline log data. Note the higher resolution available from core studies for the same interval. Note also the variety of preserved cycles, including drying upward and drying/wetting upward types.

extensive supersurface (Fig. 4). The time-span represented by the supersurface may be geologically significant, potentially from a point prior to the period of maximum aridity to the point of maximum wetness. In their study of the Page Sandstone, Havholm & Kocurek (1994) considered that a greater amount of time was incorporated within the supersurfaces and associated sabkha deposits than within the overlying and underlying aeolian deposits.

## Basin margin (proximal and distal fluvial) system

The period of maximum wetness corresponds to the maximum hinterland run-off and, for the purpose of this study, this has been equated to the maximum point of fan progradation. Because grain size decreases down the fluvial profile, this can be considered to correspond to the coarsest-grained clastics in a vertical section (e.g. the centre of cycles 6 and 7 in Fig. 5). Progressive drying upward will result in gradual back stepping and eventual abandonment of the fan. Continued drying results in desiccation and deflation of the sand grade portion of the deposits, leaving behind a lag (Facies F·).

The onset of wetting is marked by fan progradation which may be accompanied by incision as the streams regrade themselves. Progradation will typically be gradual, starting with distal deposits which become progressively more proximal upward (e.g. cycle 7 in 48/19b-7, Fig. 5). However, sharp based climatic cycles within marginal fan complexes have been observed (e.g. base of cycle 11 in 48/18b-2, Fig. 5). A tectonic origin must be considered if a sudden influx of very coarse alluvial material is encountered. Such tectonically controlled cycles are typically thicker and more asymmetric (Nichols 1987).

## Application of climatic cyclicity to understanding the depositional evolution of the Rotliegend Group

Bed-to-bed correlations, although possible on a field scale are not applicable for correlating or predicting facies on a basin scale due to the problems of bedform diachroneity. Identifying climatic trends from groups of previously correlated wells provides a composite climatic cycle curve for the preserved sediment package (Fig. 10). These curves may be correlated

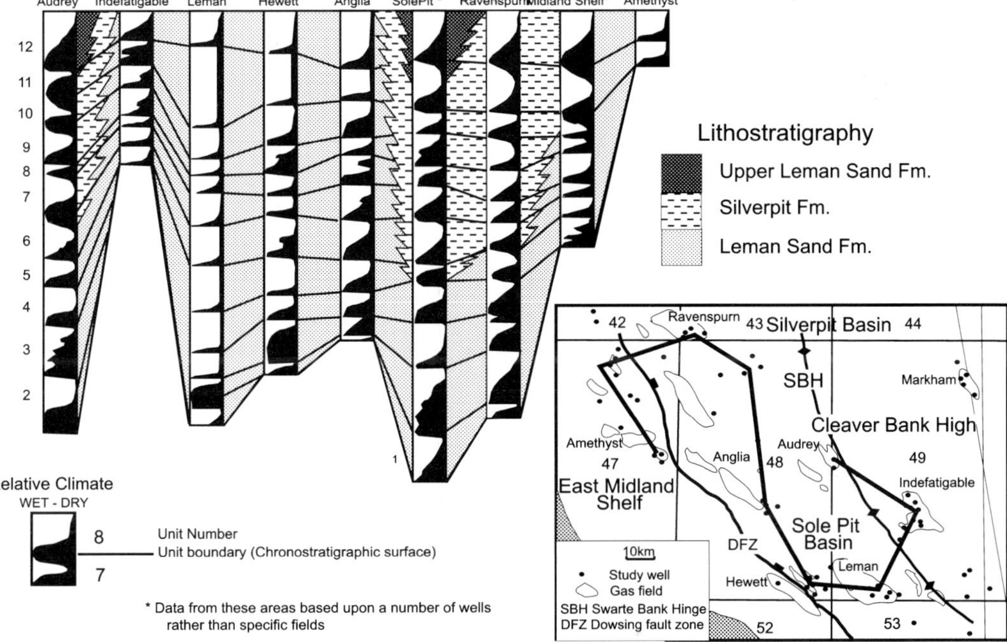


**Fig. 10.** Correlation of composite climatic curves from the sub-areas of the basin. Each curve is derived from the cross sections (Figs 3–6) and represents a summary of the relative climate cycles from each well (cf. Fig. 9). Note the varied expression of single cycles although the thickness of each cycle is typically constant, particularly within single structural blocks. Note also the unit boundaries, time lines, cross-cut the lithostratigraphic formation boundaries.

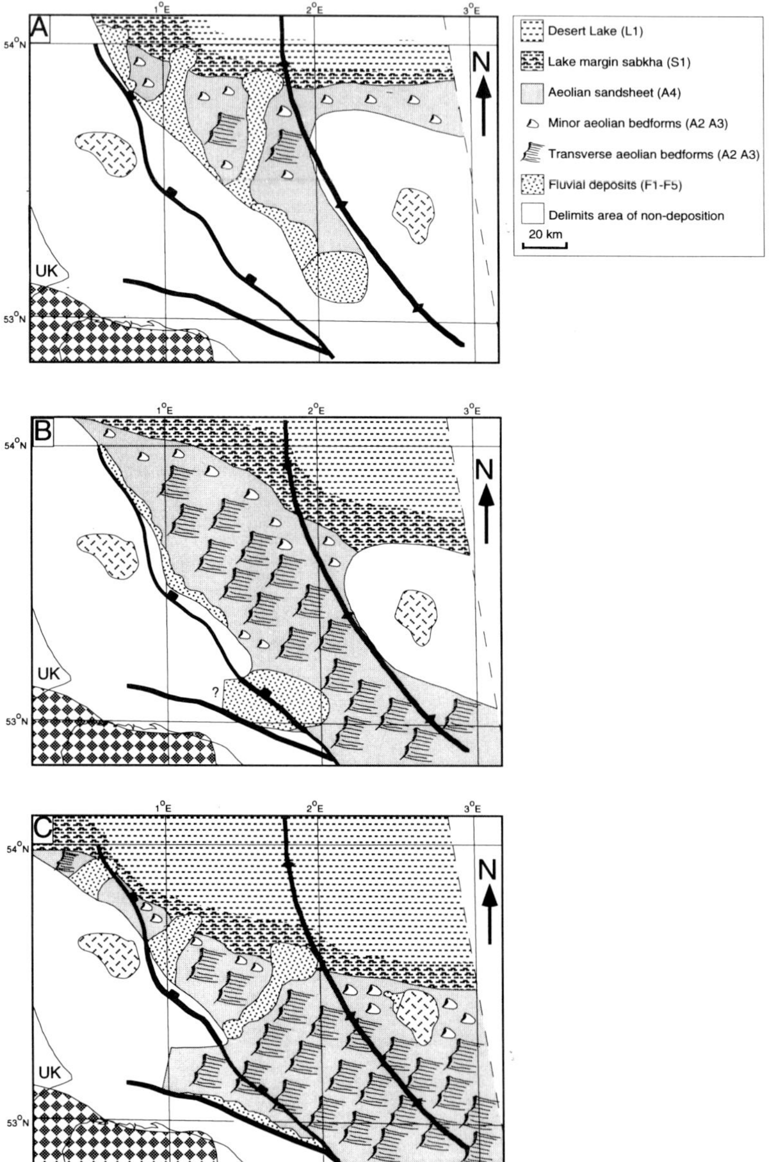



**Fig. 11.** Palaeogeographic maps for representative time intervals. Dune orientations are schematic, although WNW transport directions are predominant. (**a**) Reconstruction for climatic cycle 2. Note that fluvial deposition is confined to the Sole Pit area with aeolian deposition occurring to the North around the margin of the desert lake. (**b**) By climatic cycle 4 the main erg has been established in the south and the lake margin has migrated southward. (**c**) During cycle 7 the Indefatigable High is virtually covered by sediment derived from a large erg present over the southern portion of the area, the lake margin has migrated further south. (**d**) By cycle 10, the lake margin is at its southernmost point, the Indefatigable High is completely buried whilst the Amethyst High remains a source for fluvial sediment. A large erg covers the Sole Pit area. (**e**) At the end of the Rotliegend time interval (cycle 12), prior to the Zechstein transgression, the Amethyst High is onlapped, a large erg covers the entire area. The lake margin has migrated northward. This corresponds with the deposition of the Upper Leman sandstone Formation. (**f**) Base map of study area showing location of major structural features and gas fields referred to in the text.

between groups of wells, defining climatically driven depositional units bounded at their, variably expressed, point of maximum wetness.

From these correlations, a maximum of 12 climatically driven cycles have been identified. These cycles have been used as a chronostratigraphic framework to document the evolution of the basin fill through the Rotliegend time interval (Fig. 11). The maps include the position of a number of structural features which are believed to have influenced sediment distribution (Howell 1992; George & Berry 1993). These include a series of subsurface granite batholiths beneath the East Midland and Indefatigable Shelves (Donato & Megson 1990), the Dowsing and Hewett Faults zones and the Swarte Bank Hinge.

The evolution of the Rotliegend Basin fill through the Permian may be summarized in five intervals (Fig. 11), each one of which represents

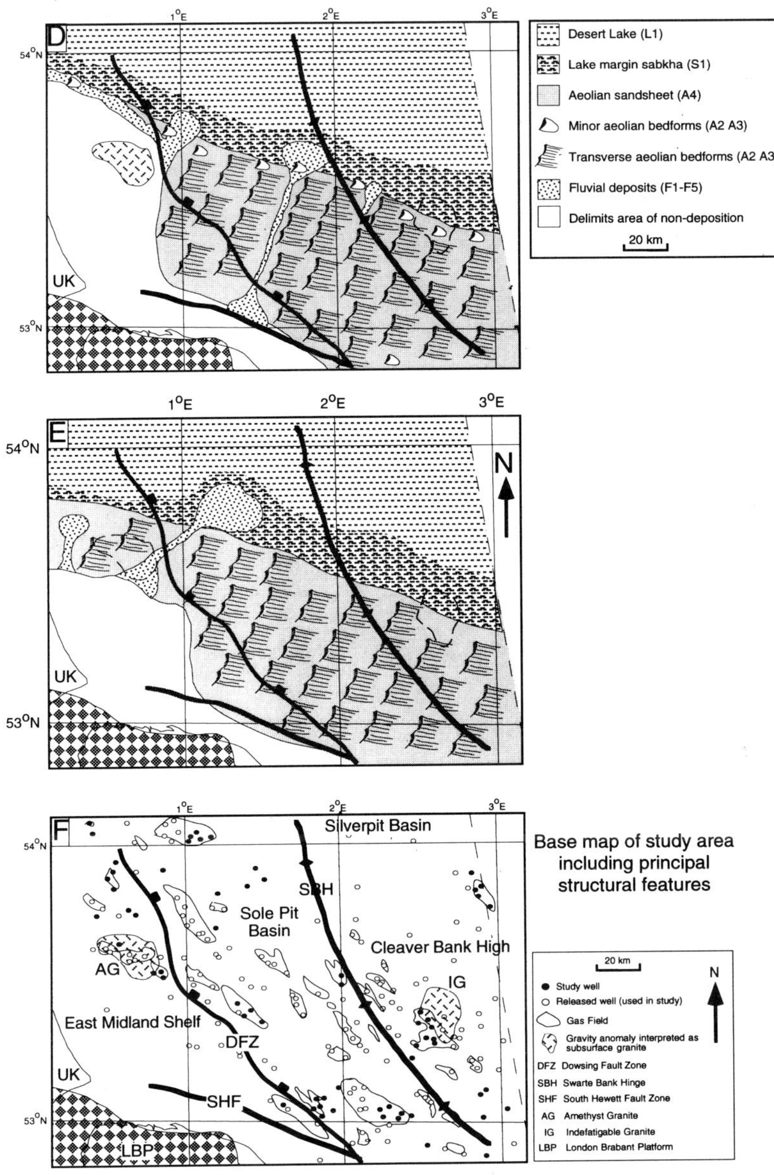


**Fig. 11.** (*Continued*).

a number of climatic cycles (units). Palaeogeographic maps for units 2 4, 7 10 and 12 are presented in Fig. 11a–e, respectively. A complete set of maps (units 1–12 inclusive) has been discussed by Howell (1992).

### Unit 1 to unit 3

During the deposition of units 1 and 2 an early stage of alluvial fan development and fluvial sheetflooding was associated with the development of northward flowing ephemeral channels (Fig. 11a). Increased aeolian activity occurred higher in the succession (unit 3), particularly in the east of the region, with the westward migration of a large erg from the Dutch Sector. Deposition was concentrated along the Sole Pit and in the north of the study area, whilst the East Midland Shelf and Indefatigable High were being eroded.

### Unit 4 to unit 5

During this period, the Silverpit lake transgressed across much of the Northern Sole Pit and North Cleaver Bank High (Fig. 11b). A significant erg was established in the Southern Sole Pit and spread eastward into the area south of the Indefatigable Shelf. A further, separate erg developed in the west of the Northern Sole Pit and Ravenspurn area.

### Unit 6 to unit 7

Onlap onto north of the East Midland Shelf began with deposition from a series of north- and northeasterly-flowing ephemeral channels (Fig. 11c). This was accompanied by a significant westward lake transgression across the Ravenspurn and the remaining Northern Sole Pit area. Aeolian/sabkha deposition occurred across the northern part of the Indefatigable Shelf. Thick accumulations of erg and erg margin deposits accumulated in the Southern Sole Pit.

### Unit 8 to unit 10

The main Sole Pit erg expanded across much of the Southern Sole Pit and the final burial of the Indefatigable Shelf occurred (Fig. 11d). The maximum southward and westward migration of the Silverpit Lake occurred across the north East Midland Shelf and Northern Sole Pit and was associated with fluvial deposition on the East Midland Shelf and in the Central Sole Pit.

### Unit 10 to unit 12

These units represent a final period of maximum aridity with deposition in large, well established aeolian bedforms across the entire south of the study area, including the Amethyst region (Fig. 11e). All fluvial activity ceased with the exception of areas located on the northeastern part of the East Midland Shelf. During this time, the northward migration (regression) of the lake margin was associated with the deposition of the Upper Leman Sandstone Formation.

## Sediment supply, accommodation and the vertical expression of preserved cycles

The recognition of climatically derived cycles permits temporal correlation of distinctly different depositional environments (Fig. 10). These correlations cross cut lithostratigraphic boundaries, such as the Silverpit–Leman Sandstone Formation boundary. The correlation panel illustrates that, although a variety of preserved expressions exist for a single cycle (drying-wetting, drying upward), individual cycles often maintain a relatively constant thickness, irrespective of depositional process. This is particularly true within individual structural blocks. This observation requires further discussion.

Two principal factors control the deposition and preservation of sediment: (1) Accommodation – if no accommodation space is created, no sediment will be preserved and migrating bedforms will simply by-pass the area. The principal process for creating accommodation space is subsidence. (2) Sediment supply – to preserve sediment, sediment supply must be greater than sediment removal. Sediment is introduced into the area by a variety of processes (aeolian, fluvial, lacustrine) within which sedimentation rates vary at different stages of the climatic cycle.

Base level is defined as a surface (e.g. sea level, fluvial equilibrium profile, water table) above which sediment will be bypassed and below which it will be deposited. Accommodation space is created beneath the base level and may be defined as 'the space available for sediment to accumulate' (Jervey 1988). If the rate of sediment accumulation exceeds the rate of accommodation creation, then any sediment deposited will lie above base level and will have a low preservation potential. It is likely to be reworked. In aeolian systems which can build large-relief bedforms a few hundred metres above base level, this reworking may take considerable time.

Accommodation creation and sediment supply can be considered against a time scale in which relative climate fluctuates cyclically for each of the principal depositional environments (Fig. 12). Within this model subsidence is treated as a constant. Time is considered in the context of absolute changes in climate, which for each environment will result in predictable but different changes in net sediment flux (the balance between sediment supply and sediment removal). These variations are provisionally modelled as symmetrical, sinusoidal curves. Sediment supply is a direct function of the depositional process (setting) and the climate (time). Periods of positive and negative budget are shown, where the supply curve is above or below the accommodation line, respectively. Idealised preserved sediment columns and the expression of the preserved climatic cycles are also shown on Fig. 12.

When considered in this manner, each of the principal depositional regimes can be studied and the internal architecture of preserved cycles can be explained. Subsidence and, consequently, preserved cycle thickness is independent of the depositional environment. Thus, the preserved thickness of individual sedimentary units is controlled largely by the rate of subsidence across the basin. Within a single episode of sedimentation, rapidly deposited aeolian bedforms may temporarily reach a thickness of several hundred metres above base level, but much of that thickness will ultimately be reworked prior to preservation. Conversely, during the same depositional episode, slowly accumulating lake deposits will eventually fill the available accommodation space such that at the end of the episode the preserved thickness from each of these very different sedimentary environments will be the same.

Although the preserved thickness of an individual sedimentary unit may be similar within a single tectonic block (irrespective of depositional process), the thickness may show considerable variation across the entire width of the basin where differential subsidence rates operate.

Within the erg environment (Fig. 12a), little sediment is moved during the wet period as the sand becomes stabilised by vegetation. Accommodation space creation outpaces sediment accumulation. As time progresses and the dry period begins, large aeolian bedforms rapidly prograde into the area and fill the space created. This space is quickly filled and exceeded and sediment is bypassed. An erosive supersurface may be formed at, or close to, the point of maximum aridity (Kocurek 1988). Alternatively,

the erg may retain a positive relief until the onset of the ensuing wet period, whence erosional modification and subsequent stabilisation will form a supersurface (Loope 1984; Talbot 1985). Either way a sandsheet is deposited above a supersurface through the subsequent wet period. The preserved depositional cycle shows a drying upward motif. The majority of sediment is deposited during a very short period of time and most of the time is encapsulated within the supersurface and the overlying sandsheet (Havholm & Kocurek 1994).

At the basin margin, within the fluvial depositional environment (Fig. 12b), net sediment flux is greatest during the wet period. Sediment supply exceeds accommodation space creation and progradation occurs. As the climate dries out abandonment and aeolian reworking occur (see earlier discussion). As the climate becomes wetter progradation occurs once more. The nature of the preserved cycle is the most variable within the semi-arid system: the cycle may be symmetrical; it may be wetting upward if the dry stage is removed by aeolian erosion or, more rarely, it may be drying upward if alluvial supply is generally low.

In the lake and marginal sabkha complex (Fig. 12c), sedimentation is usually slow. Sedimentation rates vary little during the various stages of the cycle and aggradation occurs where sediment infilling more or less matches accommodation space creation. Preserved cycles are typically symmetrical and the accommodation space is not always exceeded. When it is exceeded, it occurs at or near the dry maxima and consequently some cycles show drying upward motifs. If subsidence exceeded sediment supply, lake level would rise. Indeed, differential subsidence rates between the basin centre and margin, probably driven by oblique slip faulting (George pers. comm.), may account for unusually rapid lacustrine sedimentation rates. Such tectonically controlled lake level rises occurred throughout the Rotliegend times as the Silverpit Lake gradually transgressed the Leman Sandstone toward the south. Tectonic controls on base level may be difficult to differentiate from climatically controlled lake level fluctuations.

In all of the above scenarios, the thickness of the preserved sediment package is controlled predominantly by subsidence and accommodation creation and is largely independent of depositional process. Hence, individual cycles commonly maintain a constant thickness between depositional environments. The internal architecture of the preserved cycle however is strongly controlled by depositional environment

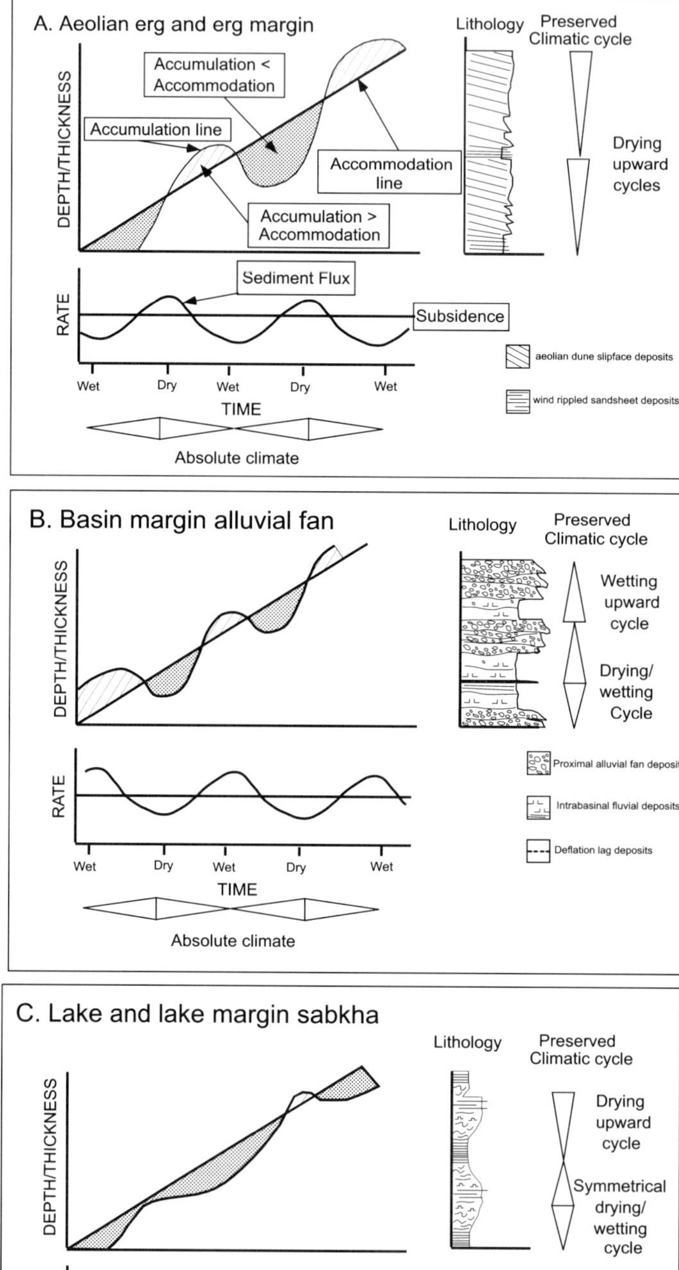


**Fig. 12.** The interaction of climate and subsidence in controlling sediment supply within the principal depositional environments. Note the varied timing of periods of high and low sediment flux and the controls upon preserved depositional cycle architecture.

and sedimentation rates and consequently cycle architecture varies greatly between different parts of the basin.

## Sequence stratigraphy and climatic cyclicity: discussion

The interplay between eustatic sea level (glacial eustasy especially) and climate is worth considering briefly. Sarnthein (1978) discussed the controls exhibited by polar ice caps upon trade wind belt size and strength. During glacial periods, expanded ice caps compress the trade wind belts and result in periods of relatively dry climate with high velocity, continuous winds. Such periods of aridity favour creation and migration of large aeolian bedforms whilst reducing fluvial activity (Glennie 1986; Kocurek *et al.* 1991).

Interglacial or pluvial periods, in contrast, are characterized by very broad, weak trade wind belts. This results in reduced aeolian bedform migration, enhanced by an increase in fluvial activity associated with increased precipitation, particularly in the hinterland.

Periods of maximum aridity, which correspond to glacial maxima, can therefore, at least hypothetically, be correlated with glacio-eustatic lowstands. Within the marine realm, such periods are characterised by sequence boundary formation on the shelf and deposition of lowstand shorelines and fan systems within the deeper marine environment (Posamentier & Vail 1988; Van Wagoner *et al.* 1990). Similarly, pluvial periods (glacial minima) correspond to glacio-eustatic highstands of sea level. An important point, however, is that the point of maximum wetness, picked as an allostratigraphic boundary here, is comparable to the top of the eustatic curve and not necessarily the maximum flooding surface. The latter occurs at the point of maximum water depth somewhere between the R infection point (the inflection point on the rising limb of the sea-level curve) and the top of the eustatic curve, which is subsequently modified by subsidence rates (Jervey 1988; Posamentier & Vail 1988). Thus, subsidence rates have no affect on climatic cyclicity whilst they partly control maximum water depth in marine systems.

Although the point of maximum aridity may correspond to the marine sequence boundary, extreme caution should be exercised in trying to recognize the standard marine systems tracts within climatic cycles (Yang & Nio 1994), especially with the high degree of variability recognized within the internal architecture of preserved cycles. Prosser (1993), using the original Brown & Fisher (1977) definition of a system tract (a linkage of contemporaneous depositional systems) proposed the term 'climatic systems tract' although she did not discuss how such systems tracts should be defined. It is proposed here that the wetting and drying portions of a single climatic cycle may be considered as systems tracts. The boundary between the wet and subsequent dry portions of the cycle is represented by a change in facies stacking patterns. Posamentier *et al.* (1988) used this as a suitable criteria for a system tract boundary (cf. the maximum flooding surface between the transgressive and highstand systems tracts). In contrast, the boundary between the top of the dry and subsequent wet portion is represented by either a deflation surface or by a change in stacking patterns. The wetting-upward systems tract and drying-upward systems tract describe a linked depositional system with time significance, but do not in any way relate to their marine counterparts. Similarly, the direct comparison of super surfaces to sequence boundary is inadvisable because of the variety of supersurfaces which may occur (Kocurek & Havholm 1994).

A further point of interest is that within an inter-montane basin, an ephemeral lake, rather than sea level, controls base level. Following an increase in climatic aridity the lake level will fall. A fall in base level would normally cause fluvial progradation. However, in this case, the reduction in precipitation will effectively shut down fluvial systems which will subsequently retrograde. Following an increase in humidity the lake level will rise and the fluvial systems will also prograde due to increased run-off. This contrasts with the typical marine continental margin scenario in which fluvial style changes from braided to meandering, and aggradation occurs during base level highstand (Shanley & McCabe 1993; Schumm 1993).

## Conclusions

The recognition and correlation of climatic cycles provides a methodology for considering semi-arid and arid depositional systems in a time stratigraphic framework. Such an approach is based upon the changes which occur in depositional processes if the climate becomes more or less arid. From this, a complete climatic cycle, bounded at the point of least climatic aridity, can be identified for different depositional settings within the basin. These can then be recognized and correlated. The application of climatic cycles to subsurface data sets is improved if groups of closely spaced wells rather than individual wells are considered. By using several wells, localised changes can be identified, and the significance of facies transitions can be more accurately assessed.

Recognition of climatic cycles from the data set presented in this paper has been used to define 12 climatically driven depositional units, that can be interpreted in a time stratigraphic manner. The identification, tracing and mapping out of these units across the various sedimentary environments encountered within the basin has resulted in palaeogeographic reconstructions that improve greatly on traditional lithostratigraphic interpretations of the SNS Basin. Units commonly maintain a constant thickness over regional scales, independent of depositional process. However, differential subsidence between various parts of the basin will ultimately control the preserved thickness of each unit. The internal architecture of the unit is strongly controlled by the depositional process and varies in a predictable manner depending upon sediment supply and depositional rate. Climatic cycles may be considered in the terms of two new systems tracts; wetting upward and drying upward. These systems tracts define linked depositional systems and are bounded by the point of maximum aridity and the point of maximum wetness, either of which may be used to define and correlate units.

The development of a sequence-stratigraphic model for the predominantly continental facies of the SNS has improved our understanding of how a series of linked depositional systems have evolved over time within a single basin. The interrelationships between each of these systems is of key importance in understanding the future hydrocarbon potential of the region. The sequence-stratigraphic model presented here has resulted in improved stratigraphic prediction and, when combined with seismic data, will provide a more precise definition of the nature of potential hydrocarbon reserves, particularly the recognition of play fairways. The recognition of laterally extensive, climatically controlled, depositional units is particularly useful in mapping out subtle interfingering between erg and fluvial facies at the basin margin, where future hydrocarbon reserves are likely to exist.

The majority of this work was undertaken as part of a regional study of the Southern Permian Basin and co-workers A. Becker, P. Turner, A. Searl, E. Edwards and G. Williams are acknowledged for help and support. Data and funding for the work came from Amerada Hess, Amoco (UK) Exploration Co., British Gas Plc, Hamilton Bros. Oil & Gas, Phillips Petroleum Co. (UK), Ranger Oil (UK), Texaco and Ultramar Exploration. S. Davies and S. Knight at the University of Liverpool commented extensively upon an early draft of the manuscript. Our thanks go to G. George, M. Sweet and K. Ziegler for their constructive reviews.

## References

AHLBRANDT, T. S. & FRYBERGER, S. G. 1982. Introduction to eolian deposits. *In*: SCHOLLE, P. A. & SPEARING, D. (eds) *Sandstone depositional environments*. American Association Petroleum Geologists Memoirs, **31**, 11–47.

ALMON, W. R. 1981. Depositional environment and diagenesis of Permian Rotliegendes sandstones in the Dutch Sector of the SNS. *In*: LONGSTAFFE, F. J. (ed.) *Clays and the Resource Geologist*. Mineralogical Association of Canada Short Course Handbooks, **7**, 119–147.

BLANCHE, G. B. 1973. The Rotliegend Sandstone Formation of the U. K. sector of the Southern North Sea Basin. *Transactions of the Institute Mineralogists and Meterologists Section B*, **82**, 85–89.

BREED, C. S. & GROW, T. 1979. Morphology and distribution of dunes in sand seas observed by remote sensing. *In*: MCKEE, E. D. (ed.) *A study of global sand seas*. USGS Professional Papers, **1052**, 253–302.

BROWN, L. F. & FISHER, W. L. 1977. Seismic stratigraphic interpretation of depositional systems: examples from Brazilian rift and pull-apart basins. *In*: PAYTON, C. E. (ed.) *Seismic Stratigraphy – Applications to Hydrocarbon Exploration*. American Association of Petroleum Geologists Memoirs, **26**, 213–248.

BULAT, J. & STOKER, S. J. 1987. Uplift determination from interval velocity STUDIES, U.K. Southern North Sea. *In*: BROOKS, J. & GLENNIE, K. W. (eds) *Petroleum Geology of North West Europe*. Graham and Trotman, London 293–305.

CLEMMENSEN, L. B., OLSEN, H. & BLAKEY, R. C. 1989. Erg-margin deposits in the Lower Jurassic Moenave Formation and Wingate Sandstone, southern Utah. *Geological Society of America Bulletin*, **101**, 759–773.

DONATO, J. A. & MEGSON, J. B. 1990. A buried granite batholith beneath the East Midland shelf of the Southern North Sea Basin. *Journal of the Geological Society, London*, **147**, 133–140.

EUGSTER, H. P. & HARDIE, L. A. 1975. Sedimentation in an ancient playa-lake complex. The Wilkins Peak Member of the Green River Formation of Wyoming. *Geological Society of America Bulletin*, **85**, 319–334.

FRYBERGER, S. G. 1993. A review of aeolian bounding surfaces, with examples from the Permian Minnelusa Formation, USA. *In*: NORTH, C. P. & PROSSER, D. J. (eds) *Characterization of Fluvial and Aeolian Reservoirs*. Geological Society, London, Special Publications, **73**, 167–198.

GEORGE, G. T. & BERRY, J. K. 1993. A new lithostratigraphy and depositional model for the Upper Rotliegend of the UK sector of the southern North Sea. *In*: NORTH, C. P. & PROSSER, D. J. (eds) *Characterization of Fluvial and Aeolian Reservoirs*. Geological Society, London, Special Publications, **73**, 291–320.

—— & ——1994. A new palaeogeographic and depositional model for the Upper Rotliegend, offshore The Netherlands. *First Break*, **12**, 147–158.

GLENNIE, K. W. 1970. *Desert sedimentary environments*. Developments in Sedimentology, **14**. Elsevier, Amsterdam.

——1972. Permain Rotliegendes of North West Europe interpreted in light of modern desert sedimentation studies. *American Association of Petroleum Geologists Bulletin*, **56**, 1048–1071.

——1983. Lower Permian Rotliegend desert sedimentation in the North Sea Area. *In*: BROOKFIELD, M. E. & AHLBRANDT, T. S. (eds) Eolian Sediments and Processes. Developments in Sedimentology, **38**. Elsevier, Amsterdam, 521–541.

——1986. Early Permian – Rotliegendes. *In*: GLENNIE, K. W. (ed.) *Intoduction to the Petroleum Geology of the North Sea*. Blackwell Scientific Publications, 63–86.

—— & BOEGNER, P. L. E. 1981. Sole Pit inversion tectonics. *In*: ILLING, L. V. & HOBSON, G. D. (eds) *The Petroleum Geology of the Continental Shelf of North West Europe*. Heyden and Son Ltd, 110–210.

HAVHOLM, K. G. & KOCUREK, G. 1994. Factors controlling aeolian sequence stratigraphy: clues from super bounding surface in the Middle Jurassic Page Sandstone. *Sedimentology*, **41**, 913–934.

HERRIES, R. G. 1993. Contrasting styles of fluvial-aeolian interaction at a downwind erg margin: Jurassic Kayenta–Navajo transition, northeastern Arizona, USA. *In*: NORTH, C. P. & PROSSER, D. J. (eds) *Characterization of Fluvial and Aeolian Reservoirs*. Geological Society, London, Special Publications, **73**, 199–218.

HILLIER, A. P. & WILLIAMS, B. P. J. 1991. The Leman field, blocks 49/26, 49/27, 49/28, 53/1, 53/2, U.K. North Sea. *In*: ABBOTTS, I. L. (ed.) *United Kingdom Oil and Gas Fields 25 Years Commemorative Volume*. Geological Society, London, Memoirs, **14**, 451–386.

HOWELL, J. A. 1992. *Sedimentology of the Rotliegend Group UK Southern North Sea*. PhD thesis, University of Birmingham, UK.

HUNTER, R. E. 1977. Basic types of stratification in small aeolian dunes. *Sedimentology*, **24**, 361–387.

JERVEY, M. T. 1988. Quantitative modelling of siliciclastic rock sequences and their seismic expression. *In*: WILGUS, C. K. *ET AL*. (eds) *Sea-level changes: an integrated approach*. Society of Economic Paleontologists and Mineralogists Special Publications, **42**, 47–70.

JOLLEY, E. J., TURNER, P., WILLIAMS, G. D., HARTLEY, A. J. & FLINT, S. 1990. Sedimentology response of an alluvial system to Neogene thrust tectonics, Atacama Desert, northern Chile. *Journal of the Geological Society, London*, **147**, 769–784.

KOCUREK, G. 1981. Significance of interdune deposits and bounding surfaces in aeolian dune sands. *Sedimentology*, **28**, 753–780.

——1988. First-order and super bounding surfaces in eolian sequences – Bounding surfaces revisited. *Sedimentary Geology*, **56**, 193–206.

—— & HAVHOLM, K. G. 1994. Eolian Sequence Stratigraphy – A Conceptual Framework. *In*: WEIMER, P. & POSAMENTIER, H. W. (eds) *Siliciclastic sequence stratigraphy. Recent developments and applications*. American Association of Petroleum Geologist Memoirs, **58**, 393–409.

—— & NIELSON, J. 1986. Conditions favourable for the formation of warm climate aeolian sand sheets. *Sedimentology*, **33**, 795–816.

——, HAVHOLM, K. G., DEYNOUX, M. & BLAKEY, R. C. 1991. Amalgamated accumulations resulting from climatic and eustatic changes, Akchar Erg, Mauritania. *Sedimentology*, **38**, 751–772.

LANGFORD, R. & CHAN, M. A. 1989. Flood surfaces and deflation surfaces within the Culter Formation and Cedar Mesa Sandstone (Permian), southeastern Utah. *Geological Society of America Bulletin*, **100**, 1541–1549.

LEE, M., ARONSON, J. L. & SAVIN, S. M. 1989. Timing and conditions of Permian Rotliegende sandstone diagenesis, Southern North Sea: K/Ar and Oxygen isotope data. *American Association of Petroleum Geologists Bulletin*, **73**, 195–215.

LOOPE, D. B. 1984. Discussion: Origin of extensive bedding planes in aeolian sandstones: A defence of Stokes' hypothesis. *Sedimentology*, **31**, 123–132.

MARIE, J. P. P. 1975. Rotliegendes stratigraphy and diagenesis. *In*: WOODLAND, A. W. (ed.) *Petroleum Geology and the Continental Shelf of N.W. Europe*. Applied Science Publishers, 205–210.

MARTIN, J. H. & EVANS, P. F. 1988. Reservoir modelling of marginal aeolian/sabkha sequences, Southern North Sea (U.K. sector). Society Petroleum Engineers, **18155**, 473–486.

NAGTEGAAL, P. J. C. 1973. Adhesion-ripple and barchan- dune sands of the recent Namib S.W. Africa and Permian Rotliegend N.W. Europe deserts. *Madoqua Series*, **2**, 5–19.

——1979. Relationship of facies and reservoir quality in Rotliegendes desert sandstones, southern North Sea region. *Journal of Petroleum Geology*, **2**, 145–158.

NICHOLS, G. J. 1987. Syntectonic alluvial fan sedimentation, southern Pyrenees. *Geological Magazine*, **124**, 121–133.

OLSEN, H., DUE, P. H. & CLEMMENSEN, L. B. 1989. Morphology and genesis of asymmetric adhesion warts – A new adhesion surface structure. *Sedimentary Geology*, **61**, 277–285.

POSAMENTIER, H. W. & VAIL, P. R. 1988. Eustatic controls on clastic deposition II – conceptual framework. *In*: WILGUS, C. K. *ET AL*. (eds) *Sea-level changes: an integrated approach*. Society of Economic Palaeontologists and Mineralogists Special Publications, **42**, 125–154.

——, JERVEY, M. T. & VAIL, P. R. 1988. Eustatic controls on clastic deposition I – conceptual framework. *In*: WILGUS, C. K. *ET AL*. (eds) *Sea-level changes: an integrated approach.* Society of Economic Paleontologists and Mineralogists Special Publications, **42**, 109–124.

PROSSER, S. D. 1993. Rift-related linked depositional systems and their seismic expression. *In*: WILLIAMS, G. D. & DOBB, A. (eds) *Tectonics and Seismic Sequence Stratigraphy.* Geological Society, London, Special Publications, **71**, 35–67.

PURVIS, K. 1989. Authigenic Zoned magnesites in the Rotliegend, Lower Permian, Southern North Sea. *Sedimentary Geology*, **65**, 307–318.

RHYS, G. H. 1975. A proposed standard lithostratigraphic nomenclature for the Southern North Sea. *In*: WOODLAND, A. W. (ed.) *Petroleum Geology and the Continental Shelf of N.W. Europe.* Applied Science Publishers, 151–163.

ROSSEL, N. C. 1982. Clay mineral diagenesis in Rotliegend aeolian sandstones of the Southern North Sea. *Clay Minerals*, **17**, 69–77.

RUBIN, D. M. & HUNTER, R. E. 1982. Bedform climbing in theory and nature. *Sedimentology*, **29**, 121–138.

SARNTHEIN, M. 1978. Sand deserts during glacial maximum and climatic optimum. *Nature*, **272**, 43–46.

SCHUMM, S. A. 1993. River response to base level change: Implications for sequence stratigraphy. *Journal of Geology*, **101**, 279–294.

SHANLEY, K. W. & MCCABE, P. J. 1993. Alluvial architecture in a sequence stratigraphic framework – a case history from the upper Cretaceous of southern Utah, U.S.A. *In*: FLINT, S. S. & BRYANT, I. D. (eds) *The geological modelling of hydrocarbon reservoirs.* International Association of Sedimentologists Special Publications, **15**, 21–56.

——1994. Perspectives on the sequence stratigraphy of continental strata. *American Association of Petroleum Geologists Bulletin*, **78**, 544–568.

SNEH, A. 1988. Permian dune patterns in North Western Europe challenged. *Journal of Sedimentary Petrology*, **58**, 44–51.

STEELE, R. P. 1985. Early Permian Rotliegendes palaeowinds of the North Sea – comments. *Sedimentary Geology*, **45**, 293–297.

STOKES, M. W. 1968. Multiple parallel-truncation bedding planes – a feature of wind deposited sandstone formations. *Journal of Sedimentary Petrology*, **38**, 510–515.

TALBOT, M. R. 1985. Major bounding surfaces in aeolian sandstones – a climatic model. *Sedimentology*, **32**, 257–265.

VAN VEEN, F. R. 1975. Geology of the Leman gas field *In*: WOODLAND, A. W. (ed.) *Petroleum Geology and the Continental Shelf of N.W. Europe.* Applied Science Pub. 223–231.

VAN WAGONER, J. C., MITCHUM, R. M., CAMPION, K. M. & RAHMANIAN, V. D. 1990. *Siliciclastic sequence Stratigraphy in well logs, cores, and outcrops.* American Association of Petroleum Geologist Methods in Exploration Series, **7**.

WALKER, R. G. 1990. Facies modelling and sequence stratigraphy. *Journal of Sedimentary Petrology*, **60**, 777–786.

YANG, C. S. & BAUMFALK, Y. A. 1994. Milankovitch cyclicity in the Upper Rotliegend Group of the Netherlands offshore. *In*: DE BOER, P. L. & SMITH, D. G. (eds) *Orbital forcing and cyclic sequences.* International Association of Sedimentologists Special Publications, **19**, 47–61.

—— & NIO, S. D. 1994. Application of high resolution sequence stratigraphy of the Upper Rotliegend in the Netherlands offshore. *In*: WEIMER, P. & POSAMENTIER, H. W. (eds) *Siliciclastic sequence stratigraphy. Recent developments and applications.* American Association of Petroleum Geologist Memoirs, **58**, 285–315.

# Compartmentalization of Rotliegendes gas reservoirs by sealing faults, Jupiter Fields area, southern North Sea


GREGORY P. LEVEILLE[1], ROB KNIPE[2], COLIN MORE[3], DAVE ELLIS[4], GRAHAM DUDLEY[5], GREG JONES[2], QUENTIN J. FISHER[2], & GARETH ALLINSON[6]

[1] *Conoco Inc., 600 North Dairy Ashford, Houston, TX 77079, USA*
[2] *Rock Deformation Research, The University of Leeds, Leeds, UK*
[3] *Conoco (UK) Limited, Aberdeen, UK*
[4] *Mobil North Sea Limited, Aberdeen, UK*
[5] *BP Exploration, Aberdeen, UK*
[6] *STATOIL (UK) Limited, London, UK*



**Abstract:** Rotliegendes gas reservoirs in the Jupiter Fields are compartmentalized by sealing faults. Significant variations in free water levels (i.e. gas–water contacts corrected for capillary pressure effects), hydrocarbon composition and pressure exist between sealed fault blocks. Displacements on many of the sealing faults are low, resulting in 'clean', porous and permeable Rotliegendes sandstones being juxtaposed across the faults. Characteristics that are important in determining the sealing potential of intra-Rotliegendes faults include (i) fault-rock cement types and volumes, (ii) intensity of cataclasis, (iii) three-dimensional continuity of the fault and associated damage zone, (iv) authigenic clay content at the time of faulting, and (v) relative orientation, timing and magnitude of later deformational events.

Detailed analysis of fault rocks from core indicates cementation is the most effective fault sealing mechanism in the Jupiter Fields. Volumetrically significant cements include salt, anhydrite and quartz. Cataclasis and deformation induced mixing of authigenic clay with fragments of framework grains also contribute to the sealing potential of some faults. Permeabilities in cemented and cataclastic fault rocks are reduced by two or more orders of magnitude compared to undeformed sandstones.

3D seismic mapping indicates NNE–SSW-trending fault zones have the highest along-strike continuity, followed by NW–SE- and E–W-trending fault zones. The NNE–SSW-trending faults also appear to have had the simplest deformational history, with NW–SE- and E–W-trending faults exhibiting evidence of movement during multiple deformational events. Static reservoir data indicate that some NNE–SSW- and NW–SE-trending faults seal over geological time, and it is expected other faults will act as seals or baffles during reservoir depletion.


Many studies have shown that sealing faults are important in trapping and compartmentalizing oil and gas reservoirs (e.g. Harding & Tuminas 1988, 1989; Hindle 1989; Allan 1989; Sassi *et al.* 1992; Jev *et al.* 1993; Smith 1966, 1980; Spencer & Larsen 1990). However, little published information exists on the sealing characteristics of faults in the Permian Rotliegendes sandstone reservoirs of the southern North Sea, even though fault-related compartmentalization has been observed in several fields. Rotliegendes fields in which faults have been reported as forming seals or barriers include the Leman (Hillier & Williams 1991), Viking (Gage 1980), Barque (Farmer & Hillier 1991*a*), Clipper (Farmer & Hillier 1991*b*), and Indefatigable Fields (Pearson *et al.* 1991).

In this paper, we describe the results of a detailed study of faulting and fault sealing in the Jupiter Fields, which consist of five Rotliegendes gas accumulations: Ganymede, Callisto, Europa, Sinope and Thebe (Fig. 1). The objective of the study was to determine the impact of sealing faults on appraisal, development and field management activities. In this paper, we present evidence that the Jupiter Fields are compartmentalized by sealing faults, relate fault zone development to the structural history of the area, and characterise the sealing potential of faults within the Rotliegendes reservoirs. Regional implications of this work are also described.

In order to characterize the sealing potential of faults it is necessary to have an integrated understanding of their geometry and petrophysical properties. Seismic and core data from several North Sea oil and gas fields have revealed that faults commonly interpreted as


*From Ziegler, K., Turner, P. & Daines, S. R. (eds), 1997, Petroleum Geology of the Southern North Sea: Future Potential*, Geological Society Special Publication No. 123, pp. 87–104.

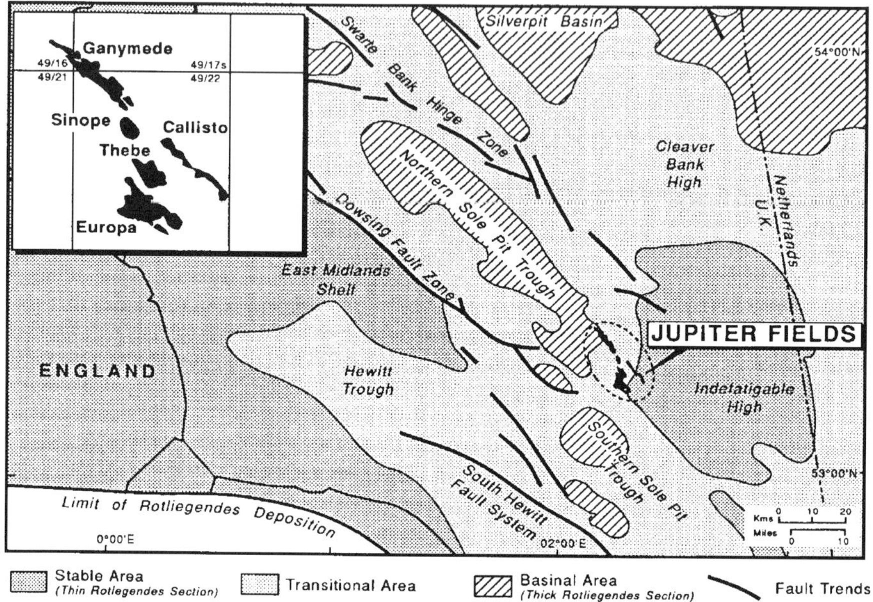


**Fig. 1.** Regional Rotliegendes structural features map. The Jupiter Fields are located between the Sole Pit Trough to the west and the Indefatigable High to the east in the UK sector of the Southern North Sea (see inset for detailed map of the fields).

single slip planes on seismic data are actually composed of a zone of large- and small-displacement faults and fractures which are termed fault 'damage zones'. Damage zones have been recognized by field geologists in outcrops for many years (Engelder 1974; Aydin & Johnson 1978; Edwards *et al.* 1993; Antonellini & Aydin 1995), but are seldom identified in seismic interpretations. Core data and outcrop studies indicate that damage zones can extend for several hundreds of metres away from what appears to be a single fault plane on seismic data.

Damage zones represent the accommodation of strain in the volume surrounding large faults, and are the products of fault growth, propagation and linking sequences. Damage-zone geometries and internal fault-population distributions can vary as a result of several factors, such as variable lithological distribution in the subsurface or local variations in the geometry of the fault zone. Damage-zone geometries may also change as a result of fault-growth processes over time.

The petrophysical properties of fault rocks encountered in the depth range of most oil and gas reservoirs are largely determined by the amount of cementation, mechanical grain re-arrangement, grain fracturing, frictional grain-

boundary sliding, and cataclastic flow that has occurred (Knipe 1993). These processes control pore sizes and pore geometries, and thereby determine the porosity, permeability and sealing capacity (i.e. capillary pressure) of fault rocks. Relationships between fault rock types and changes in petrophysical properties have been documented by several workers (Pittman 1981; Knipe 1992; Antonellini & Aydin 1994), but more work is needed in this area to develop predictive models for fault sealing.

Characteristics of sealing faults in the Jupiter Fields have been investigated at seismic, macroscopic and microscopic scales. The data base utilized in this work consisted of 17 wells, 1282 m of conventional core, and a 3D seismic survey. From the 17 available wells, five key wells (49/17–10, 49/22–9z, 49/22–12, 49/22–13 and 49/22–15) were selected for detailed analysis. Analysis performed on these wells included (i) 'logging' of structural features in cores, (ii) thin-section analysis of fault rocks and cements, (iii) scanning electron microscopy (SEM), backscattered scanning electron microscopy (BSEM), analytical transmission electron microscopy (ATEM), and cathodoluminescene (CL), and (iv) permeability and mercury intrusion capillary pressure measurements. In addition, static pressure data from wireline formation tester tools were utilized along with 3D seismic mapping to determine the distribution of compartments in the Rotliegendes. No dynamic reservoir data were available for use in this study since production from the Jupiter Fields had not commenced at the time the study was completed.

## Geological history

The geological history of the Jupiter Fields is complicated as a result of several tectonic events (Fig. 2). The oldest deformation directly observed in the Jupiter Fields area is related to the latest Carboniferous Variscan Orogeny,

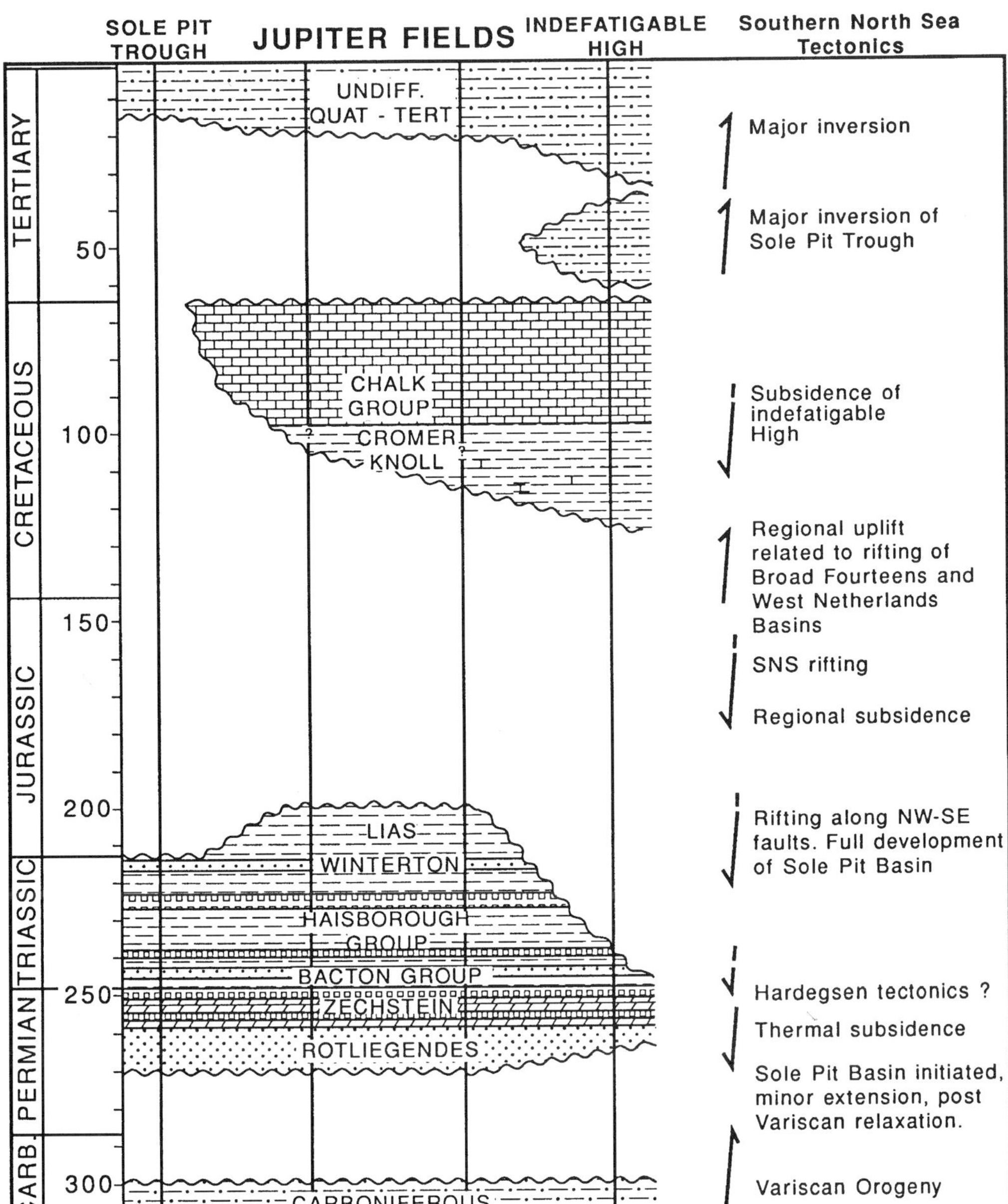


**Fig. 2.** Stratigraphy and associated tectonic events in the Jupiter Fields area. Upward-pointing arrows indicate uplift/inversion events, downward-pointing arrows subsidence/extension.

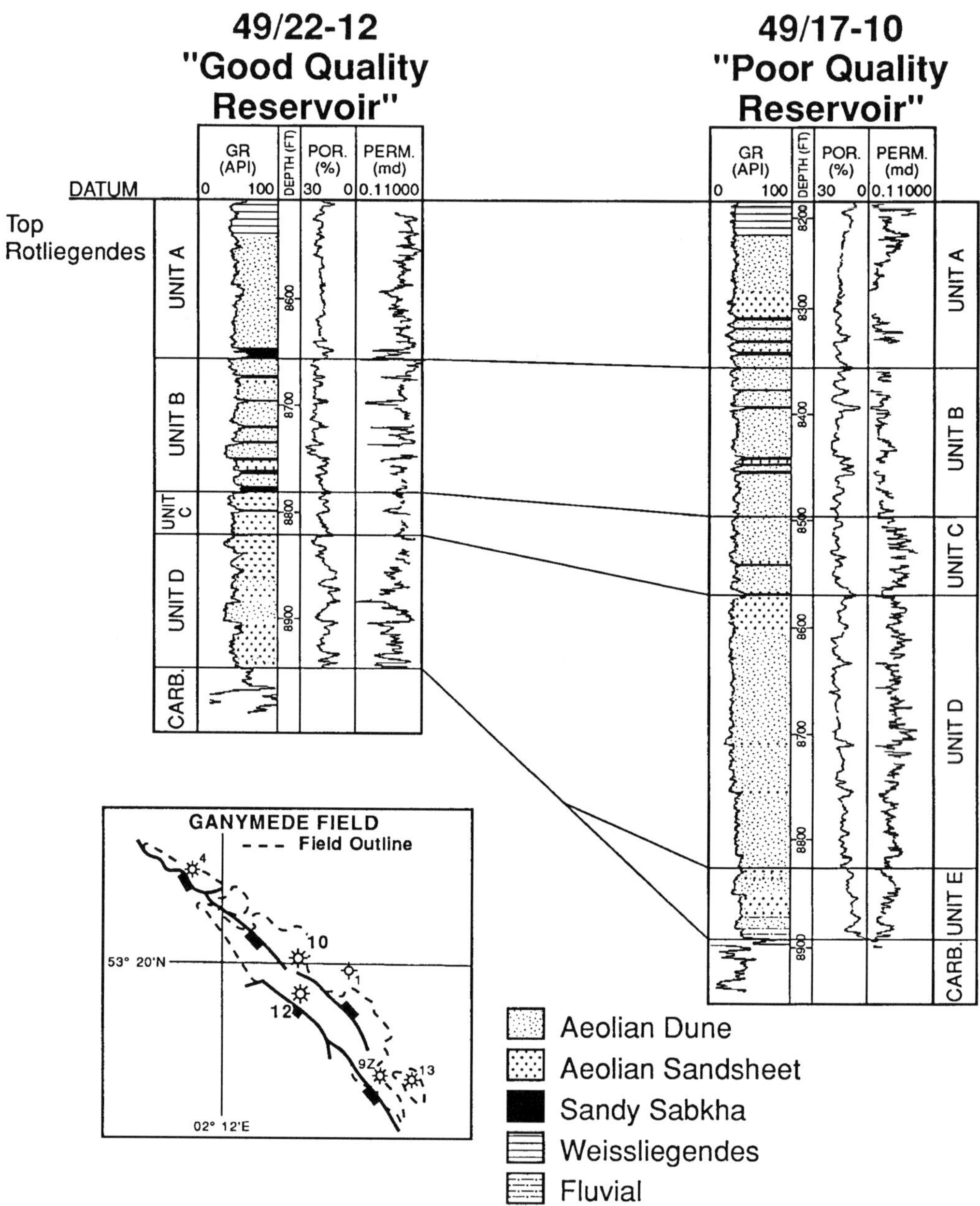



**Fig. 3.** Petrophysical and depositional facies data from the 49/22–12 and 49/17–10 wells in the Ganymede Field.

which resulted in formation of the regional Saalian Unconformity (Glennie 1990). This event was followed by a period of extension and regional subsidence during which the Lower Permian Rotliegendes Group, Upper Permian Zechstein Supergroup, and Triassic strata were deposited. This phase of regional subsidence was followed by accelerated rifting during the Jurassic, which was associated with regional extension throughout northwest Europe (Ziegler 1988). Extension was terminated by regional uplift in the Late Jurassic to Early Cretaceous, which produced the 'Cimmerian' unconformity. A return to subsidence in the Cretaceous provided accommodation space for deposition of the Cromer Knoll and Chalk Groups (Van Hoorn 1987). Deposition of these units was followed by compression and fault inversion in the latest Cretaceous, with episodic inversion and subsidence events occurring throughout the Tertiary (Van Hoorn 1987; Alberts & Underhill 1991).

Thermal-history modelling based on apatite fission track data indicate maximum burial of the Rotliegendes in the Jupiter Fields area occurred in the Late Jurassic to Early Cretaceous (see fig. 9 in Leveille *et al.* this volume). Burial was terminated by uplift associated with 'Cimmerian' regional inversion. This was followed by another phase of burial in the Cretaceous, with further uplift in the latest Cretaceous and Tertiary. The average amount of net uplift for the Jupiter Fields since maximum burial in the Late Jurassic to Early Cretaceous is estimated at between 900 and 1100 m based on the apatite fission track data, regional trends in Bunter Shale velocities and Carboniferous vitrinite reflectance data.

Diagenesis in Rotliegendes sandstones varies considerably between fault blocks in the Jupiter Fields. These variations resulted from differences in structural position and pore fluid chemistry during the main phase of diagenesis in the Jurassic to Early Cretaceous (Leveille *et al.* this volume). Fault blocks that were structurally low during this period contain large volumes of authigenic illite (11–14 wt%), along with quartz, dolomite, anhydrite and siderite. Fault blocks that were structurally high contain lesser amounts of illite (5–8 wt%), as well as berthierine/chlorite, kaolin, quartz, dolomite, anhydrite and siderite.

## Evidence for fault sealing

Gas accumulations in the Jupiter Fields are contained in sandstones of the Lower Permian Rotliegendes Group. In the vicinity of the fields, the Rotliegendes is between 75 and 250 m thick and is composed mainly of aeolian dune and aeolian sandsheet facies, with minor amounts of sandy sabkha and fluvial facies (Fig. 3). Average porosities in the sandstones range from 12 to

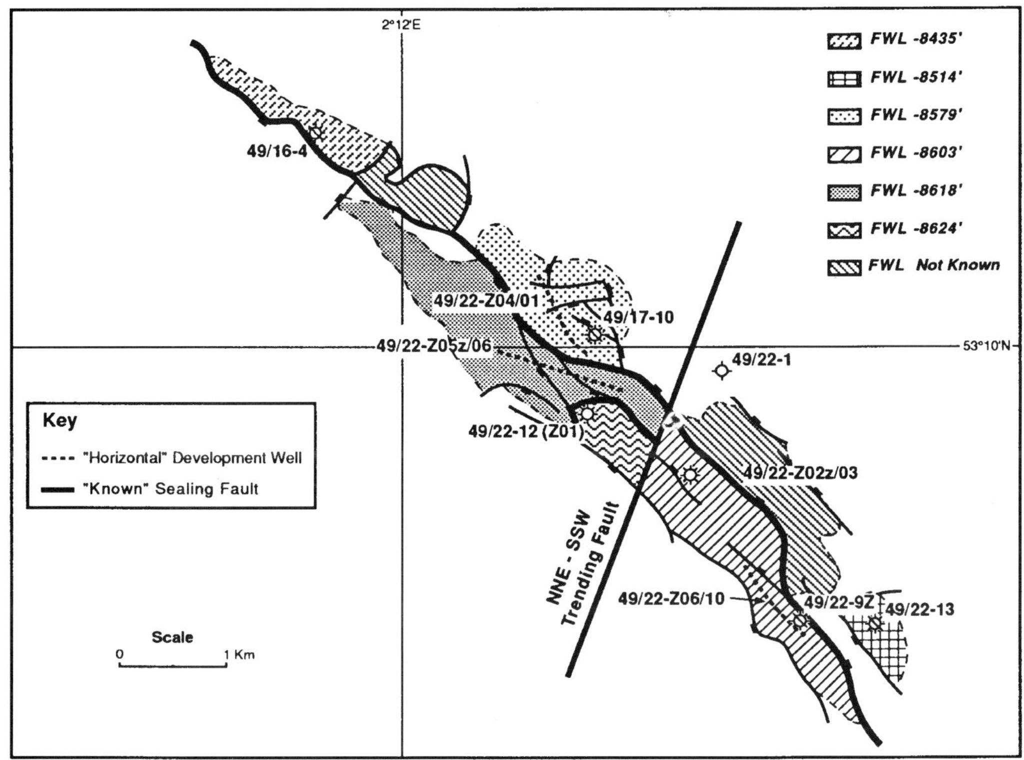


**Fig. 4.** Map showing differences in free water levels (FWL) in the Ganymede Field and the location of known sealing faults.

21%, and geometric mean permeabilities vary between 2 and 200 md.

Faulted dip closures at top Rotliegendes form the primary trap for the Jupiter Fields gas accumulations. Maximum gas column heights for the five accumulations vary from 15 m to over 200 m. Many of the gas accumulations are internally faulted, with some faults being sealing based on differences in free water levels, hydrocarbon compositions and pressures between fault blocks. As displacements on many of the sealing faults are less than the thickness of the Rotliegendes, fault sealing occurs where Rotliegendes sandstones are juxtaposed across the faults (i.e. intra-Rotliegendes sealing faults).

The presence of sealing faults is best documented in the Ganymede Field, where data from six exploration and appraisal wells, and four development wells define several fault sealed compartments. Fault sealing also controls separation between the Thebe and Sinope discoveries, and may compartmentalize the Europa gas accumulation.

The Ganymede Field is an elongate, faulted, high-relief dip closure at top Rotliegendes level.

Six different free water levels have been encountered by wells drilled to date in the field (Fig. 4). These have been used to define six reservoir compartments, four of which require faults to seal where Rotliegendes sandstones are juxtaposed across the faults. The maximum offset between free water levels in adjacent fault sealed compartments is a few tens of metres.

The differences in free water levels in the Ganymede Field are comparable to differences observed in the Indefatigable Field (see fig. 2 in Pearson *et al.* 1991). Larger differences in free water levels have been reported for the Barque Field (up to 50 m; see fig. 1 in Farmer & Hillier 1991*a*) and the Clipper Field (up to 115 m; see figs 1 and 5 in Farmer & Hillier 1991*b*).

Wireline formation tester pressure data acquired prior to first production in the Ganymede Field indicate pressure differences of up to 35 PSIa exist between adjacent fault blocks (Fig. 5). Pressure differences are observed in the gas-filled portion of the Rotliegendes and also in the underlying aquifer. Well control and 3D seismic mapping indicate the differences in pressure are due to intra-Rotliegendes sealing faults.

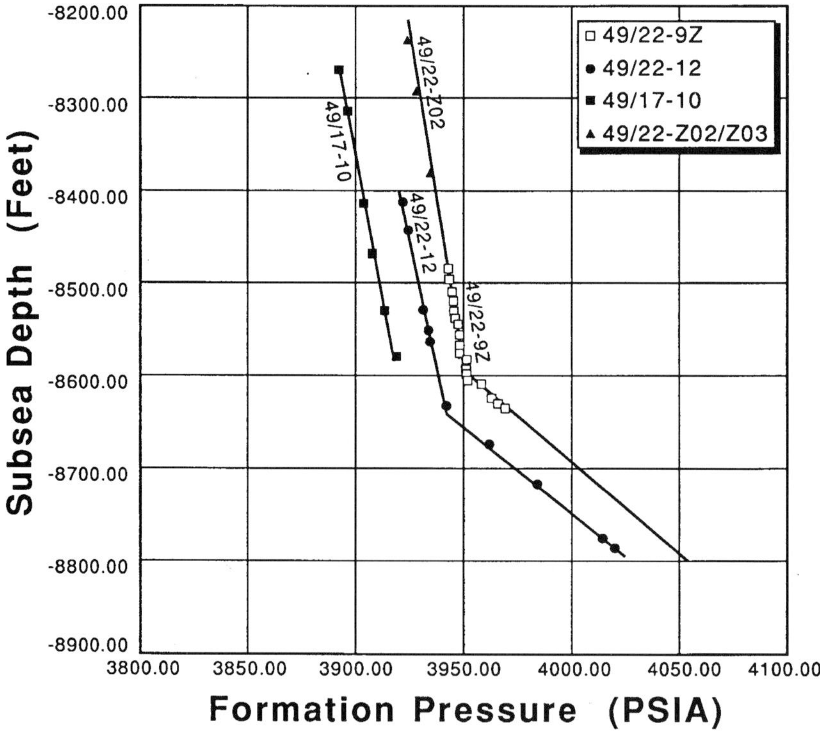


**Fig. 5.** Wireline formation tester data from the Rotliegendes in the Ganymede Field. Notice pressure differences between wells in both the gas filled portion of the Rotliegendes and the underlying aquifer (well locations shown in Fig. 4).

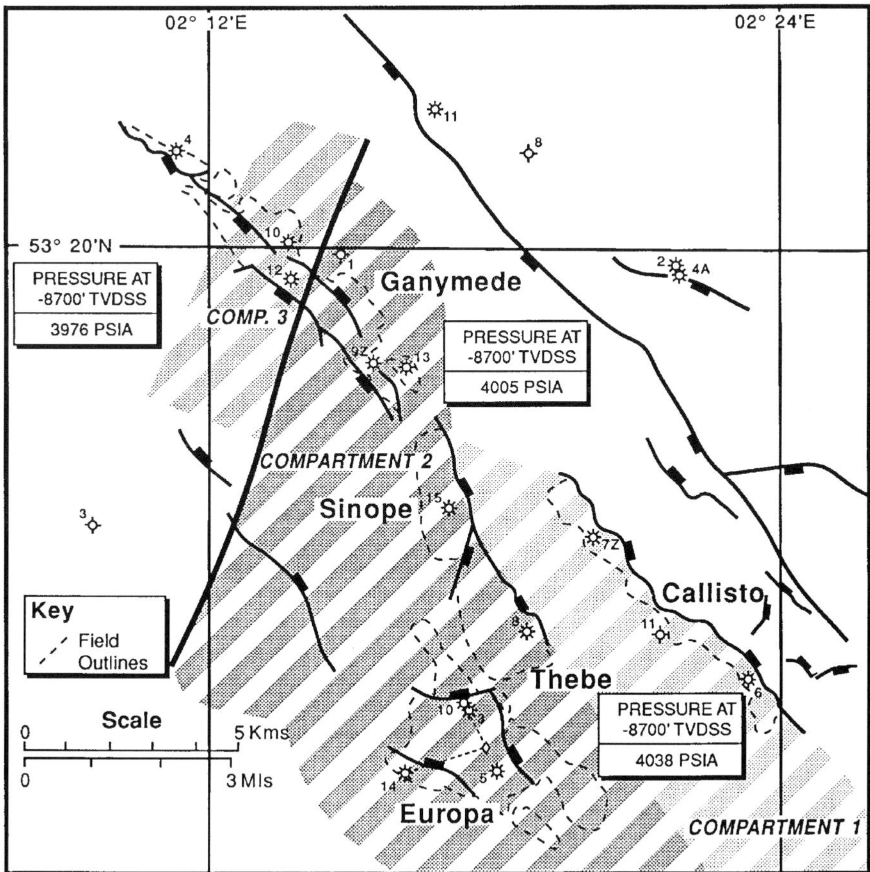


**Fig. 6.** Map showing aquifer compartments, major faults and gas accumulations in the Jupiter Fields area.

Three aquifer pressure compartments have been identified in the Jupiter Fields area, two of which underlie parts of the Ganymede Field (Fig. 6). The size of the aquifer compartments are large, with some of the compartments covering several tens of square kilometres. The boundaries between compartments appear to be sharp, and are interpreted as sealing faults. 3D seismic mapping indicates the compartments are sealed by a combination of intra-Rotliegendes sealing faults and faults that juxtapose the Rotliegendes against the overlying impermeable Zechstein evaporites or underlying mostly impermeable Carboniferous section.

Hydrocarbon composition data from the Ganymede Field provides further evidence that the field is compartmentalised by sealing faults. Hydrocarbon compositions are different on the two sides of a major fault running NW–SE down the centre of the field (Fig. 7). Wells drilled to the west of the fault have recovered gas with a lower density and lower condensate yield than wells drilled on the east side of the fault. The presence of different fluid compositions across this fault is consistent with data that indicate gas charging occurred at different times on either side of the fault (Leveille *et al.*, this volume).

## Characteristics of sealing faults

In this section we discuss the characteristics of intra-Rotliegendes faults in the Jupiter Fields area at three different scales; seismic (tens of metres to kilometres), macroscopic ('core scale': millimetre to tens of metres) and microscopic (less than one millimetre). Observations regarding the sealing potential of faults based on these characteristics are reviewed in the next section.

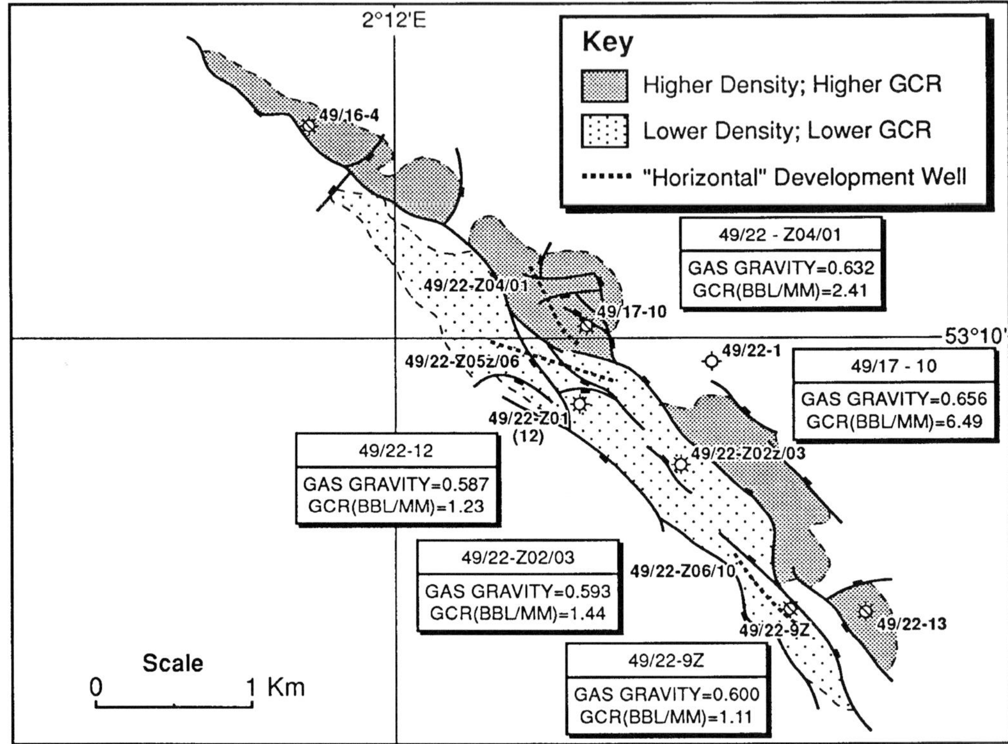


**Fig. 7.** Map showing differences in hydrocarbon composition in the Ganymede Field. Hydrocarbon compositions shown in boxes are from DST and PVT data. Wireline formation tester pressure measurements from 49/22–13 well suggest hydrocarbon compositions similar to 49/17–10.

## Seismic-scale characteristics of faults

Three main fault trends are present at top Rotliegendes level in the vicinity of the Jupiter Fields. These are a NW–SE trend, an E–W trend, and a NNE–SSW trend (Fig. 8). Differences in the geometries of faults of each of these trends have implications for their sealing potential.

The dominant fault trend in the Jupiter Fields area is NW–SE. This trend is present over much of the Southern North Sea and is commonly referred to as the 'Gas Basin' trend. Faults of this trend are believed to have formed as a result of reactivation of older structural features that originally formed during Late Caledonian and/or Variscan tectonic events (Glennie 1990; Ziegler 1990; Coward 1993; Oudmayer & de Jager 1993). The post-Rotliegendes deformational history of many of these faults is believed to have been complicated, with faults of this trend controlling Rotliegendes thickness trends in the Lower Permian, diagenesis in the Jurassic to Early Cretaceous, and inversion movements in the Cretaceous and Tertiary (Leveille et al. this volume).

Characteristics of the NW–SE-trending faults include high length-to-throw ratios, throw reversals along strike, and complex fault plane geometries. These characteristics are believed to have resulted from repeated reactivation during the several tectonic events that affected the Jupiter Fields area. These characteristics also suggest that a component of oblique slip was associated with some of the tectonic events, which is consistent with structural models for the evolution of the southern North Sea (Oudmayer & de Jager 1993; Van Hoorn 1987; Albert & Underhill 1991).

Some faults with a NW–SE trend are known to be sealing in the Jupiter Fields area (see Fig. 4). Faults with this trend have also been identified as sealing faults in other Rotliegendes gas fields (see fig. 2 in Pearson et al. 1991). Seismic scale characteristics that contribute to the sealing potential of faults with this trend include their continuity along strike and their propensity to link with other faults.

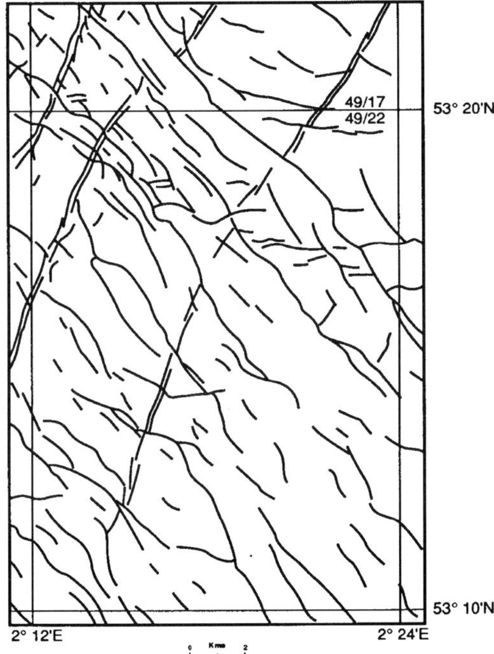


**Fig. 8.** Line drawing of fault pattern at top Rotliegendes level. Fault pattern is based on 3D seismic mapping and top Rotliegendes seismic time images.

E–W-trending faults are much less common in the Jupiter Fields area than NW–SE-trending faults, being confined mostly to broad zones in the northeastern quadrant of the study area (Fig. 8). Some of the E–W-trending faults exhibit reversals in the sense of displacement along strike and display complex fault plane geometries, suggesting multiple phases of deformation similar to the NW–SE-trending faults. Faults of this trend are therefore believed to be relatively old features that have been reactivated several times.

E–W-trending faults generally have lower length to throw ratios and are less continuous along strike than NW–SE-trending faults. None of the E–W-trending faults are known to be sealing in the Jupiter Fields area, but this may be more a function of their relative scarcity and the position of wells drilled to date rather than a true reflection of their sealing potential.

The other significant fault trend present at top Rotliegendes is NNE–SSW. Faults of this trend are subtle but highly continuous features, with strike lengths of tens of kilometres but only limited vertical offset. These faults are difficult to image without 3D seismic data, and therefore have not often been identified on 2D seismic interpretations.

An interesting characteristic of faults with a NNE–SSW trend is the absence of significant vertical offset despite strike lengths that exceed tens of kilometres. Vertical offset is generally only a few tens of metres, close to the resolution of seismic data. The faults commonly occur as paired, subparallel fault systems with a low-relief graben situated between the bounding faults (Fig. 9). The width of the downthrown central block is generally several tens of metres to a few hundreds of metres. Spacing between the fault bounded grabens is variable, but is usually between 1 to 4 km.

The timing of movement on the NNE–SSW-trending faults is believed to be relatively recent since fault geometries are simple and faults of this trend are continuous across other structures. The mechanism by which these paired fault systems formed is not completely understood at this time, although it is believed these features may have resulted from reactivation of older basement lineaments (Caledonian–Variscan) due to oblique shear during Tertiary inversion events. Additional study is needed, however, to determine how these 'paired' fault systems with a downthrown central block formed.

Although vertical offset on NNE–SSW-trending faults is usually insignificant, some faults with this trend are sealing. A good example of this occurs in the Ganymede Field, where a NNE–SSW-trending fault zone compartmentalizes both the gas filled portion of the Rotliegendes and the underlying aquifer (see Figs 4 and 5). Reservoir quality in wells located on either side of the fault zone is excellent, with geometric mean permeabilities of tens to hundreds of millidarcies. Permeability reductions of many orders of magnitude must therefore have occurred within the fault zone to have created a seal capable of compartmentalising the gas reservoir and aquifer over geological time.

Seismic scale characteristics of the NNE–SSW trending faults that contribute to sealing potential include their extremely continuous nature and their occurrence in fault pairs bounding a downthrown central block. The occurrence of these faults in pairs may be particularly important in determining sealing potential, since it effectively creates a zone tens to hundreds of metres wide across which fluid flow is impaired.

### Macroscopic-scale characteristics of faults

Macroscopic-scale characteristics of faults in the Jupiter Fields were investigated by studying faults and fractures present in cores from the

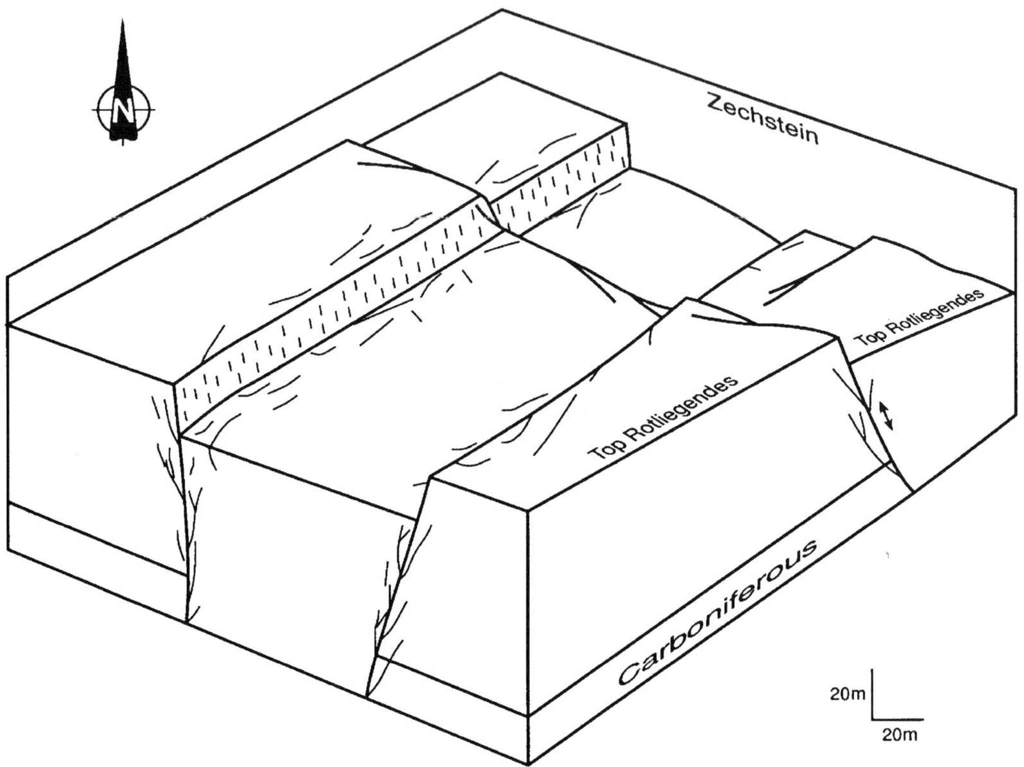



**Fig. 9.** Schematic block diagram showing characteristic geometry of NNE–SSW-trending faults. Diagram depicts NNE–SSW-trending fault cutting NW–SE-trending fault that had both extensional and later reverse movements (i.e. reactivated).

Rotliegendes. None of the cored wells cut a known sealing fault, but several of the wells do penetrate seismically mappable and/or subseismic faults.

One of the more interesting wells for studying macroscopic scale characteristics of faults is the 49/22–9z well. This well penetrates a major NW–SE trending fault zone in the southern part of the Ganymede Field (see Fig. 3 for location). The well entered an upthrown, gas-filled fault block at the Rotliegendes level, after drilling through a fault which juxtaposes Zechstein salts against Rotliegendes sandstones (Figs 10 and 11). About 80 m of the Rotliegendes section is 'faulted out' in the well. Present-day vertical offset across the fault is about 120 m, but may have been considerably greater prior to Cretaceous and Tertiary inversion events (see Leveille *et al.*, this volume).

Conventional core was acquired in the 49/22–9z well from approximately 2 m below the top of the Rotliegendes to just below the contact between the Rotliegendes and underlying

Carboniferous strata. The cored section does not include the main slip surface of the fault zone, but does contain the section in which the damage zone associated with the main slip surface should occur.

The distribution and characteristics of faults and fractures in the cored section are shown in Fig. 12. Several well-defined clusters of faults and fractures are present in the upper part of the cored section adjacent to the main slip surface. The clusters of deformation represent areas of strain accommodation (i.e. small faults) in the region around the main fault (i.e. the damage zone). Core scale characteristics of the damage zone observed in the 49/22–9z well indicate that damage zones contribute to the sealing capacity of faults in the Rotliegendes. The width of the damage zone encountered in the well is estimated at several tens of metres (correcting for the wellbore and fault geometries). Within this zone, numerous small faults and fractures are present that degrade reservoir quality by reducing permeability. The

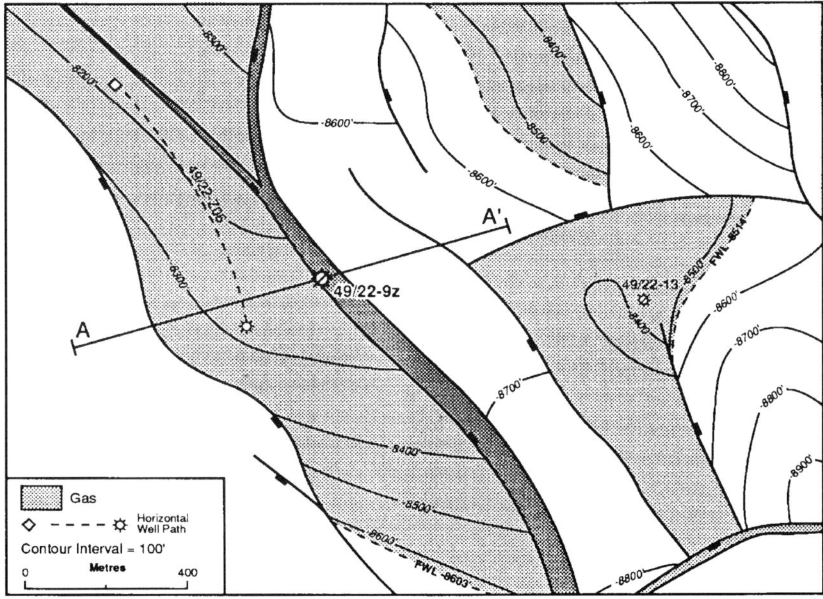


**Fig. 10.** Top Rotliegendes depth structure map of the southern part of the Ganymede Field. Location of structural cross section in Figure 11 shown by line A–A'.

frequency of faults and fractures decreases away from the main slip surface, as does the amount of offset on the faults. Faults in the core exhibit complex, anastomosing geometries that are probably typical of the 3D distribution of structural features within the damage zone on a large scale (Fig. 13). These characteristics suggest that structural features within the damage zone can form significant impediments to fluid flow, which along with the main

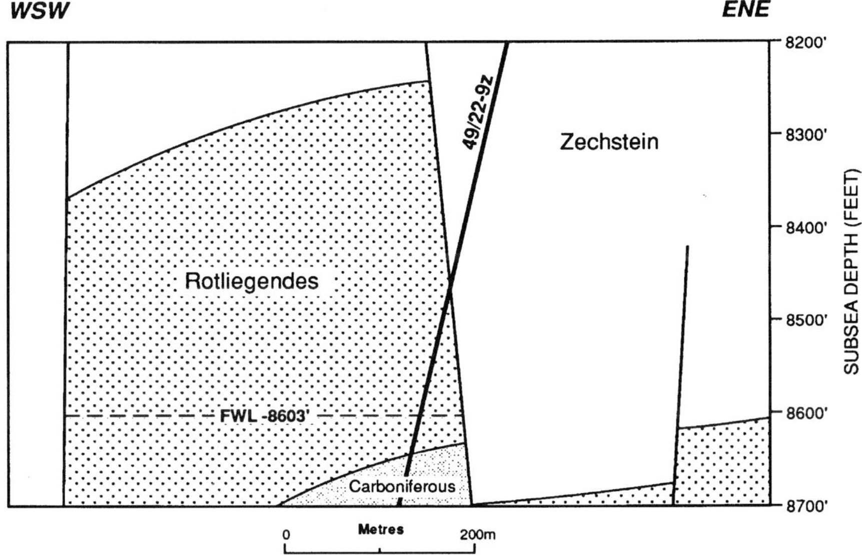


**Fig. 11.** Structural cross-section showing the geometry of the 49/22–9z well in the southern part of the Ganymede Field (location of cross-section is shown in Fig. 10).

slip surface, can substantially diminish transmissivity across the fault zone and thereby compartmentalize the reservoir.

### Microscopic-scale characteristics of faults

The types of deformation features commonly observed in Rotliegendes cores from the Jupiter Fields area include disaggregation zones, cataclastic faults, and fractures. Many of the faults and fractures have been cemented by salt, anhydrite and/or quartz. No open faults or fractures were observed in any of the cored sections.

Disaggregation zones formed in the Jupiter Fields during shallow burial and tend to have similar reservoir properties (i.e. porosity, permeability, and capillary pressure characteristics) to the surrounding undeformed 'host' sandstones. Reservoir properties were not significantly altered in disaggregation zones for two reasons. Firstly, disaggregation zones developed at shallow burial depths where confining stress levels were low and the sandstones were uncemented. Deformation therefore occurred by independent particulate flow rather than by the cataclastic processes of grain fracturing and frictional grain sliding which are more prone to damaging reservoir properties. Secondly, the disaggregation zones formed before the onset of authigenic clay growth, so the 'host' sandstones were essentially clay-free. Therefore, formation of the disaggregation zones did not result in deformation-induced mixing of phyllosilicates with framework grains, which is a process that often results in a reduction in reservoir properties due to the replacement of macroporosity with microporosity.

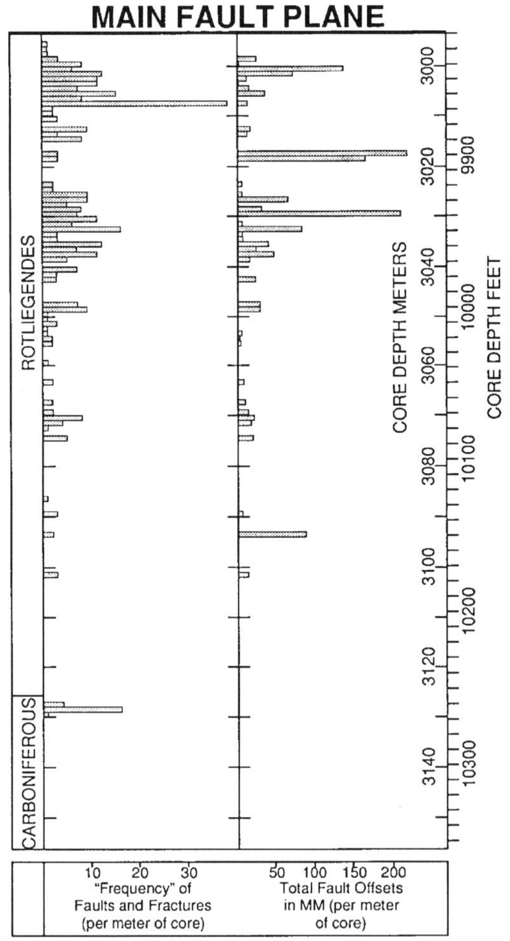


**Fig. 12.** Structural 'log' of the cored interval from the 49/22–9z well showing the frequency of faults and fractures in the core (number of faults and fractures per metre of core) and the cumulative amount of offset measured on faults in the core (per metre of core).

**Fig. 13.** Portion of the cored Rotliegendes interval in the 49/22–9z well. Dark-coloured portions of the core are undeformed sediment, white portions are deformation bands.

Cataclastic faults in the Jupiter Fields area formed at deeper burial depths than disaggregation zones and can be divided into three types; cataclastic faults (*sensu stricto*), cemented cataclastic faults and phyllosilicate-rich cataclastic faults. Differentiation of cataclastic faults into these three types is based on differences in the principal processes responsible for the reduction of reservoir properties. In cataclastic faults (*sensu stricto*) grain fracturing and friction sliding resulted in a decrease in grain size and grain sorting leading to porosity collapse. These processes led to a reduction in pore size and hence diminished reservoir properties. Cemented cataclastic faults formed by the same processes but were also subject to the extensive growth of new minerals within and directly adjacent to the fault during and soon after deformation. Phyllosilicate-rich cataclasites are those in which the pore structure was partly controlled by the presence of fine-grained phyllosilicates. This type of fault rock forms as a result of the deformation-induced mixing of authigenic clays with framework grains and fragments produced by cataclasis.

Fault rocks in which cataclasis (*sensu stricto*) is the main process responsible for a reduction in reservoir properties generally develop in clean sandstones with a low phyllosilicate content (<10%). If the phyllosilicate content is higher than about 10%, then the deformation induced mixing of phyllosilicates with framework grains becomes an important process in reducing reservoir properties. In the Jupiter Fields, the Rotliegendes contains negligible detrital clay, but in places has moderate to high abundances of authigenic clay. This has resulted in the formation of phyllosilicate-rich cataclastic rocks in some areas. This type of fault rock is particularly common in cores from parts of the Jupiter Fields where early precipitation of authigenic berthierine/chlorite occurred, but may also be present in other areas (see Leveille *et al.*, this volume, for a discussion of authigenic clay distribution).

Cataclastic faults in the Jupiter Fields area appear to be capable of reducing reservoir properties sufficiently to form effective seals. Cataclasites observed in cores exhibit a wide range of microstructures from weakly lithified to well lithified cataclastic fault rocks (Fig. 14a and b). Sealing capacity increases with the intensity of lithification, with well lithified cataclasites having permeabilities up to three orders of magnitude lower than surrounding 'host' sandstones.

Cemented faults and fractures exhibit the largest reductions in reservoir properties of any of the deformation features present within the

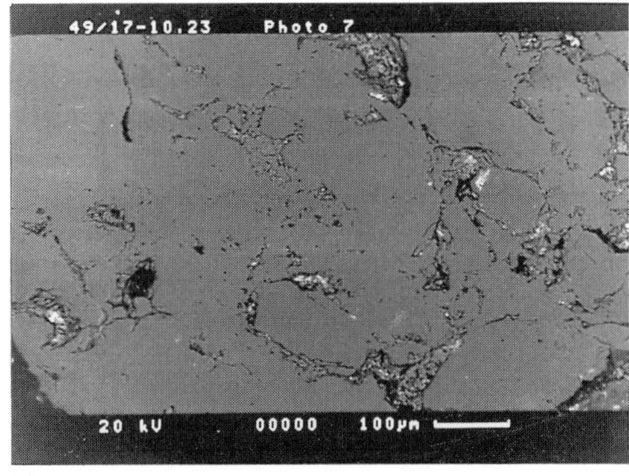

(a)

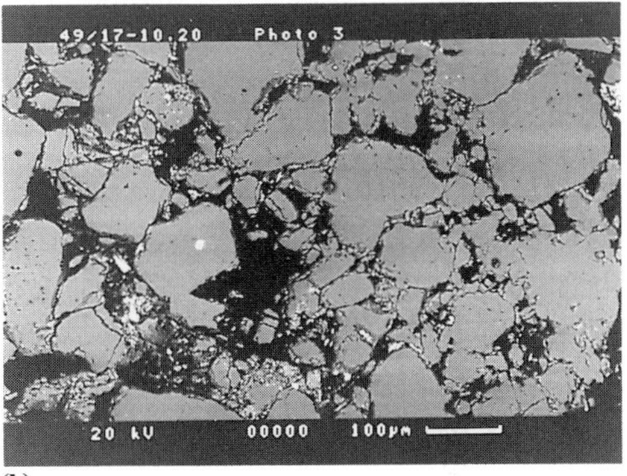

(b)

**Fig. 14.** BSEM micrographs of: (**a**) very well lithified cataclastic fault, (**b**) poorly lithified cataclasite.

Jupiter Fields area. Volumetrically significant cements include salt, anhydrite and quartz (Fig. 15a and b). Cements are generally restricted to individual faults and fractures, with surrounding 'host' sandstones being essentially cement free. Permeability reductions of over three orders of magnitude have been measured in cemented fault rocks.

Salt and anhydrite cements are present in a large number of the faults and fractures in the Jupiter Fields area. The exact mechanism for salt and anhydrite precipitation within the faults and fractures is not known. However, it seems likely that these cements were ultimately derived from pore fluids that originated within the overlying Zechstein evaporites (Fig. 16). Saline-rich Zechstein pore fluids are believed to have moved downwards into the Rotliegendes during fault dilation events due to their greater density (Leveille *et al.* this volume). Precipitation of halite and anhydrite may then have occurred as a result of changes in temperature, pressure, and/or mixing with *in situ* Rotliegendes pore fluids. Alternatively, precipitation may have taken place during fault reactivation as Zech-

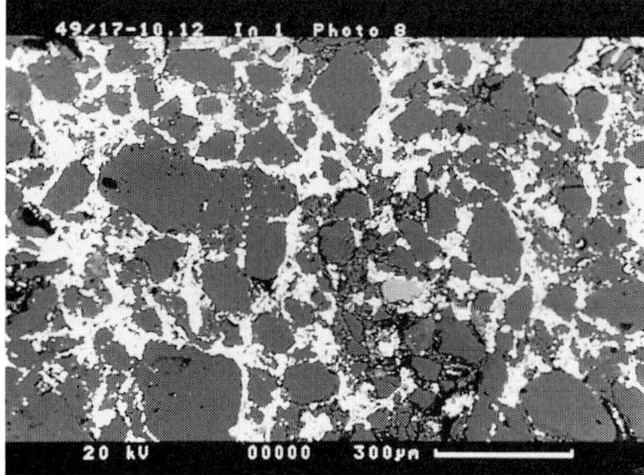

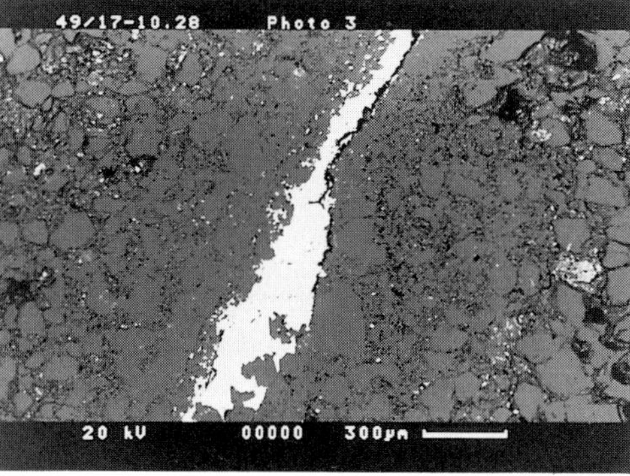

**Fig. 15.** (a) Photomicrograph of fault in Rotliegendes showing pervasive halite cementation (bright areas). (b) Anhydrite cemented fault (anhydrite bright areas). Note early cataclastic fault rock was fractured and subsequently filled with anhydrite cement.

stein derived brines within the Rotliegendes were expelled along a decreasing pressure/temperature gradient. Evidence that faults acted as both conduits and seals at different times due to episodic reactivation of faults is present in some cores (Fig. 15b).

Quartz cement is present in a number of the cataclastic faults, although it is generally absent, or present in only small amounts within undeformed Rotliegendes sandstones. Quartz cement was probably produced locally within the Rotliegendes as a result of pressure solution within fault zones and surrounding 'host' sandstones, although the possibility of precipitation from Carboniferous-derived pore fluids that moved up faults can not be excluded. It is believed that preferential cementation of faults occurred because cataclastic processes produced fresh surfaces which acted as ideal nucleation sites for silica. The scarcity of quartz cement in undeformed portions of the Rotliegendes is considered to be a result of the detrital quartz

grains being covered with authigenic clay and/or hematite coatings, which suppressed quartz overgrowth nucleation. This conclusion is supported by the observation that quartz cements occur as isolated euhedral crystals rather than syntaxial overgrowths.

Permeability values for the fault types present in the Jupiter Fields area are summarized in Fig. 17. Permeability measurements were made with a specially constructed water-flow permeameter capable of measuring permeability to 0.0001 mD. The lowest permeabilities occur in salt and anhydrite cemented faults ($<0.007$ mD), with strongly lithified cataclasites that contain quartz cement being the next best seals, followed by phyllosilicate-rich fault rocks and finally less lithified cataclasites. Permeability reduction factors for all fault types are generally two or more orders of magnitude compared to the surrounding 'host' sandstones.

The permeability values obtained for fault rocks in cores are believed to underestimate the sealing potential of fault zones as a whole for two reasons. Firstly, although some wells cut major faults in the Rotliegendes, the main slip surface of a large fault has never been cored. Since deformation tends to be concentrated on the main slip surface of a large fault, the permeability values measured from small faults in the damage zone adjacent to the main slip surface probably underestimate the overall sealing capacity of the fault zone. The second reason for suggesting that the measured permeabilities overestimate the true permeabilities arises from the difficulty of accurately measuring permeabilities in salt cemented fault rocks. Given the solubility and mobility of salt, even with the careful sampling techniques utilized, dissolution of salt can result in permeability measurements that are higher than the true permeability of the fault rocks. For these reasons, it is believed lower permeability values than those measured are possible for faults within the Jupiter Fields area.

## Sealing potential of fault zones

The sealing potential of fault zones in the Jupiter Fields area can be related to the deformational and cementation histories of the faults and to the timing of diagenetic events within the Rotliegendes as a whole. Critical attributes in determining the sealing capacity of intra-Rotliegendes faults include the (i) three dimensional continuity of the faults and associated damage zones, (ii) intensity of cataclasis, (iii) cement types and volumes, (iv) authigenic clay content at time of

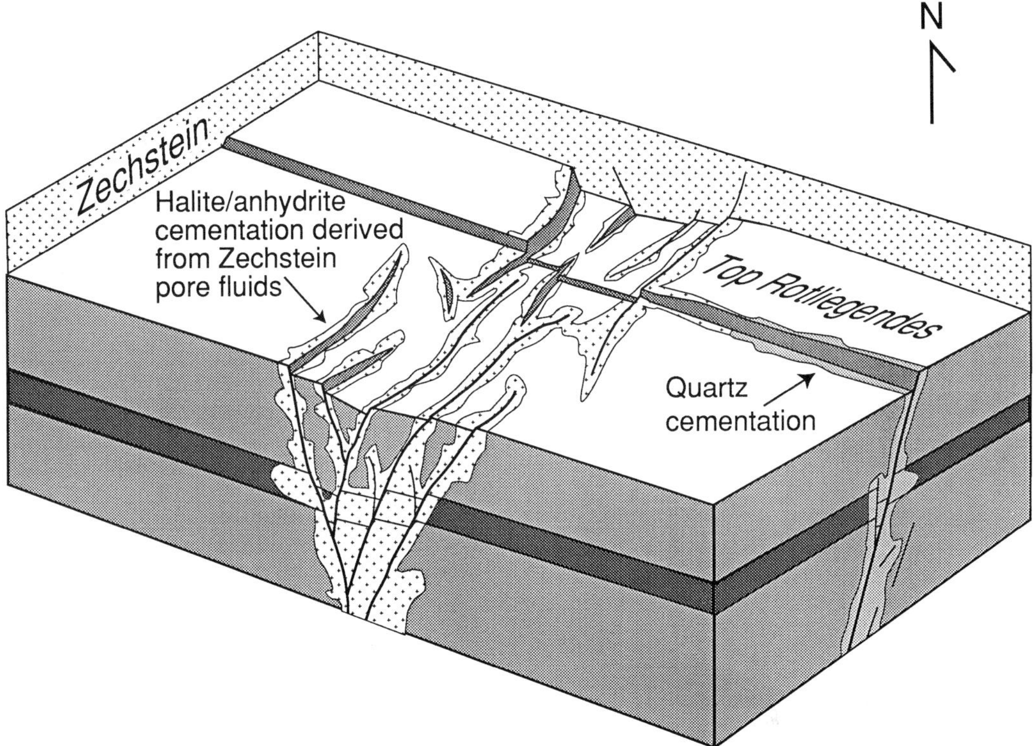



**Fig. 16.** Schematic block diagram depicting the formation of salt and anhydrite cements from saline, high-density pore fluids derived from the overlying Zechstein Supergroup.

faulting, and (v) relative orientation, timing and magnitude of later deformational events. Characteristics of the NW–SE-, E–W- and NNE–SSW-trending faults with regard to these attributes are summarized in Table 1.

From the evidence of the 3D seismic mapping, NNE–SSW-trending faults have the highest along-strike continuity, followed by NW–SE-trending faults, with E–W-trending faults generally being the least continuous. Map-view geometries of the NNE–SSW-trending faults also tend to be simpler than those of the other two fault trends, which commonly are composed of numerous fault segments connected by steps and ramps, and which often form complicated anastomosing patterns. Therefore, from the perspective of fault plane continuity, faults with a NNE–SSW trend are likely to have the highest sealing capacity.

Identification of variations in the intensity of cataclasis associated with faults of different trends cannot be directly measured from the available well database. However, because faults with a NW–SE trend, which tend to have the greatest amount of present day vertical offset,

show evidence of repeated reactivation, and were active during maximum burial (Leveille *et al.* this volume), it is believed that fault zones with this trend will have experienced the most intense cataclasis.

Quantitative data on cement types and volumes suggest that: (i) NW–SE trending faults generally contain large amounts of quartz cement as well as salt and anhydrite cements, and (ii) quartz cement may be less common in NNE–SSW-trending faults, which contain relatively larger volumes of salt and anhydrite cements. The presence of a large amount of quartz cement in the NW–SE-trending faults and little quartz cement in the NNE–SSW-trending faults is consistent with the observation that the former fault trend formed at a deeper structural level, where the supply of dissolved silica resulting from local pressure solution was higher. No direct data on cement types and volumes are available for the E–W-trending faults, but it is thought these would likely have similar characteristics to the NW–SE-trending faults given the similarities in deformational histories.

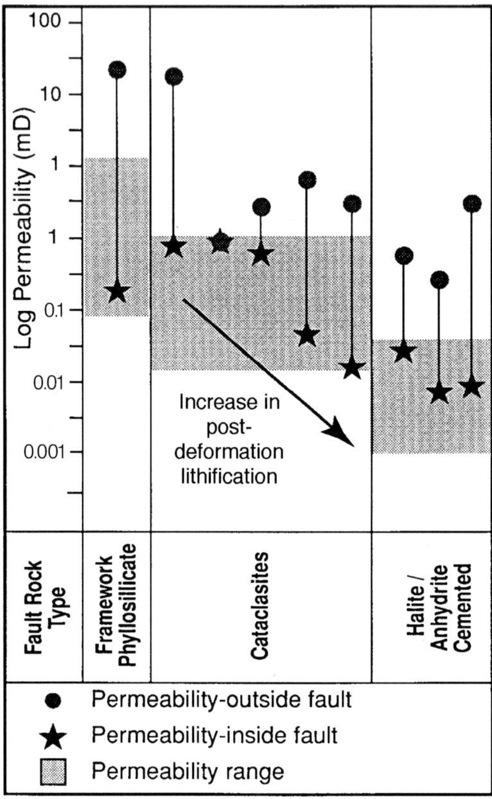



**Fig. 17.** Graph showing permeability measurements for cemented, cataclastic and phyllosilicate-rich fault rocks in Ganymede field.

The importance of phyllosilicate-rich fault rocks in fault zones of various trends is dependent on the volume of authigenic clay present at the time of fault movement, with clay contents greater than about 10% required to form fault rocks of this type. As many of the NW–SE-trending fault zones show evidence of early movement (Leveille *et al.* this volume), some of these fault zones may not contain phyllosilicate-rich fault rocks since movement occurred prior to formation of authigenic clays. However, because fault zones of all trends generally show evidence for movement after the main phase of authigenic clay formation in the Jurassic to Early Cretaceous, the distribution of phyllosilicate-rich fault rocks in the Jupiter Fields area is thought to be mostly dependent on 'area-wide' variations in authigenic clay abundance.

As mentioned previously, many of the NW–SE- and E–W-trending faults show evidence of having been reactivated. Movement on faults of these trends is likely to have occurred during both extensional and compressional events, some of which had components of oblique slip. Core evidence suggests inversion related faulting produced reverse faults and also reactivated pre-existing extensional faults. Reactivation resulted both in breaching of fault seals formed during previous tectonic events and formation of new seal rocks. As a consequence of this relationship, the types of fault rocks formed during the most recent period of movement on a fault are believed to largely control the sealing capacity of faults with NW–SE and E–W trends.

Based on the characteristics summarized in Table 1 the sealing potential of faults in the Jupiter Fields area can be compared and ranked (Table 2). Faults with the greatest sealing potential are believed to be the NNE–SSW-trending faults, with continuous NW–SE- and E–W-trending faults that contain salt/anhydrite cements also being highly effective seals. Sealing potential of faults that have been reactivated is greatest when reactivation was accompanied by salt/anhydrite cementation, although reactivated faults that experienced salt cementation prior to reactivation may retain seal capacity because of

Most of the clay found in the phyllosilicate-rich fault rocks is authigenic, with berthierine/chlorite being especially common in the faults analysed in core. Detailed diagenetic studies indicate that almost all of the authigenic clay formed prior to maximum burial of the Rotliegendes in the Late Jurassic to Early Cretaceous (Leveille *et al.* this volume). These studies also indicate that clay volumes vary dramatically between fault blocks, with clay volumes in some fault blocks being relatively low (4–8%), whereas in other fault blocks they are considerably higher (10–14%).

**Table 1.** *Characteristics of NW–SE-, E–W- and NNE–SSW-trending faults that are important in determining sealing potential of intra-Rotliegendes faults*

| Fault trend | Fault continuity | Intensity of cataclasis | Cement type | Clay content | Reactivation |
|---|---|---|---|---|---|
| NW–SE | Moderate–high | High | Quartz & salt/anhy. | Low–high | High |
| E–W | Low–moderate | Moderate–high | Quartz & salt/anhy. | Low–high | Moderate–high |
| NNE–SSW | High | Low | Salt/anhydrite | Low–high | Low |

**Table 2.** *Relative sealing potential of intra-Rotliegendes faults in the Jupiter Fields area*

**High-sealing potential**

- NNE–SSW-trending faults, especially in sandstones with high authigenic clay content.
- NW–SE- & E–W-trending faults not reactivated after salt/anhydrite cements formed.
- NW–SE- & E–W-trending faults that were subject to intensive cataclasis and were not reactivated.

**Low sealing potential**

- Faults reactivated after salt/anhydrite cementation.
- Disaggregated and cataclastic faults with little or no cementation.

the ability of salt to rapidly 'heal' any breaches generated. The seal capacity of NW–SE- and E–W-trending faults that contain cataclastic, quartz cemented fault rocks will tend to be greatest where they have not been reactivated. Cataclastic faults with little or no cementation have the poorest sealing potential.

## Conclusions

(1) Sealing faults divide Rotliegendes gas accumulations in the Jupiter Fields area into compartments with different free water levels, hydrocarbon compositions and pressures.

(2) Cementation is probably the most effective fault seal mechanism. Volumetrically significant cements include salt, anhydrite and quartz. Cataclastic processes and deformation induced mixing of authigenic clay also contribute to fault seal potential.

(3) Some NW–SE- and NNE–SSW-trending fault zones are known to be sealing, and it is believed E–W-trending fault zones may seal in places.

(4) The sealing potential of individual fault zones is related to their deformational and cementation histories, and to the timing of diagenetic events within the Rotliegendes as a whole. The deformational histories of many fault zones are complicated as a result of the numerous tectonic events that have affected the Jupiter Fields area. Differences in diagenetic and cementation histories between fault zones are mostly related to timing of fault movement relative to depth of burial and pore fluid flow.

(5) Characteristics that enhance the sealing potential of faults include (i) fault plane continuity, (ii) abundant salt and anhydrite cementation, (iii) intense cataclasis, (iv) fault movement after growth of moderate to large volumes of authigenic clay in the 'host' sandstones and (v) little or no reactivation subsequent to seal formation.

## Implications

(1) Appraisal and development programmes for Rotliegendes gas accumulations should be designed with the knowledge that free water levels can vary considerably between fault blocks.

(2) In order to mitigate against the possibility of poor hydrocarbon recovery, development wells should be planned so as to maximise drainage from possible fault sealed compartments. Utilization of high angle and/or horizontal wells that penetrate multiple fault blocks may be a cost-effective way to limit the negative effects of fault sealing. Wells of this type were used in the Ganymede Field development to improve gas recovery in highly faulted areas of the field.

(3) Sealing faults may enhance gas recovery in some areas by limiting water influx.

We wish to thank all of our colleagues at Conoco, The University of Leeds, Mobil, BP and Statoil who contributed to our understanding of fault sealing in the Jupiter Fields. Early versions of this paper were improved as a result of review by K. Ziegler, P. Herrington, and R. Gaupp. Thanks also to K. Ziegler, S. Daines and P. Turner who organized the 'Petroleum Geology of the Southern North Sea: Future Potential' meeting.

## References

ALBERTS, M. A. & UNDERHILL, J. R. 1991. The effect of Tertiary structuration on Permian gas prospectivity, Cleaver Bank area, southern North Sea, UK. *In:* SPENCER, A. M. (ed.) *Generation, Accumulation, and Production of Europe's Hydrocarbons.* Special Publications of the European Association of Petroleum Geoscientists, 1, 161–173.

ALLAN, U. S. 1989. Model for hydrocarbon migration and entrapment within faulted structures. *American Association of Petroleum Geologists Bulletin,* **73,** 803–811.

ANTONELLINI, M. & AYDIN, A. 1994. Effect of faulting on fluid flow in porous sandstones: petrophysical properties. *American Association of Petroleum Geologists Bulletin,* **78,** 355–377.

—— & ——1995. Effect of faulting on fluid flow in porous sandstones: geometry and spatial distribution. *American Association of Petroleum Geologists Bulletin,* **79,** 642–671.

AYDIN, A. & JOHNSON, A. M. 1978. Development of faults as zones of deformation bands and as slip surfaces in sandstone. *Pure and Applied Geophysics,* **116,** 931–942.

COWARD, M. P. 1993. The effect of Late Caledonian and Variscan continental escape tectonics on basement structure, Palaeozoic basin kinematics and subsequent Mesozoic basin development in NW Europe. *In*: PARKER, J. R. (ed.) *Petroleum Geology of Northwest Europe: Proceedings of the 4th Conference.* Geological Society, London, 1095–1108.

EDWARDS, H. E., BECKER, A. D. & HOWELL, J. A. 1993. Compartmentalisation of an aeolian sandstone by structural heterogeneities: Permo-Triassic Hopeman Sandstone, Moray Firth, Scotland. *In*: NORTH, C. P. & PROSSER, D. J. (eds) *Characterization of Fluvial and Aeolian Reservoirs.* Geological Society, London, Special Publications, **73**, 339–365.

ENGELDER, J. T. 1974. Cataclasis and the generation of fault gouge. *Geological Society of America Bulletin*, **85**, 1515–1522.

FARMER, R. T. & HILLIER, A. P. 1991a. The Barque Field, Blocks 48/13a, 48/14, UK North Sea. *In*: ABBOTTS, I. L. (ed.) *United Kingdom Oil and Gas Fields, 25 Years Commemorative Volume.* Geological Society, London, Memoirs, **14**, 395–400.

——1991b. The Clipper Field, Blocks 48/19a, 48/19c, UK North Sea. *In*: ABBOTTS, I. L. (ed.) *United Kingdom Oil and Gas Fields, 25 Years Commemorative Volume.* Geological Society, London, Memoirs, **14**, 417–423.

GAGE, M. 1980. A review of the Viking Gas Field. *In*: HALBOUTY, M. T. (ed.) *Giant Oil and Gas Fields of the Decade 1968–1978.* American Association of Petroleum Geologists Memoirs, **30**, 39–55.

GLENNIE, K. W. 1990. Rotliegend sediment distribution: a result of late Carboniferous movements. *In*: HARDMAN, R. F. P. & BROOKS, J. (eds) *Tectonic Events Responsible for Britain's Oil and Gas Reserves.* Geological Society, London, Special Publications, **55**, 127–138.

HARDING, T. P., & TUMINAS, A. C. 1988. Interpretation of footwall (lowside) fault traps sealed by reverse and convergent wrench faults. *American Association of Petroleum Geologists Bulletin*, **72**, 738–757.

—— & ——1989. Structural interpretation of hydrocarbon traps sealed by basement normal blocks and at stable flank of foredeep basins and rift basins. *American Association of Petroleum Geologists Bulletin*, **73**, 812–840.

HILLIER, A. P. & WILLIAMS, B. P. J. 1991. The Leman Field, Blocks 49/26, 49/27, 49/28, 53/1, 53/2, UK North Sea. *In*: ABBOTTS, I. L. (ed.) *United Kingdom Oil and Gas Fields, 25 Years Commemorative Volume.* Geological Society, London, Memoirs, **14**, 451–458.

HINDLE, A. D. 1989. Downthrown traps of the NW Witch Ground Graben, UK North Sea. *Journal of Petroleum Geology*, **12**, 405–418.

JEV, B. I., KAARS-SIJPESTEIJN, C. H, PETERS, M. P. A. M., WATTS, N. L. & WILKIE, J. T. 1993. Akaso Field, Nigeria: use of integrated 3-D seismic, fault slicing, clay smearing, and RFT pressure data on fault trapping and dynamic leakage. *American Association of Petroleum Geologists Bulletin*, **77**, 1389–1404.

KNIPE, R. J. 1992. Faulting processes and fault seal. *In*: LARSEN, R. M. (ed.) *Structural and Tectonic Modelling and its Application to Petroleum Geology.* NPF, Stavanger, 325–342.

——1993. The influence of fault zone processes and diagenesis on fluid flow. *American Association of Petroleum Geologists Studies in Geology*, **36**, 135–154.

LEVEILLE, G. P., PRIMMER, T. J., DUDLEY, G., ELLIS, D. & ALLINSON, G. 1997. Diagenetic controls on reservoir quality in Rotliegendes sandstones, Jupiter Fields, Southern North Sea. *This volume.*

OUDMAYER, B. C. & DE JAGER, J. 1993. Fault reactivation and oblique slip in the Southern North Sea. *In*: PARKER, J. R. (ed.) *Petroleum Geology of Northwest Europe: Proceedings of the 4th Conference.* Geological Society, London, 1281–1290.

PEARSON, J. F. S., YOUNG, R. A. & SMITH, A. 1991. The Indefatigable Field, Blocks 49/18, 49/19, 49/23, 49/24, UK North Sea. *In*: ABBOTTS, I. L. (ed.) *United Kingdom Oil and Gas Fields, 25 Years Commemorative Volume.* Geological Society, London, Memoirs, **14**, 443–450.

PITTMAN, E. D. 1981. Effect of fault-related granulation on porosity and permeability of quartz sandstones, Simpson Group (Ordovician), Oklahoma. *American Association of Petroleum Geologists Bulletin*, **65**, 2381–2387.

SASSI, W., LAVERA, S. E., & CALINE, B. P. R. 1992. Reservoir compartmentalisation by faults in Cormorant BLK. IV, U. K. northern North Sea. *In*: LARSEN, R. M. (ed.) *Structural and Tectonic Modelling and its Application to Petroleum Geology.* NPF, Stavanger, 355–364.

SMITH, D. A. 1966. Theoretical considerations of sealing and non-sealing faults. *American Association of Petroleum Geologists Bulletin*, **50**, 363–374.

——1980. Sealing and non-sealing faults in the Louisiana Gulf Coast. *American Association of Petroleum Geologists Bulletin*, **64**, 145–172.

SPENCER, A. M. & LARSEN, V. B. 1990. Fault traps in the Northern North Sea. *In*: HARDMAN, R. F. P. & BROOKS, J. (eds) *Tectonic Events Responsible for Britain's Oil and Gas Reserves.* Geological Society, London, Special Publications, **55**, 281–298.

VAN HOORN, B. 1987. Structural evolution, timing and tectonic style of the Sole Pit Inversion. *In*: ZIEGLER, P. A. (ed.) *Compressional Intra-Plate Deformations in the Alpine Foreland. Tectonophysics*, **137**, 239–284.

ZIEGLER, P. A. 1988. *Evolution of the Arctic–North Atlantic rift system.* American Association of Petroleum Geologists Memoirs, **43**.

——1990. *Geological Atlas of Western and Central Europe (2nd edition).* Shell Internatinale Petroleum Maatschappij B.V., The Hague.

# Diagenetic controls on reservoir quality in Permian Rotliegendes sandstones, Jupiter Fields area, southern North Sea


GREGORY P. LEVEILLE[1], TIM J. PRIMMER[2], GRAHAM DUDLEY[2],
DAVID ELLIS[3] & GARETH J. ALLINSON[4]

[1] *Conoco Inc., 600 North Dairy Ashford, Houston, TX 77079, USA*
[2] *BP Exploration, Aberdeen, UK*
[3] *Mobil North Sea Limited, Aberdeen, UK*
[4] *STATOIL (UK) Limited, London, UK*



**Abstract:** Reservoir quality in Lower Permian Rotliegendes Group sandstones in the Jupiter Fields varies considerably between fault blocks due to differing degrees of diagenetic alteration. Average porosities in individual fault blocks range from 12 to 21%, and geometric mean permeabilities vary between 2 and 200 md. Rotliegendes stratigraphy and sedimentology varies little across the area, and therefore does not strongly influence reservoir quality.

The two most important diagenetic processes were the formation of authigenic illite and burial compaction. Illite growth mainly occurred during the Middle Jurassic to Early Cretaceous, coincident with maximum burial of the Rotliegendes. Fault blocks with good reservoir quality generally contain (1) relatively low illite abundances, (2) low-potassium illite, (3) increasing illite abundances with depth, (4) illite ages that decrease with depth and (5) comparatively low amounts of burial compaction. Fault blocks with poor reservoir quality all exhibit (1) high illite abundances, (2) high-potassium illite, (3) no consistent trend in illite abundances or age with depth and (4) high amounts of burial compaction. These observations are believed to indicate that reservoir quality was preferentially preserved in fault blocks that were structurally high during the main phase of illite growth. Factors contributing to the preservation of reservoir quality in structurally high fault blocks included differences in pore fluid chemistry, less burial compaction, and, in places, early hydrocarbon charging. Reservoir quality in undrilled fault blocks can be predicted by reconstructing the structural geometries present during the Mid-Jurassic to Early Cretaceous.


Large differences in reservoir quality exist in the Jupiter Fields area due to varying degrees of diagenetic alteration. These differences impact flow rates, recovery factors, economic value, optimal well designs, and development options. Understanding what processes control reservoir quality is therefore critical to effective exploration, appraisal, and development in this area.

The Jupiter Fields consist of five gas accumulations (Ganymede, Callisto, Europa, Sinope and Thebe) located in Blocks 49/16, 49/17s, and 49/22 in the UK sector of the Southern North Sea (Fig. 1). The gas accumulations are contained in sandstones of the Lower Permian Rotliegendes Group. In the vicinity of the fields, the Rotliegendes is comprised mainly of aeolian dune and sandsheet facies, with minor amounts of sandy sabkha and fluvial facies. These facies were deposited under arid climatic conditions in the interior of a long-lived, mainly dry erg.

The Jupiter Fields provide an excellent opportunity to study diagenetic controls on reservoir quality in Rotliegendes Group sandstones for three reasons. Firstly, the Rotliegendes exhibits little stratigraphic variation across the study area, thus eliminating the problem of distinguishing sedimentological and diagenetic controls on reservoir quality. Secondly, reservoir quality variations are very large and can be related to specific diagenetic processes. Thirdly an excellent data base consisting of 17 wells, 1282 m of conventional core, and a 3D seismic survey is available.

Diagenesis has been shown to have been important in determining reservoir quality in the Rotliegendes in many parts of northwest Europe (Glennie *et al.* 1978; Seemann, 1979; Arthur *et al.* 1986). Several studies describing regional aspects of diagenesis have been published (e.g. Robinson *et al.* 1993; Gaupp *et al.* 1993; Lee *et al.* 1989). There have, however, been few detailed, field-specific diagenetic studies reported in the literature. Field-specific diagenetic studies that have been published include work by Turner *et al.* (1993) on the Ravenspurn North Field and a study by Lahann *et al.* (1993) on the Vanguard Field.

This paper presents the results of a comprehensive study into diagenetic controls on reservoir quality in the Jupiter Fields. The paper


*From* Ziegler, K., Turner, P. & Daines, S. R. (eds), 1997, *Petroleum Geology of the Southern North Sea: Future Potential,* Geological Society Special Publication No. 123, pp. 105–122.

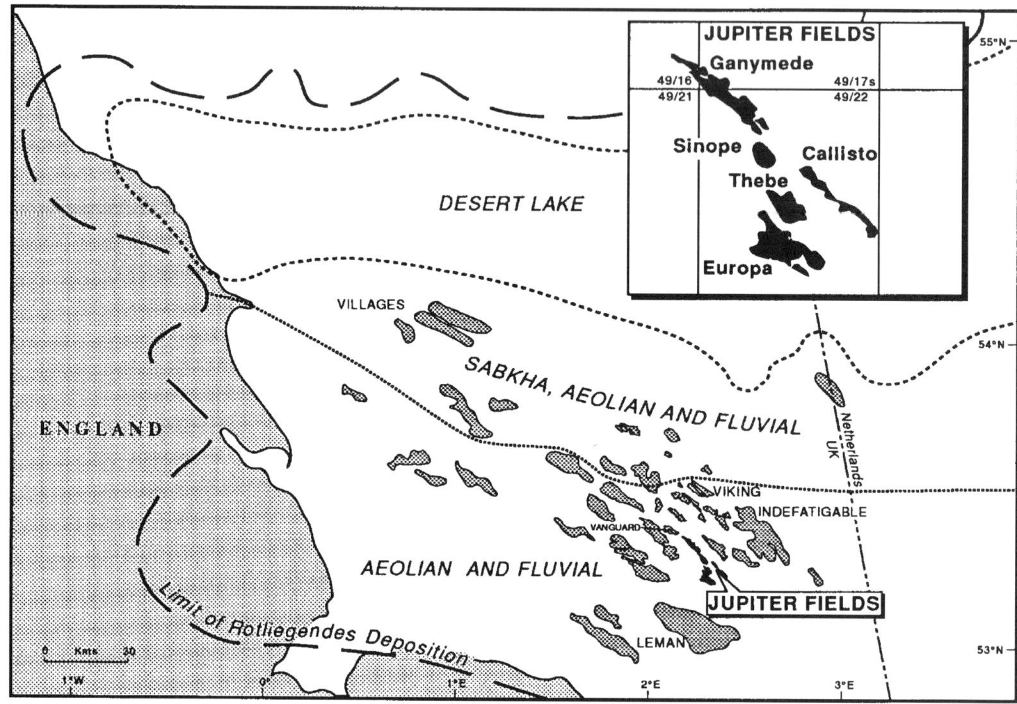


**Fig. 1.** Palaeogeographic map for the Rotliegendes Group showing the location of the Jupiter Fields (solid black; detailed map in inset) and other Rotliegendes gas fields in the UK sector of the Southern North Sea.

describes how diagenetic processes created the large variations in reservoir properties, and relates the diagenetic history to the overall geological history of the area. Observations on how reservoir quality might be predicted prior to drilling are also presented.

## Reservoir quality variations

Reservoir quality in the vicinity of the Jupiter Fields varies considerably between fault blocks. This is most pronounced in the Ganymede Field which is divided into a good quality fault block and a poor quality fault block by a fault running NW–SE down the centre of the field (Fig. 2). The six wells drilled to date on the good quality side of the field have encountered average porosities between 18 and 21%, with geometric mean permeabilities ranging from 38 to 200 mD. Porosities in the five wells on the poor quality side of field vary between 12 and 16%, and permeabilities range from 1.7 to 2.5 mD.

The differences in reservoir quality in the Ganymede Field can be shown to be unrelated to stratigraphic variations since similar facies assemblages with similar textures and detrital mineralogies are present on both the good quality and poor quality sides of the field. In both areas the Rotliegendes consists mostly of aeolian dune and aeolian sandsheet deposits, with lesser amounts of sandy sabkha and fluvial units (Fig. 3). The internal stratigraphy of the Rotliegendes is also the same in both areas, with five regionally correlatable time stratigraphic units identified (Reservoir Units A, B, C, D, E). Type logs from the good- and poor-quality sides of the Ganymede Field showing the stratigraphic correlations and differences in reservoir properties are provided in Fig. 4.

Rotliegendes sandstones in the Jupiter Fields area have sublithic to subarkosic compositions, similar to those reported in other parts of the Southern North Sea (e.g. Glennie 1972). Monocrystalline quartz is the dominant detrital mineral with significant amounts of polycrystalline quartz and lesser amounts of rock fragments and feldspars. The sandstones are generally fine or medium grained and well to moderately sorted. Detrital and infiltrated clay content is low in all facies except the sandy sabkha deposits.

Aeolian dune sandstones usually exhibit the best reservoir properties of any facies in a given area regardless of the overall level of diagenesis

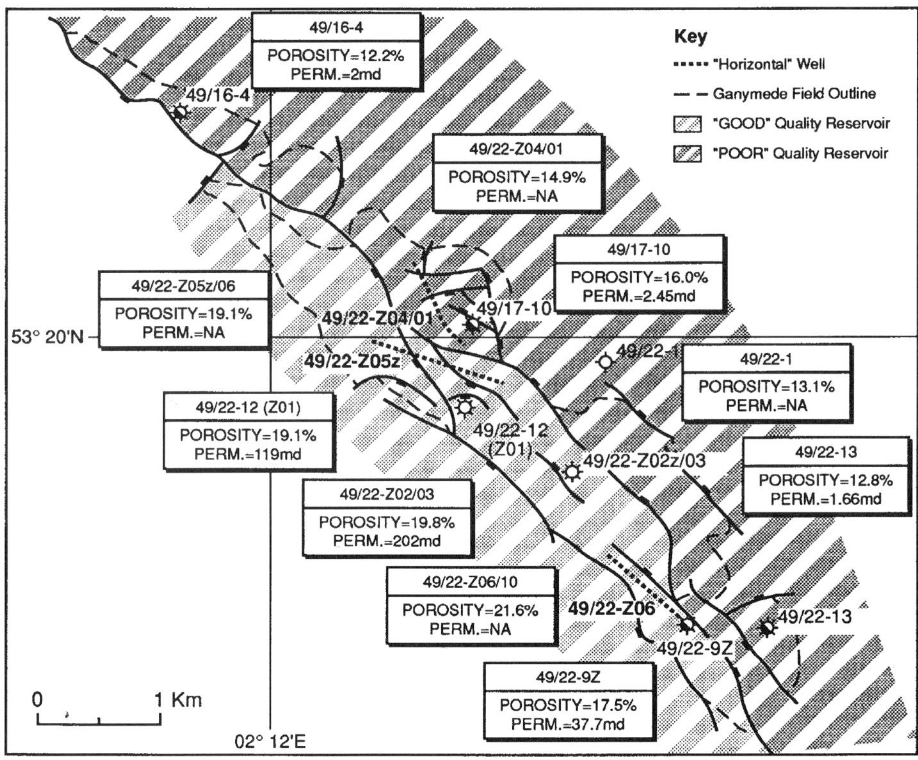


**Fig. 2.** Map showing variations in Rotliegendes Group reservoir quality in the Ganymede Field.

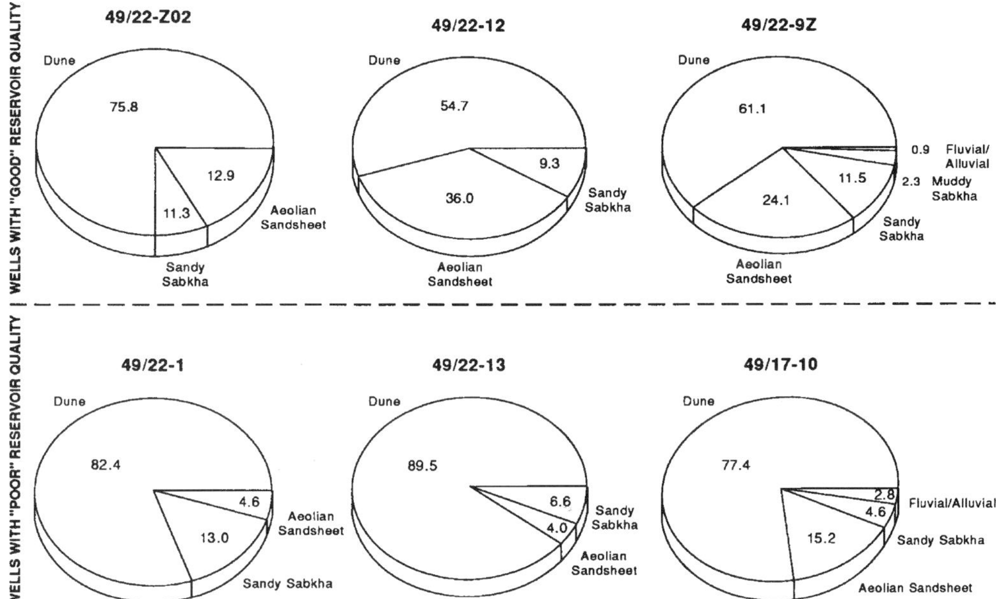


**Fig. 3.** Facies distribution by percentage for wells in the Ganymede Field.

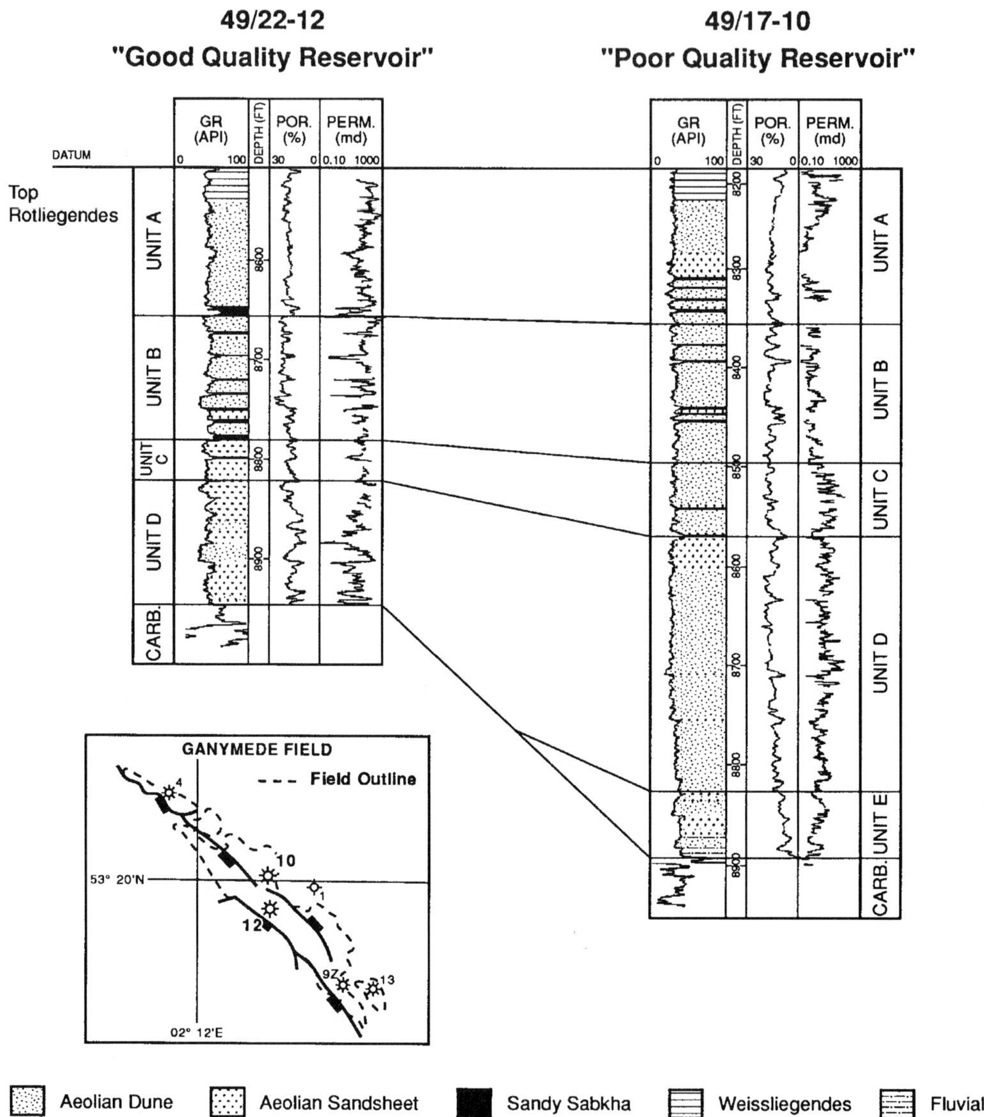



**Fig. 4.** Petrophysical and depositional facies data from the 49/22–12 and 49/17–10 wells. Note the very large differences in reservoir quality, especially permeability, between the two wells (see Fig. 2 for well locations).

that has occurred. This is due to the better sorting of aeolian dune sandstones compared to sandstones of other facies. Aeolian sandsheet deposits generally have reservoir properties almost as good as the dune sandstones, with fluvial and sandy sabkha deposits having the poorest properties.

Sieve analysis and visual grain size estimates indicate that the upper part of the Rotliegendes (Reservoir Units A and B) is, in general, finer grained than the lower part of the Rotliegendes (Reservoir Units C D and E) throughout the Jupiter Fields area. Similar differences in grain size are present in the V-Fields just west of the Jupiter Fields (Grant 1992). This difference in average grain size is an important secondary control on reservoir quality in areas where the overall amount of diagenetic alteration is large (e.g. poor quality area of the Ganymede Field). In these areas permeability is usually lower in

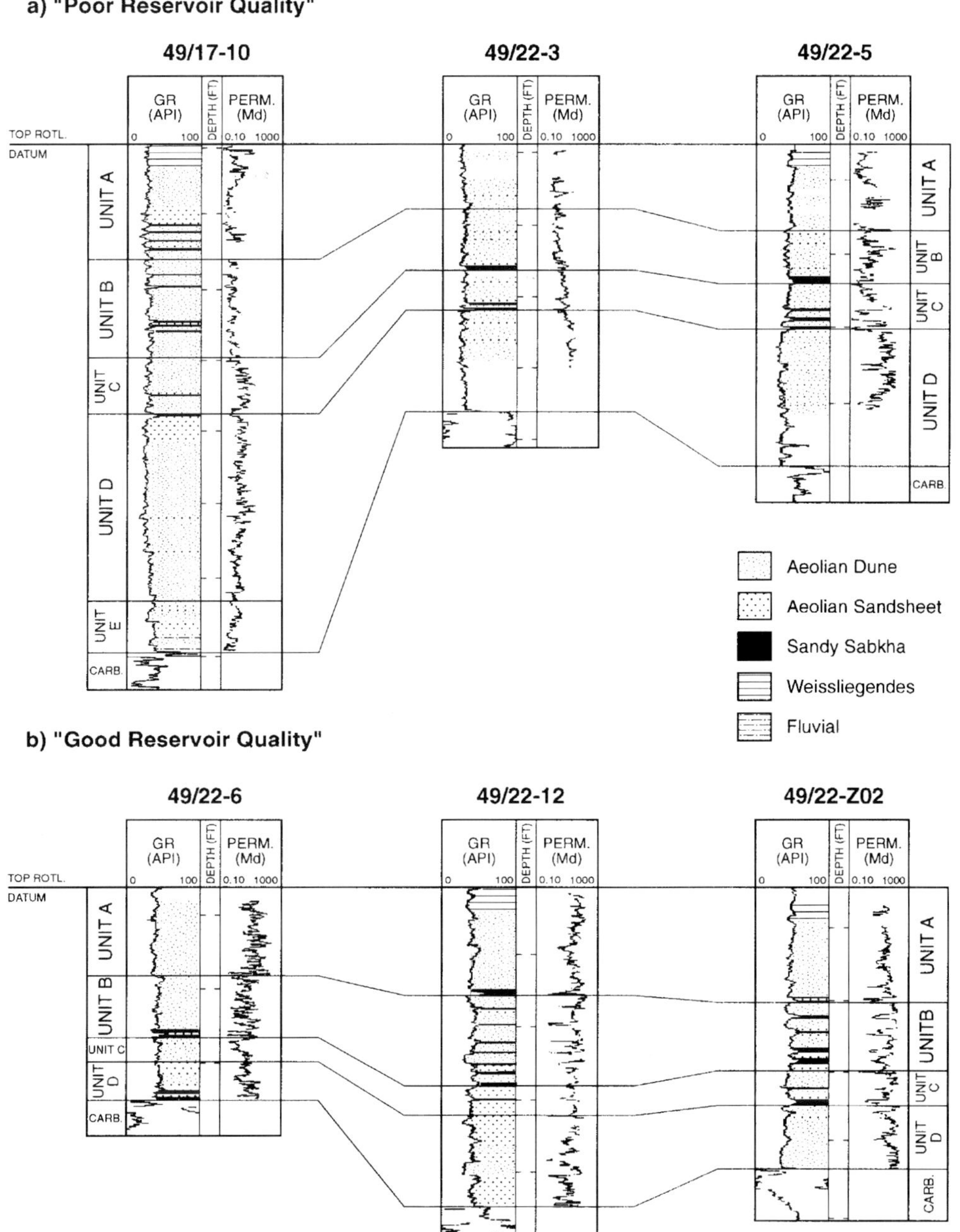



**Fig. 5.** Core permeability data from Jupiter Fields wells. (**a**) Wells with overall poor reservoir quality; (**b**) wells with overall good properties.

the upper, finer-grained part of the Rotliegendes than in the lower, coarser-grained section (Fig. 5a). This is due to the smaller pore throats in the finer-grained sandstones being more easily 'bridged' by authigenic clays. In areas where the overall amount of diagenetic alteration is low (e.g. good quality side of the Ganymede Field), no discernable, consistent differences in reservoir quality exist between the upper and lower parts of the Rotliegendes (Fig. 5b).

## Diagenesis

### *Analytical techniques*

Diagenetic controls on reservoir quality were investigated in detail on specially selected sample sets from eight wells (49/16–4, 49/17–10, 49/22–5, 6 7z, 9z, 12 and 13). All of the samples analysed were aeolian grainflow sandstones for which grain size, porosity and permeability measurements were available. Aeolian grainflow sandstones were utilized exclusively to minimize the problem of contamination of authigenic illite samples with detrital illite and to eliminate ambiguity with regard to controls on diagenesis that might be introduced by mixing data from several facies. Likewise, grain-size measurements were made in order to ensure systematic biases were not introduced as a result of utilizing unrepresentative sample sets from one or more wells.

Analysis performed on all of the samples included (i) standard thin section modal and textural analysis and (ii) X-ray diffraction, thermogravimetric, evolved water analysis (XRD/TG/EWA) for clay mineral quantification (following techniques described by Thornley & Primmer 1995). Other analysis performed on some samples included (i) K–Ar dating of authigenic illite, and (ii) combined backscattered scanning electron microscopy/cathodoluminescence (BSEM/CL) analysis of quartz cement. Apatite fission-track analysis was also carried out on four wells (49/17–10, 49/22–5, 49/22–6 and 49/22–9z) to quantify burial histories. In addition, analytical transmissive electron microscopy (ATEM) and whole rock X-ray fluorescence (XRF) analysis were available for samples from the 49/17–10 and 49/22–12 wells.

### *Diagenetic sequence and petrology*

A generalized diagenetic sequence for the Jupiter Fields is shown in Fig. 6. The diagenetic sequence is broadly similar to that reported for the Rotliegendes in other parts of the UK sector of the Southern North Sea (e.g. Robinson *et al.* 1993).

Illite is by far the most important authigenic mineral both in terms of volume and impact on reservoir quality. Illite abundance determined by XRD/TG/EWA ranges from 1 to 17 wt%. Illite occurs as characteristic filamentous grain rimming and pore bridging overgrowths (Fig. 7). Where illite is present in moderate to high abundances (>10%), it reduces permeabilities by as much as two orders of magnitude compared to permeabilities in sandstones with low illite abundances.

Other volumetrically important authigenic minerals include chlorite (0–7 wt%) and to a lesser extent kaolin (0–3 wt%). These clays do not, however, significantly degrade reservoir quality. Dolomite, quartz, anhydrite, and siderite cements are also present but are volumetrically unimportant. Quartz cement abundances are usually less than 1.5% by volume. Microcrystalline dolomite is locally abundant near the top of the Rotliegendes, but decreases in abundance rapidly with depth, a feature noted in other Rotliegendes reservoirs (Sullivan *et al.* 1990).

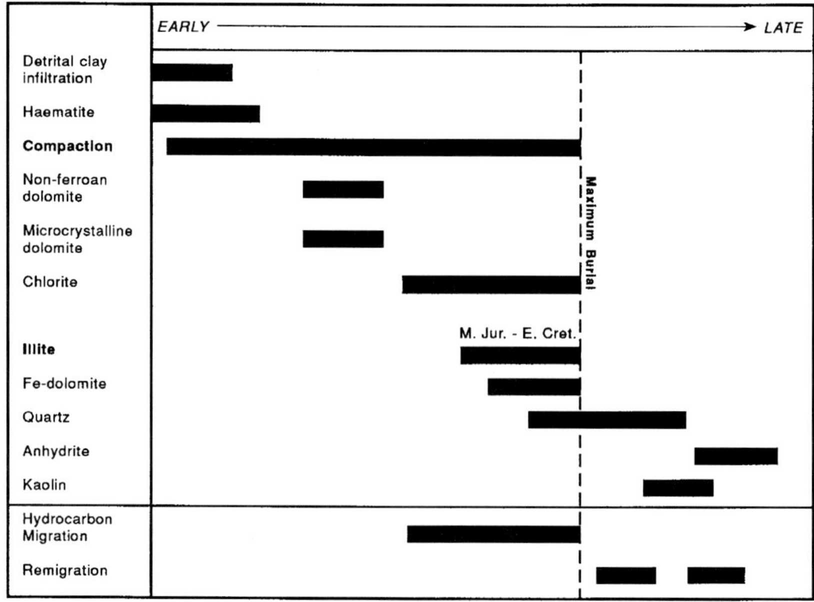


**Fig. 6.** Diagenetic sequence for the Jupiter Fields.

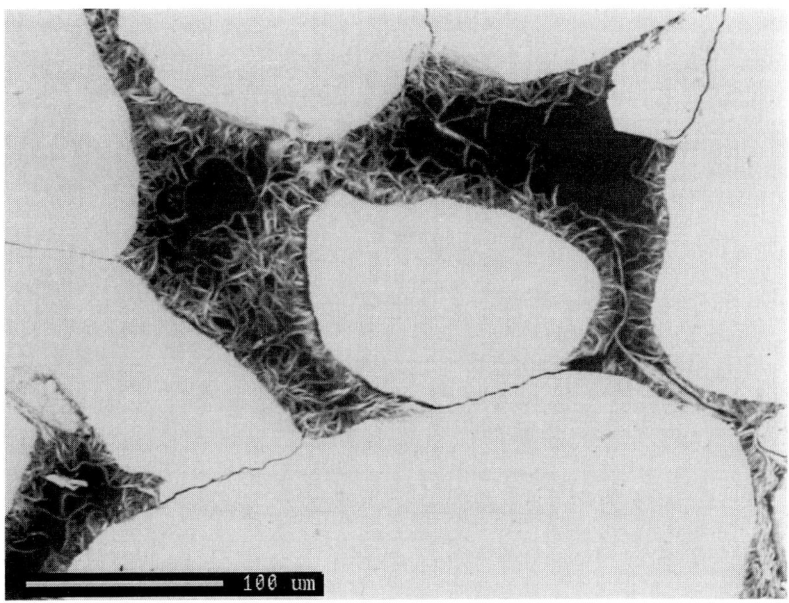


**Fig. 7.** Backscattered electron micrograph showing extensive development of illite as both a pore lining and pore bridging diagenetic phase (49/17–10 well; 2562.3 m core depth).

## Illite K–Ar ages

The most pragmatic interpretation of K–Ar ages for authigenic illite is that they represent the average age of the period over which the bulk of the illite grew. This interpretation is supported by calculations which show that authigenic illite is unlikely to lose radiogenic argon or have its argon 'clock' reset under diagenetic conditions (Hamilton *et al.* 1989). Each K–Ar age is therefore related to the average age of the main growth phase of illite. However, it is not possible to determine directly the age at which growth started or finished.

K–Ar dating of authigenic illite was done on 39 samples from the eight wells included in the diagenetic study in order to determine the timing of illite formation. Separation and analysis techniques utilized were broadly similar to those described by Robinson *et al.* (1993) and were designed to avoid contamination with other potassium-bearing minerals. All of the ages determined lie in the range from 177 to 122 Ma (Mid-Jurassic to Early Cretaceous), except for two analyses from the 49/16–4 well which gave ages of 88 and 104 Ma (Early to Late Cretaceous). These ages are broadly consistent with regional age distributions in the Southern North Sea (e.g. Lee *et al.* 1989; Robinson *et al.* 1993, Ziegler 1992). The K–Ar ages are interpreted as indicating that the main phase of illite growth occurred from the Mid-Jurassic to Early Cretaceous.

## Geological and burial history

The geological history of the Jupiter Fields area is summarized in Fig. 8. The Rotliegendes Group unconformably overlies Carboniferous coals and organic shales which are the principal source rocks for the gas accumulations in the Southern North Sea. The unconformity between the Carboniferous and the Rotliegendes Group resulted from uplift and erosion associated with the Variscan Orogeny (Glennie 1990). The Variscan Orogeny was followed by an extended period of regional subsidence during which the Lower Permian Rotliegendes reservoir sandstones, Upper Permian Zechstein evaporites and Triassic strata were deposited. Regional subsidence in the Southern North Sea gave way to accelerated rifting in the Jurassic associated with regional extension throughout northwest Europe (Ziegler 1988). This was terminated by uplift associated with the Late Jurassic to Early Cretaceous 'Cimmerian' event. A return to gradual subsidence in the Cretaceous provided accommodation space for deposition of the Cromer Knoll and Chalk Groups. Deposition

    G. P. LEVEILLE *ET AL.*

## JUPITER FIELDS

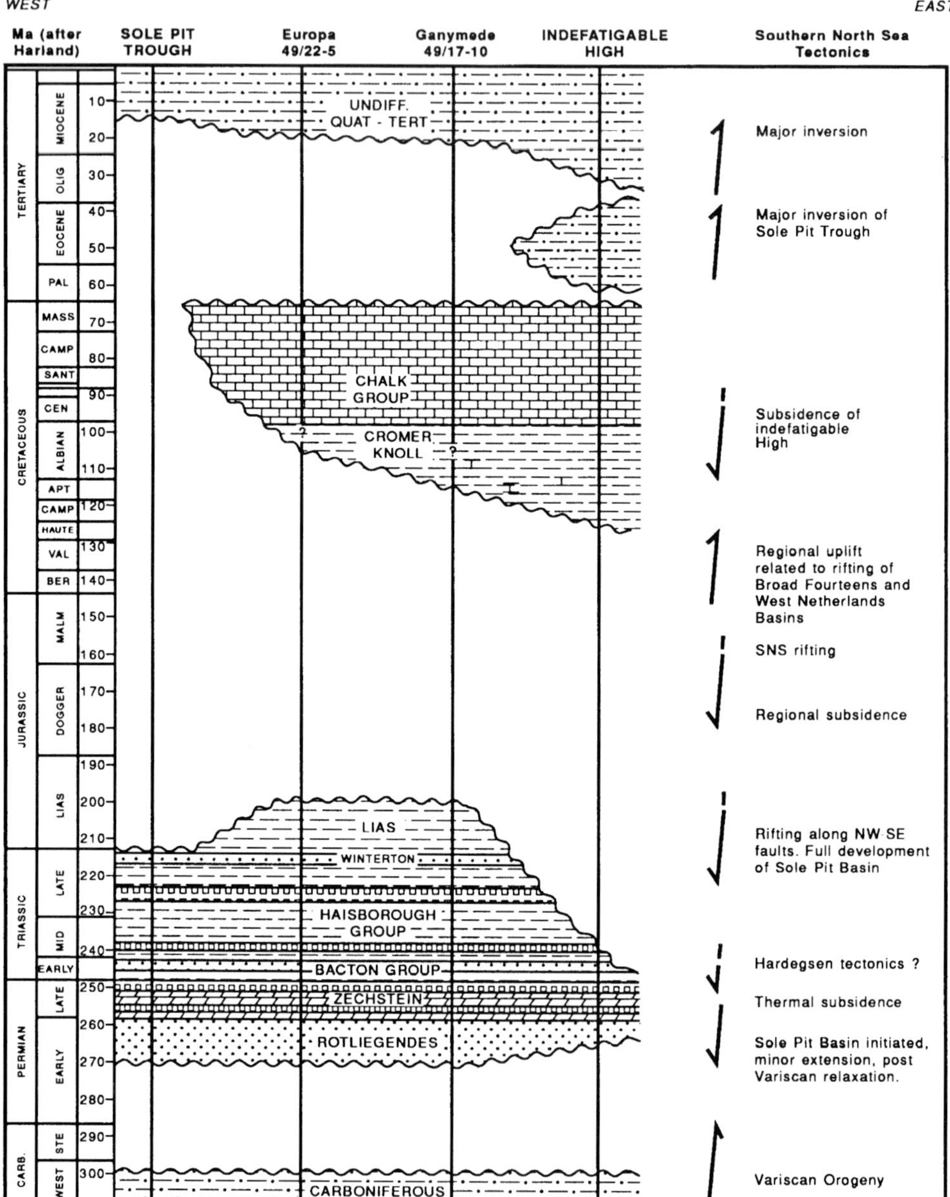

**Fig. 8.** Stratigraphy and associated tectonic events in the Jupiter Fields area.

of these units was followed by renewed inversion in the Late Cretaceous and Tertiary. The youngest deposits in the area are Tertiary and Quaternary clastic sediments.

A thermal history representative of the Jupiter Fields area was generated using apatite fission-track data in order to determine the relationship between timing of illite growth and burial history. These data indicate maximum burial to temperatures around 120°C occurred in the Late Jurassic to Early Cretaceous (Fig. 9). Burial was terminated by uplift associated with 'Cimmerian' regional inversion which resulted in cooling of 15–30°C. This was followed by

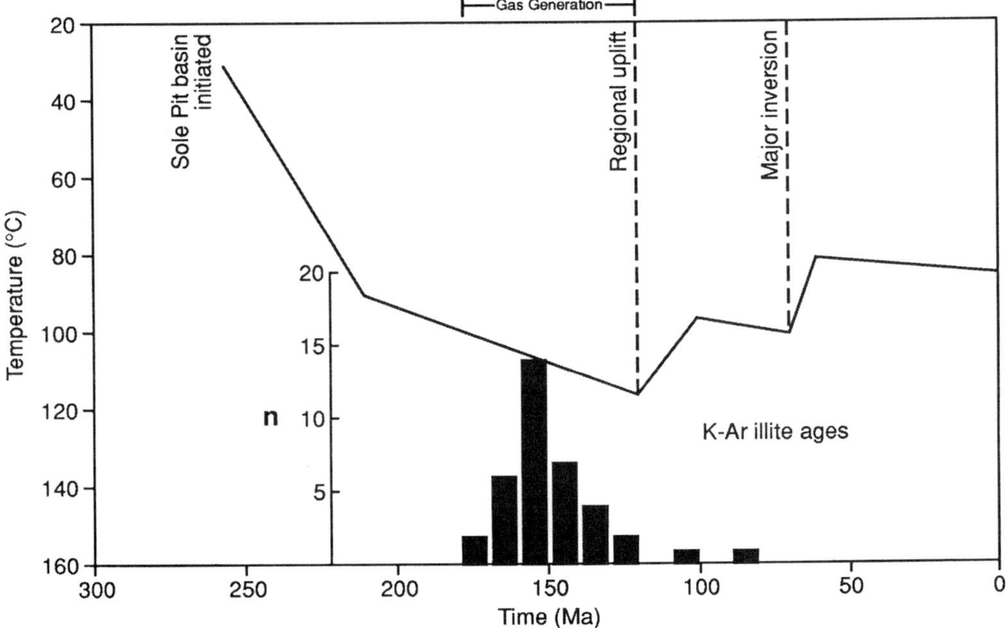



**Fig. 9.** Thermal history reconstructed from regional tectonic information, well data, and apatite fission-track analysis from four wells showing good correspondence between onset of 'rifting', temperatures of burial in excess of 100°C and illite K–Ar ages (black bars).

burial to temperatures between 100–110°C in the Late Cretaceous, with further uplift in the Tertiary.

Combining the burial history information with the illite age data suggests that illite growth primarily occurred during the period when the Rotliegendes was approaching its maximum burial depth in the Mid-Jurassic to Early Cretaceous (Fig. 9). Source rock maturation modelling indicates that Carboniferous source rocks underlying the Rotliegendes were generating large volumes of gas at this time. The age of illite growth in the Jupiter Fields area also coincided with a time of accelerated rifting in the Southern North Sea. A similar linkage between accelerated rifting and illite growth in the Village Fields area has been noted by Robinson *et al.* (1993), who suggested a phase of regional pore water movement associated with increased tectonic activity may have been responsible for creating changes in pore fluid chemistry which led to illite formation.

## Diagenesis and reservoir property variations

The large, abrupt variations in reservoir quality in the Jupiter Fields area are due to differing degrees of authigenic illite precipitation and burial compaction between fault blocks. Physical differences between fault blocks with good and poor reservoir quality are summarized in Table 1.

### Variations in illite abundance

Illite abundance varies considerably between fault blocks in the Jupiter Fields area. In areas with good reservoir quality, average illite abundances in wells range from 5 to 8 wt%, whereas in areas with poor reservoir properties, average abundances range from 11 to 14 wt%. The 3–9% greater amounts of illite found in areas of poor properties can be shown by empirical studies (Pallatt *et al.* 1984) and numerical modelling (Cade *et al.* 1994) to be capable of reducing permeabilities by two orders of magnitude in sandstones with textures and mineralogies similar to those in the Jupiter Fields area. Determining the factors responsible for the large variations in illite abundance between fault blocks is therefore fundamental to understanding controls on reservoir properties.

Illite in the Jupiter Fields area evidently formed as a true 'cement' in essentially clay-free sandstones. Variations in illite abundance must therefore be related to differences in

**Table 1.** *Differences in physical properties between areas of good and poor reservoir quality*

|                                        | Good quality           | Poor quality          |
| -------------------------------------- | ---------------------- | --------------------- |
| Average porosity                       | 17–21%                 | 12–16%                |
| Geometric mean permeability            | 38–200 mD              | 1–5 mD                |
| Illite abundance (wt%)                 | 4.7–8.3%               | 10.5–13.9%            |
| Illite abundance depth trend           | Increasing with depth  | None                  |
| Illite K–Ar age trend                  | Decreasing with depth  | None                  |
| Average illite chemical composition    | Low-potassium illite   | High-potassium illite |
| Relative compactional porosity loss    | Lower                  | Higher                |

physical and chemical conditions during diagenesis. The most likely causes for the variations in abundance are (i) differences in potassium ion concentration in formation fluids, (ii) differences in temperature and (iii) inhibition of illite growth due to hydrocarbon charging.

Although illite is a commonly observed authigenic mineral in sandstones, little published work exists to illustrate a consistent understanding of how differences in formation fluid chemistry (i.e. potassium ion concentration) and temperature control illite abundance. Recent experimental work (Primmer *et al.* 1993) has shown that there appears to be a consistent relationship between potassium ion concentration, illite chemical composition and illite abundance in the temperature range over which illite growth occurred in the Jupiter Fields (100–120°C). In this temperature range, pore fluids with low potassium ion concentrations will produce illite with low potassium content, whereas pore fluids enriched in potassium will produce high-potassium illite. In contrast, geochemical modelling has shown that differences in temperature in this temperature range will not produce measurable variations in illite chemical composition (Warren & Fritz 1989). A correlation between illite chemical composition and illite abundance is also apparent, with low potassium illites usually occurring in relatively small quantities (<5 wt%), probably due to a restricted supply of potassium, whereas high potassium illites can occur in much greater quantities (usually >10 wt%). Because of these relationships between illite chemical composition, illite abundance and potassium ion concentration in pore fluids, it should be possible to determine if variations in pore fluid chemistry were important in controlling differences in illite abundance in the Jupiter Fields by analysing the chemical composition of illites.

Analytical transmissive electron microscopy (ATEM) was performed on approximately 30 samples from the 49/17–10 and 49/22–12 wells in order to determine illite chemical composition. The 49/17–10 well is located in an area of poor reservoir quality and contains a large amount of illite (14 wt%), whereas the 49/22–12 well is located in an area of good reservoir quality and contains much less illite (7 wt%). The ATEM results indicate that illites in the 49/17–10 well are much more potassium rich than illites in the 49/22–12 well, with potassium concentrations in the 49/17–10 well averaging 1.9 moles of potassium per unit formula versus 1.3 moles in the 49/22–12 well (pure muscovite = 2.0 moles). Since the temperature range over which illite formed in the Jupiter Fields was not large enough to cause the observed variations in illite chemistry, this relationship is interpreted as indicating that differences in pore fluid chemistry existed between areas of poor reservoir quality (i.e. 49/17–10 well) and areas of good reservoir quality (i.e. 49/22–12 well) during illite growth.

The differences in illite chemical composition observed in ATEM data are supported by XRD data, which can be utilised to distinguish, in a broad sense, illites of different chemical compositions. Work by numerous authors (Srodon *et al.* 1986, 1992; Eberl & Srodon 1988) has established a formal relationship between illite composition and XRD character (peak position, peak shapes, relative intensities, etc.). For the purposes of this study, the width of the illite 001 peak at half maximum height above background (the illite 'crystallinity' parameter) has been utilised as a semi-quantitative indicator of illite composition. In this context, illites with low potassium, relatively low alumina contents have broad 001 peaks, whereas illites with high potassium, relatively high alumina contents have narrow 001 peaks.

Analysis of XRD peak width data indicate that the relationship between illite chemical composition and illite abundance observed in ATEM data from the 49/17–10 and 49/22–12 wells is consistent throughout the Jupiter Fields area. In areas of high illite abundance (i.e. poor reservoir quality), high-potassium illite is present, while in areas of low illite abundance (i.e. good reservoir quality), the potassium concentration in the illite is lower (Fig. 10). The potassium concentrations indicated by XRD peak width data for good and poor reservoir quality rocks do overlap more than the ATEM data would suggest, which could either reflect real variations in illite chemistry or result from limits in the accuracy of the XRD peak width technique.

In addition to the differences in illite chemical composition and reservoir quality noted above, there are discernible trends in illite abundance and illite K–Ar ages with depth in individual wells that can be correlated with differences in reservoir quality (Fig. 11). In areas with good reservoir quality, illite abundance generally increases with depth and illite K–Ar ages decrease with depth (i.e. 49/22–12, 49/22–9z and to a lesser extent 49/22–6 and 49/22–7z wells). In areas with poor reservoir quality, these trends are not present and variations in abundance and K–Ar age usually show no clear trend (i.e. 49/17–10, 49/22–5, 49/22–13, 49/16–4 wells).

One possible explanation for the trends observed in wells with good reservoir quality is that hydrocarbon charging occurred coincident with illite formation. If this were the case, one would expect illite to be least abundant at the top of the section where hydrocarbons accumulated earliest, with abundances increasing with

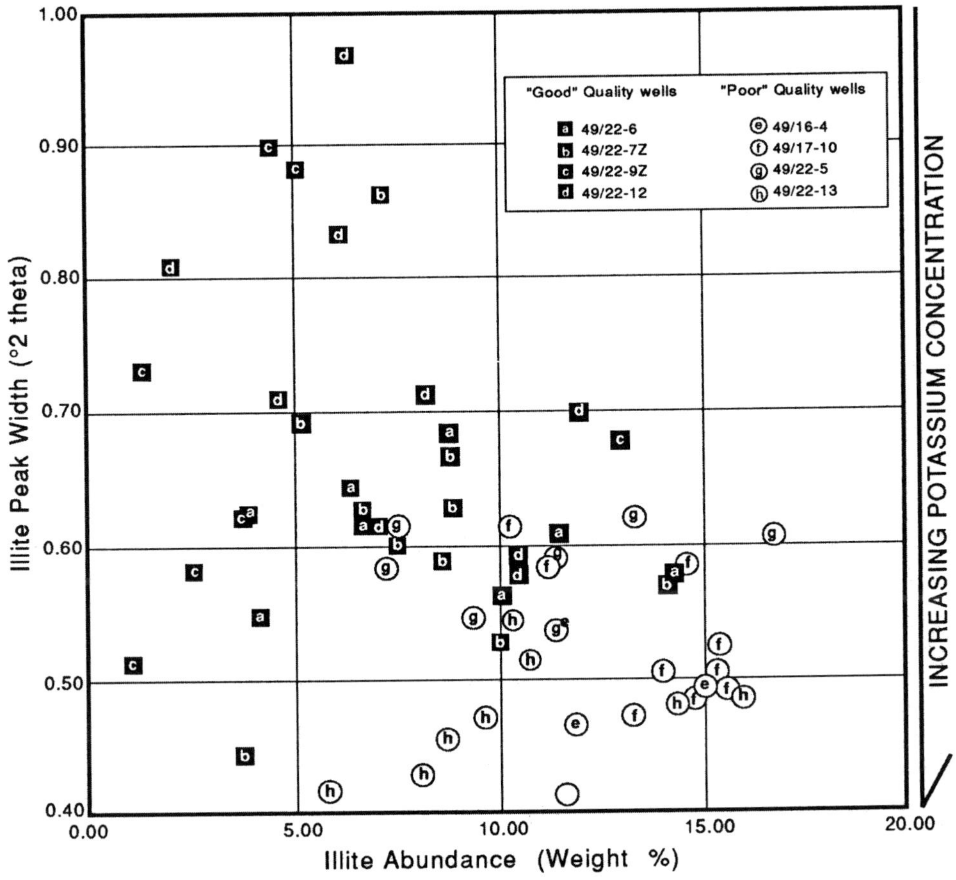


**Fig. 10.** Illite chemical composition from XRD peak width data plotted against illite abundance. Note that wells with good reservoir quality contain mostly low-potassium illites, whereas wells with poor reservoir quality contain only high-potassium illites.

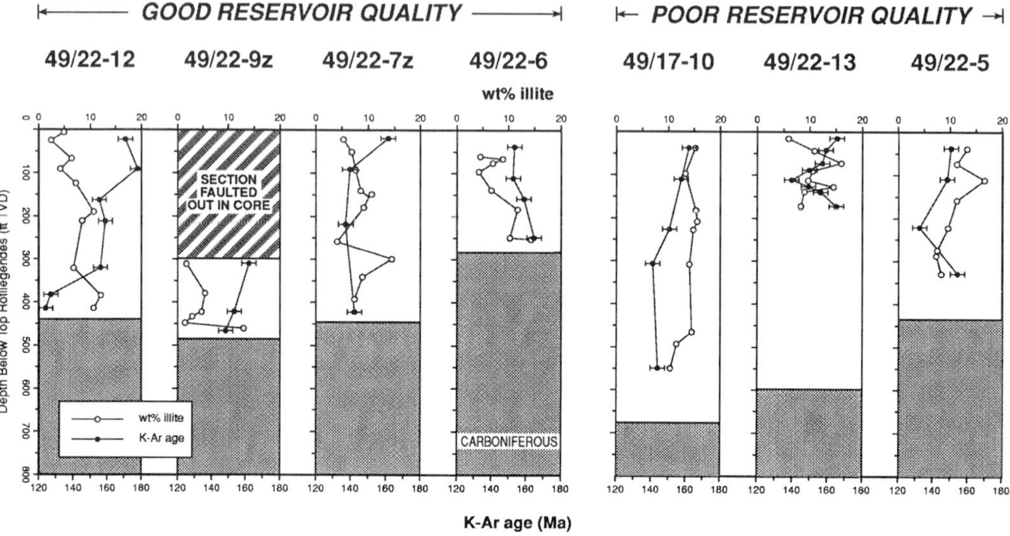



**Fig. 11.** Trends in illite K–Ar age and abundance with depth below the top of the Rotliegendes.

depth as a result of the downward movement of the hydrocarbon water contact with time. One would also expect illite K–Ar ages to decrease with depth for the same reasons, since the average time available for illite growth would be greater deeper in the reservoir ahead of the downward moving hydrocarbon front. A similar explanation for decreasing illite K–Ar ages with depth in some Rotliegendes reservoirs has previously been suggested by Lee *et al.* (1985).

### Compaction

The amount of compactional porosity loss (as defined by Lundegard 1992) that has occurred in the Rotliegendes is generally higher in fault blocks with poor reservoir properties than in fault blocks with good properties. This relationship is independent of the amount of framework grain dissolution that has occurred and is also independent of present day depth of burial. As the total amount of authigenic cements (e.g. quartz) is very low throughout the area, and does not vary significantly between areas with good and poor porosity, cementation also does not appear to be an important control on differences in compaction. These differences are therefore considered to be the result of past variations in burial history not expressed by the current structural configuration of the area.

Estimating absolute differences in past burial depths is difficult given the complex structural history of the Southern North Sea and the imprecise 'tools' available to make these kinds of estimates. Sand compaction models (using techniques described in Baldwin & Butler 1985) do, however, suggest that considerable structural relief, on the order of many hundreds of meters, must have existed between fault blocks with good and poor properties sometime during burial. Since some of the fault blocks which exhibit the greatest amount of compaction are those at the shallowest present day depths, significant inversion across local structural features must have occurred subsequent to deepest burial. This is consistent with the structural history of the Jupiter Fields area, with well documented inversion events having occurred in the Cretaceous and Tertiary (Walker & Cooper 1987; Van Hoorn 1987; Alberts & Underhill 1991).

## Controls on reservoir quality variations

It is believed that the relationships between reservoir quality variations, trends in illite abundance, and differences in compaction observed in the Jupiter Fields area (Table 1) resulted from reservoir quality being preferentially preserved in fault blocks that were structurally high during the main phase of diagenesis in the Middle Jurassic to Early Cretaceous. Processes thought to have contributed to the preservation of reservoir quality in structurally high fault blocks and to the destruction of reservoir quality in adjacent structural lows include (i) less burial compaction in structurally high areas, (ii) differences in pore fluid chemistry between high and low blocks, and (iii) hydrocarbon charging of early formed structural traps coincident with illite growth (Fig. 12).

Evidence for differences in maximum burial depths between areas of good and poor reservoir quality comes from the previously described petrographic measurements of compactional porosity loss. Areas with good reservoir quality all exhibit less compaction and therefore are interpreted to have been buried less deeply than the poor quality areas. Direct evidence for inversion related uplift is also present in some of the poor quality areas based on seismic interpretation and fault geometries observed in cores.

Evidence for differences in pore fluid chemistry between structurally high fault blocks and adjacent structural lows comes from illite chemical composition data. Specifically, the presence of low-potassium illite in areas that were paleostructural highs is believed to indicate potassium ion concentration in pore fluids was relatively low in structurally high areas compared to concentrations in structurally low areas, which contain only high-potassium illite. The segregation of pore fluids between structurally high and low areas is postulated to have occurred as a result of halokinetically driven

pore fluid movement in a structurally complex setting due to differences in fluid density. That is, potassium-rich fluids, which would have been highly saline, would have moved downwards towards structural lows due to their greater density, while low-potassium fluids, which would have had relatively low densities due to low concentrations of dissolved solids, would have moved upwards towards structurally high areas. Similar halokinetically driven 'gravity' fluid flow has been reported around salt domes in the Gulf of Mexico (Hanor 1988).

There are undoubtedly several ways in which pore fluids of different chemical compositions could have been generated and kept segregated within the Rotliegendes. The most straightforward and preferred interpretation is that the pore fluids were derived from different sources. Low potassium pore fluids in sublithic to subarkosic sandstones similar to those in the Rotliegendes are commonly formed during diagenesis as a result of dissolution of potassium bearing detrital phases like K-feldspar (e.g. see Warren 1987; Warren & Curtis 1989). To form potassium-rich pore fluids in these same sandstones, an external potassium-rich source, such as potassium salts, would be required. Based on the distribution of lithologies in the Rotliegendes and adjacent units, it is believed the low-potassium illite present in areas of good

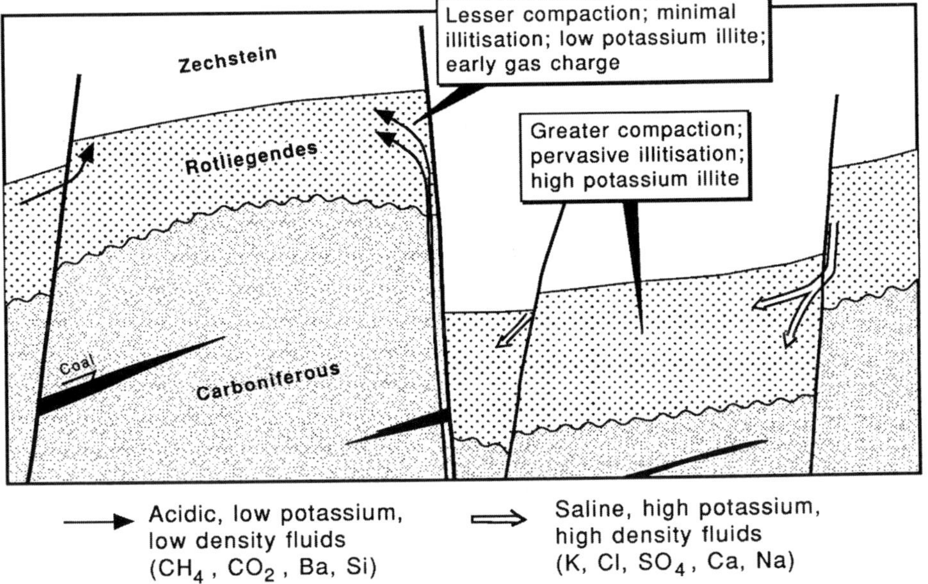


**Fig. 12.** Schematic cross section depicting the differences in (1) structural elevation, (2) pore fluid migration pathways and (3) timing of gas charge that controlled variations in reservoir quality. Structurally high areas suffered less compaction and were less prone to illite precipitation due to the predominance of low potassium pore fluids and, in places, early gas charge.

reservoir quality formed from *in-situ* Rotliegendes pore fluids and/or Carboniferous pore fluids. The presence of high-potassium illite in areas of poor reservoir quality is believed to be due to the influx of 'significant' amounts of Zechstein derived pore fluids which had a high potassium content from the dissolution of evaporitic potassium salts.

Direct evidence for the influx of pore fluids from the Zechstein includes the presence of salt and anhydrite cements in numerous faults and fractures in the Rotliegendes (Leveille *et al.* this volume). These cements are believed to have formed from dense, saline-rich pore fluids that moved downward into the Rotliegendes during fault dilation events. The presence of large volumes of Carboniferous-sourced gas in the Rotliegendes supports the postulated upward movement of fluids from the Carboniferous.

Evidence of early hydrocarbon charge includes the trends in illite abundance and age seen in some areas with relatively good reservoir quality (especially 49/22–12 and 49/22–9z wells). The trends in illite abundance and age are believed to have resulted from the progressive filling of early formed structural traps with hydrocarbons. Gas was almost certainly the primary hydrocarbon phase to fill the early formed traps because of the overwhelming volume of gas prone source rocks present in the underlying Carboniferous section. Residual oil staining does, however, suggest some oil may also have been generated. The residual oil staining typically occurs in dark brown bands and is interpreted to have formed from oil rims on early gas accumulations.

Additional evidence for hydrocarbon charging having inhibited illite growth comes from variations in the abundance of authigenic chlorite. Chlorite is usually found in relatively high abundances in areas that are interpreted as having been structurally high during diagenesis (good reservoir quality wells; 49/22–6, 49/22–7z, 49/22–9z, 49/22–12). A trend of decreasing chlorite abundance with depth is also commonly present in these wells. Chlorite is usually absent or present in very small amounts in wells which were presumably structurally low during diagenesis (49/16–4, 49/17–10, 49/22–5, 49/22–13). These relationships are believed to be related to redox reactions associated with migration of hydrocarbon bearing fluids out of the Carboniferous, with the reactions being similar to those described by Surdam *et al.* (1993) for other red sandstones (i.e. hematite stained). This interpretation is supported by whole-rock XRF measurements that indicate similar iron contents are present in all sandstones regardless of chlorite content, suggesting the iron in the chlorite was derived by reduction of hematite in a closed system. The trend of decreasing chlorite content with depth may have resulted from low density, reducing fluids being concentrated at the top of early-formed structural highs.

## Reservoir quality prediction

Predicting reservoir quality distribution in the Jupiter Fields area requires the ability to reconstruct structural geometries during the main phase of diagenetic alteration in the Mid-Jurassic to Early Cretaceous. Determining the structural configuration present during this period is difficult however because structural geometries were significantly modified by Late Cretaceous and Tertiary inversion related deformation events. The most striking example of this occurs in the Ganymede Field where the structurally highest portion of the field around the 49/17–10 well has poor reservoir properties and shows evidence of having been deeply buried, whereas the structurally lower area around the 49/22–12 well has excellent reservoir properties and was evidently structurally high during the main phase of diagenesis (see Fig. 2 for well locations).

Decoupling of the post-Permian section on the Upper Permian Zechstein salts further hinders reconstruction of paleostructural geometries in the Jupiter Fields. Empirical observations and modelling experiments indicate that the post Permian section has been effectively decoupled from the Rotliegendes level by plastic deformation of the Zechstein salts (Hooper *et al.* in press). This greatly diminishes the utility of conventional methods of predicting paleostructural geometries, such as cross section balancing techniques.

Methods that have been successfully utilised to predict paleostructural configuration, and therefore reservoir quality, include (i) analysis of fault geometries and fault trends and (ii) isopaching of the Rotliegendes stratigraphic section. With regard to analysis of fault geometries, fault blocks that show clear indications of having been significantly inverted have been found to usually contain poor reservoir properties. Determining whether a fault block has been inverted from seismic data has its pitfalls, however, since fault plane geometries are often poorly resolved and other structural indicators of inversion are commonly ambiguous. Extrapolation of fault trends, and associated reservoir quality trends, is likewise a possible approach, but one that needs to be used with caution.

A potentially more robust method for predicting reservoir quality distribution that seems to work well in the Jupiter Fields area is to construct a 'high-resolution' interval isopach map of the Rotliegendes section. This method works well because there is a direct relationship between Rotliegendes stratigraphic thickness and local variations in reservoir quality. In areas of good reservoir quality, the Rotliegendes is much thinner than in nearby wells with poor reservoir quality (Fig. 13).

The best examples of the relationship between reservoir thickness and reservoir quality occur in the Ganymede Field. In this field, the 49/22–12 and 49/17–10 wells are less than one kilometre apart but are separated by the major fault that runs down the centre of the field (see Fig. 2 for well locations). The Rotliegendes in the 49/22–12 well exhibits excellent reservoir properties (geometric mean permeability of 120 mD) and is relatively thin (approximately 134 m; 439 ft), whereas in the 49/17–10 well it has poor reservoir properties (geometric mean permeability of 2 mD) and is relatively thick (approximately 212 m; 695 ft).

The correlation between reservoir quality variations and thickness differences is believed to be due to structural features having controlled both diagenesis and depositional thickness trends (Fig. 14). Structural features evidently influenced thickness trends in the Rotliegendes by controlling palaeotopography and syndepositional differential subsidence. Structural deformation during Rotliegendes deposition in the Early Permian

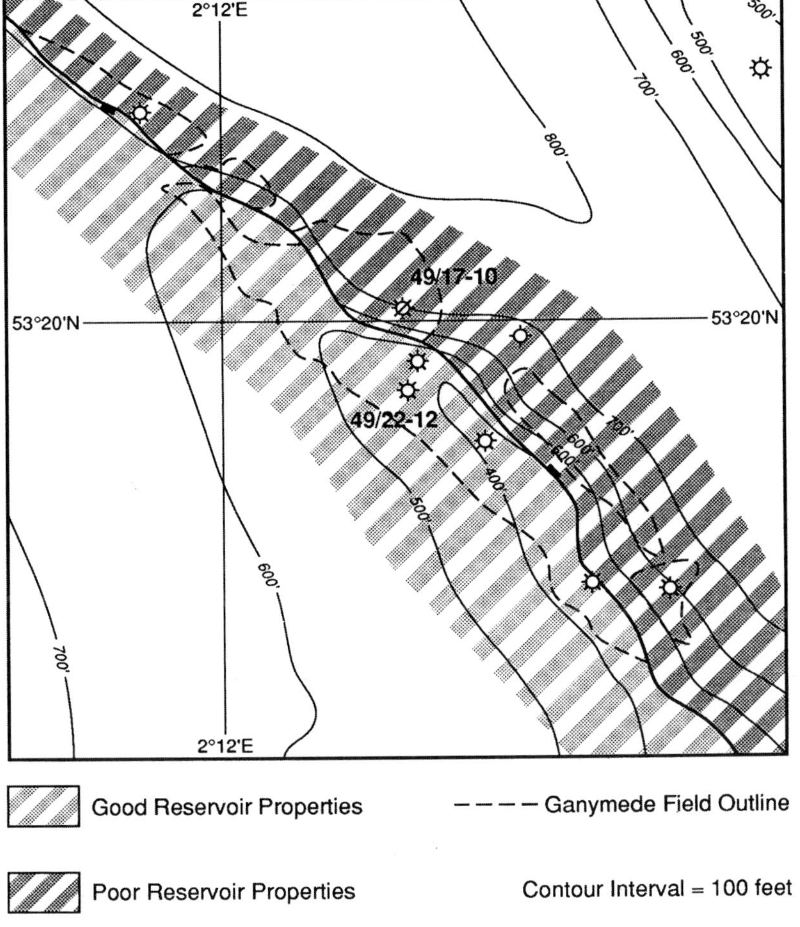


**Fig. 13.** Rotliegendes interval isopach map of the Ganymede Field area showing the correlation between reservoir quality and reservoir thickness.

**Latest Carboniferous**

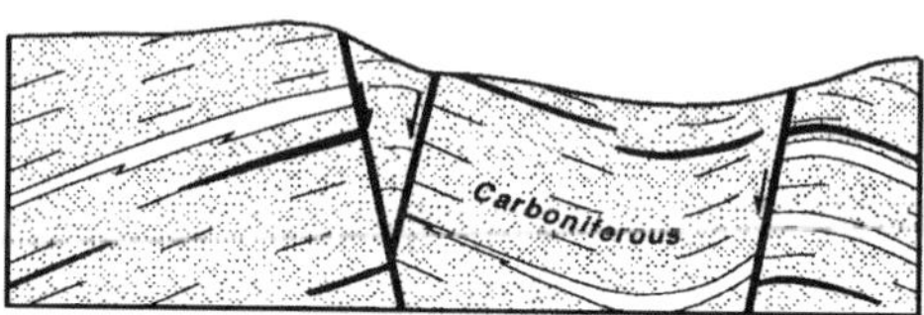



- Post Variscan Orogeny relaxation and extension (oblique extension)

- Structural elements influence topography

**Permian - Triassic**

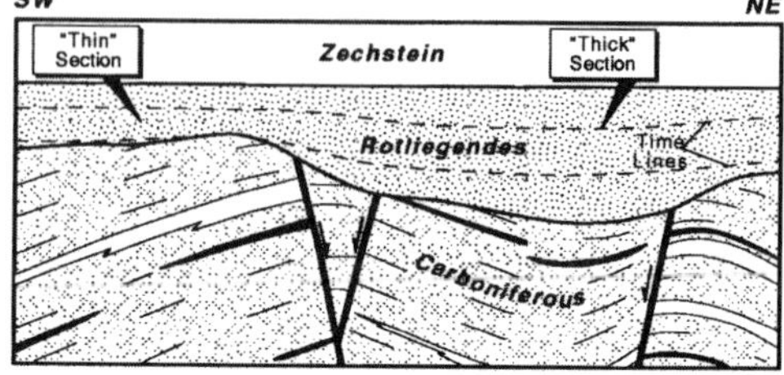



- Initiation of deposition in 'Southern Permian Basin'

- In-filling of 'Carboniferous topography' by Rotliegendes deposition

- Early differential subsidence followed by regional thermal subsidence

**Jurassic - Early Cretaceous**

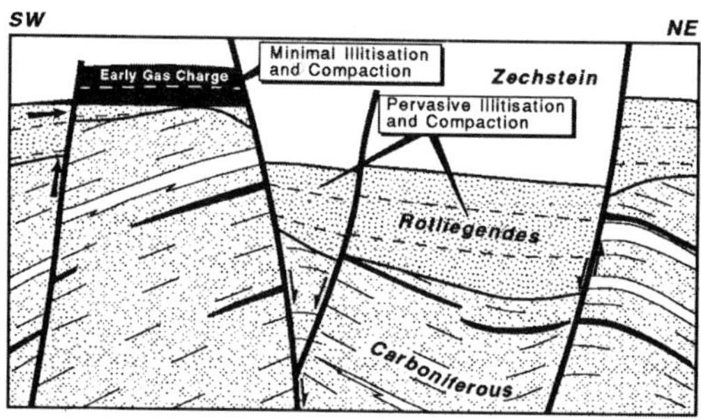



- Major rifting (oblique extension)

- 'Deep' burial of Rotliegendes; maturation of Carboniferous source rocks; large scale fluid movement; charging of early structural traps in Rotliegendes

- Formation of authigenic illite in Rotliegendes

**Cretaceous - Tertiary**

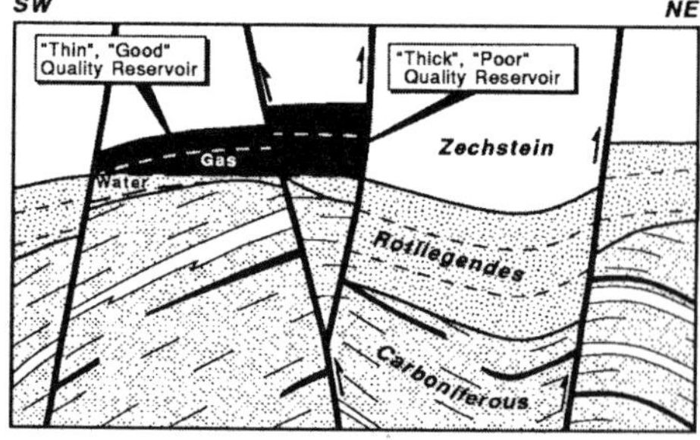



- Inversion and uplift (oblique compression)

- Remigration of gas due to structural deformation

- Cessation of significant diagenesis and source rock maturation

**Fig. 14.** A series of NE–SW schematic cross-sections through the Ganymede Field showing how structural features controlled both Rotliegendes thickness trends and reservoir quality variations. The southwestern fault block in the Ganymede Field was structurally high during the main phase of diagenesis, and therefore contains good reservoir properties, while the northeastern fault block was structurally low and contains poor properties.

was likely mostly extensional, possibly with some component of strike-slip movement (George & Berry 1994). Thin Rotliegendes sections were deposited over 'upthrown' fault blocks, while thick sections were deposited in adjacent lows. Extension associated with regional subsidence continued from the Late Permian through to the Jurassic, during which a phase of accelerated rifting occurred (Fig. 9). Thus, when the main phase of illite growth occurred in the Jurassic to Early Cretaceous, the fault blocks that had been 'structurally high' during Rotliegendes deposition were still structurally high, and were therefore less damaged by diagenesis than structurally low areas.

The relationship between local variations in thickness and reservoir quality distribution is believed to have significance beyond the Jupiter Fields area. An example of another area where this relationship seems to be present is the area around the Viking Fields (see Fig. 1 for location). In the Viking Fields area, a fault block with poor reservoir properties and a thick Rotliegendes section exists adjacent to fault blocks with good reservoir quality and thinner sections.

## Conclusions and observations

Significant conclusions and observations regarding the geology of the Jupiter Fields include the following:

(1) Large variations in reservoir quality exist. These variations are primarily controlled by differing degrees of diagenetic alteration. The two most important diagenetic processes were the formation of authigenic illite and compaction.

(2) The formation of authigenic illite mainly occurred during the Mid-Jurassic to Early Cretaceous (K–Ar ages of 177–122 Ma). During this period the Rotliegendes was approaching its maximum burial depth.

(3) Fault blocks that were structurally high during the Mid-Jurassic to Early Cretaceous were less compacted than structurally low areas. The structurally high fault blocks also contain less illite due to differences in pore fluid chemistry. Gas charge of early formed structural traps occurred coincident with illitization and further inhibited illite formation.

(4) Predicting reservoir quality distribution requires the ability to reconstruct structural geometries during the main phase of diagenetic alteration in the Mid-Jurassic to Early Cretaceous. Decoupling of the post-Permian section on the Zechstein salts limits the utility of cross section restoration techniques. Because a direct relationship exists between local variations in Rotliegendes thickness and reservoir quality due to structural control of deposition and diagenesis, a 'high resolution' Rotliegendes isopach map can be used to predict reservoir quality distribution.

(5) The presence of dramatic differences in reservoir quality over short distances in the Jupiter Fields suggest that overlooked exploration and development potential may be present in the 'mature' Rotliegendes gas play in the Southern North Sea. Areas previously discounted because of poor results from one or more wells might yet be found to contain fields with good reservoir properties. Similarly, surprises may be waiting in untested parts of fields and discoveries where the degree of diagenetic alteration varies across structural features.

We wish to thank our co-workers at Conoco, Mobil, BP, and Statoil who contributed to our understanding of controls on reservoir quality in the Jupiter Fields area. Special thanks goes to Quentin Fisher of the University of Leeds who provided ATEM and XRF data, and contributed many insights regarding diagenetic processes. Early versions of this paper were improved as a result of review by R. Lahann, Q. Fisher, K. Ziegler, R. Gaupp, G. Mound, and an anonymous reviewer. Thanks also to SURRC, Eurotrack, and D. Thornley (University of Reading) who did analytical work utilized in the study.

## References

ALBERTS, M. A. & UNDERHILL, J. R. 1991. The effect of Tertiary structuration on Permian gas prospectivity, Cleaver Bank area, Southern North Sea, UK. *In:* SPENCER, A. M. (ed.) *Generation, accumulation, and production of Europe's hydrocarbons.* Special Publications of the European Association of Petroleum Geoscientists, Oxford University Press, Oxford, **1**, 161–173.

ARTHUR, T. J., PILLING, D., BUSH, D. & MACCHI, L. 1986. The Leman Sandstone Formation in UK Block 49/28. Sedimentation, diagenesis, and burial history. *In:* BROOKS, J., GOFF, J. C. & VAN HOORN, B. (eds) *Habitat of Palaeozoic gas in N.W. Europe.* Geological Society, London, Special Publications, **23**, 251–266.

BALDWIN, B. & BUTLER, C. O. 1985. Compaction curves. *American Association of Petroleum Geologists Bulletin*, **69**, 622–629.

CADE, C. A., EVANS, I. J. & BRYANT, S. L. 1994, Analysis of permeability controls: a new approach. *Clay Minerals*, **29**, 491–501.

EBERL, D. D. & SRODON, J. 1988. Ostwald ripening and interparticle diffraction effects for illite crystals. *American Mineralogist*, **73**, 1335–1345.

GAUPP, R., MATTER, A., PLATT, J., RAMSEYER, K. & WALZEBUCK, J. 1993. Diagenesis and fluid evolution of deeply buried Permian (Rotliegende) gas reservoirs, northwest Germany. *American Association of Petroleum Geologists Bulletin*, **77**, 68–80.

GEORGE, G. T. & BERRY, J. K. 1994. A new palaeogeographic and depositional model for the Upper Rotliegend, offshore The Netherlands. *First Break*, **12**, 147–158.

GLENNIE, K. W. 1972. Permian Rotliegendes of northwest Europe interpreted in light of modern desert sedimentation studies. *American Association of Petroleum Geologists Bulletin*, **56**, 1048–1071.

——1990. Rotliegend sediment distribution: a result of late Carboniferous movements. *In* HARDMAN, R. F. P. & BROOKS, J. (eds) *Tectonic Events Responsible for Britain's Oil and Gas Reserves.* Geological Society, London, Special Publications, **55**, 127–138.

——, MUDD, G. & NAGTEGAAL, P. J. C. 1978. Depositional environment and diagenesis of Permian Rotliegendes sandstones in Leman Bank and Sole Pit areas of the UK southern North Sea. *Journal of the Geological Society, London*, **135**, 25–34.

GRANT, N. 1992. The impact of horizontal wells in the North Valiant Field. *Extended abstracts from PETEX conference*, London.

HAMILTON, P. J., KELLY, S. & FALLICK, A. E. 1989. K–Ar dating of illite in hydrocarbon reservoirs. *Clay Minerals*, **24**, 215–232.

HANOR, J. 1988. *Origin and migration of subsurface sedimentary brines.* SEPM Short Course, **21**.

HOOPER, R. J., MORE, C., KANTOROWICZ, J., HANLEY, D. & MCDONALD, N. in press. Analysis of salt-related structures in the U.K. southern North Sea. *Proceedings from modern developments in section validation techniques conference.* London, in press.

LAHANN, R. W., FERRIER, J. A. & CORRIGAN, S. 1993. Reservoir heterogeneity in the Vanguard Field, UKCS. *In*: ASHTON, M. (ed.) *Advances in Reservoir Geology.* Geological Society, London, Special Publications, **69**, 33–56.

LEE, M., ARONSON, J. L. & SAVIN, S. M. 1985. K–Ar dating of time of gas emplacement in Rotliegendes sandstones, Netherlands. *American Association of Petroleum Geologists Bulletin*, **69**, 1381–1385.

——, —— & ——1989. Timing and conditions of Permian Rotliegende Sandstone diagenesis, southern North Sea: K/Ar and oxygen isotope data. *American Association of Petroleum Geologists Bulletin*, **73** 195–215.

LEVEILLE, G. P., KNIPE, R., MORE, C., ELLIS, D., DUDLEY, G., JONES, G., FISHER, Q. & ALLINSON, G. 1997. Compartmentalization of Rotliegendes gas reservoirs by sealing faults, Jupiter Fields area, southern North Sea. *This volume.*

LUNDEGARD, P. D. 1992. Sandstone porosity loss – a "big picture" view of the importance of compaction. *Journal Sedimentary Petrology*, **62**, 250–260.

PALLATT, N., WILSON, M. J. & MCHARDY, W. J. 1984. The relationship between permeability and the morphology of diagenetic illite in reservoir rocks. *Journal of Petroleum Technology*, **36**, 2225–2227.

PRIMMER, T. J., WARREN, E. A., SHARMA, B. K. & ATKINS, M. P. 1993. The stability of experimentally grown clay minerals: implications for modelling the stability of neoformed diagenetic clay minerals. *In*: MANNING, D. A. C., HALL, P. L. & HUGHES, C. (eds) *Geochemistry of Clay–Pore Fluid Interactions.* Mineralogical Society, London, 163–180.

ROBINSON, A. G., COLEMAN, M. L. & GLUYAS, J. G. 1993. The age of illite cement growth, Village Fields area, Southern North Sea: evidence from K–Ar ages and $^{18}O/^{16}O$ ratios. *American Association of Petroleum Geologists Bulletin*, **77**, 68–80.

SEEMANN, U. 1979. Diagenetically formed interstial clay minerals as a factor in Rotliegende sandstone reservoir quality in the Dutch sector of the North Sea. *Journal of Petroleum Geology*, **1**, 55–62.

SRODON, J., ELSASS, F., MCHARDY, W. J. & MORGAN, D. J. 1992. Chemistry of illite-smectites inferred from TEM measurements of fundamental particles. *Clay Minerals*, **27**, 137–158.

——, MORGAN, D. J., ESLINGER, E. V., EBERL, D. D. & KARLINGER, M. R. 1986. Chemistry of illite/smectite and end-member illite. *Clay Minerals*, **34**, 368–378.

SULLIVAN, M. D., HAZELDINE, R. S. & FALLICK, A. E. 1990. Linear coupling of carbon and strontium isotopes in Rotliegende Sandstone, North Sea: evidence for cross-formational flow. *Geology*, **18**, 1215–1218.

SURDAM, R. C., JIAO, Z. S. & MACGOWAN, D. B. 1993. Redox reactions involving hydrocarbons and mineral oxidants: a mechanism for significant porosity enhancement in sandstones. *American Association of Petroleum Geologists Bulletin*, **77**, 1509–1518.

THORNLEY, D. M. & PRIMMER, T. J. 1995, Thermogravimetry/evolved water analysis (TG/EWA) combined with XRD for improved quantitative whole-rock analysis of clay minerals in sandstones. *Clay Minerals*, **30**, 27–38.

TURNER, P., JONES, M., PROSSER, D. J., WILLIAMS, G. D. & SEARL, A. 1993. Structural and sedimentological controls on diagenesis in the Ravenspurn North gas reservoir, UK Southern North Sea. *In*: PARKER, J. R. (ed.) *Petroleum Geology of Northwest Europe: Proceedings of the 4th Conference.* The Geological Society, London, 771–785.

VAN HOORN, B. 1987. Structural evolution, timing and tectonic style of the Sole Pit inversion. *Tectonophysics*, **137**, 239–284.

WALKER, I. & COOPER, W. G. 1987. The structural and stratigraphic evolution of the northeast margin of the Sole Pit Basin. *In*: BROOKS, J. & GLENNIE, K. (eds) *Petroleum geology of northwest Europe.* Graham and Trotman, 263–275.

WARREN, E. A. 1987. The application of a solution-mineral equilibrium model to the diagenesis of Carboniferous sandstones, Bothamsall oilfield, East Midlands, England. *In*: MARSHALL, J. D. (ed.) *Diagenesis of Sedimentary Sequences.* Geological Society, London, Special Publications, **36**, 55–69.

—— & CURTIS, C. D. 1989. The chemical composition of authigenic illite within two sandstone reservoirs as analysed by ATEM. *Clay Minerals*, **24**, 137–156.

—— & FRITZ, B. 1989. The application of a solid solution model to illite precipitation in sandstones. *Abstracts of the Mineralogical Society "Stability of Minerals" meeting*, December 1989, London.

ZIEGLER, K. 1992. Authigenic illites in the Rotliegend Leman Sandstone, Southern North Sea: K/Ar dating, oxygen and hydrogen isotopes. *Abstracts 31st Annual Meeting BSRG*, Southampton.

ZIEGLER, P. A. 1988. *Evolution of the Arctic–North Atlantic rift system.* American Association of Petroleum Geologists Memoir, **43**.

# Probing the lower limits of a fairway: further pre-Permian potential in the southern North Sea

NICK CAMERON[1] & TOM ZIEGLER[2]

[1] *GeoMark Research, Inc. (European Division), 'Chalkyfold', Grove Lane, Chesham, Bucks HP5 3QQ, UK*
[2] *PGS Nopec (UK) Ltd, PGS House, Mayfield Road, Walton on Thames, Surrey KT12 5PL, UK*

**Abstract:** Although the southern North Sea has become a major centre of hydrocarbon production, the lower limit of the fairway has not been established. We show, using a review of the pre-Permian source-rock potential of the southern North Sea, that the regional geology will permit exploration below the present base of the fairway. Seismic mapping indicates that this portion of the stratigraphic column is associated with multiple objectives. Two very different examples demonstrate the range of opportunities. The first example is a conventional structural trap located on the northeast flank of the London–Brabant Ridge in Quadrant 53. Reservoirs are the Westphalian to Devonian succession. Source-rocks are lacustrine horizons within the local Westphalian succession and the Carboniferous to Devonian section to the northeast of the South Hewett Fault. The second example presents a new objective, a stratigraphic trap to the ESE of a high located in Quadrants 42 and 43 and identified for the first time by this study. The anticipated reservoir is Dinantian carbonate progradational wedge derived from the adjacent high. Namurian and Dinantian source-rocks are predicted. Until opportunities of these types are evaluated, the lower limit of the Southern North Sea fairway will remain undefined and ultimate reserves estimates will continue to be poorly constrained.

The southern North Sea (SNS) is defined as the region south of the crest of the Mid North Sea High and to the north of the crest of the London–Brabant Ridge. This area is illustrated in Fig. 1.

The objective is to present the geological and geophysical evidence for new pre-Permian opportunities in the SNS. The time intervals examined are the Westphalian and Namurian outside the Quadrants 43 and 44 region, and the pre-Namurian succession throughout the area. A two-step investigative procedure was adopted. Firstly, source-rock possibilities, derived from regional geological and global source rock abundance considerations, were reviewed to determine whether new opportunities could exist. Source horizons investigated were the Namurian, Dinantian, Devonian and Silurian. Westphalian oil-prone source-rock possibilities were also examined. Then, regional seismic lines were inspected for structural and stratigraphic opportunities dependent on the source-rock findings.

The world's major hydrocarbon provinces are characterized by extended geological histories, the presence of multiple source-rocks, a range of reservoirs and a wide variety of structural styles. As exploration progresses in the SNS, it is evident that this basin shares these features. The SNS fairway may be considered as presently consisting of three tiers. In this paper these are termed the upper, middle and lower tiers. The upper tier is represented by the mainly early Cretaceous-reservoired oil and gas fields of the onshore and offshore Netherlands (Bodenhausen & Ott 1981; Roelofsen & de Boer 1991). Total reserves amount to approximately 0.6 BBOE (Leckie *et al.* 1995). The Bunter and Rotliegend fields constitute the middle tier. This, the dominant fairway, has ultimate recoverable reserves of some 150 trillion cubic feet (TCF) of gas (Glennie 1996). The lower tier is formed by the Upper Carboniferous gas fields of the Untied Kingdom Quadrants 43/44 region and the adjacent quadrants in The Netherlands. 6 TCF have been discovered since the initial discoveries in the mid-1980s (Leckie *et al.* 1995).

We will demonstrate that the SNS is capable of hosting new plays within the lower tier of the fairway and also in an entirely new basal tier positioned in the pre-Namurian section. Whilst previous papers (Tubb *et al.* 1986; Fraser & Gawthorpe 1990) have examined some aspects of the potential that we will describe, their significance appears to have been overlooked. This may have been because the lower tier had not then taken on fairway status and the basal tier was at the concept stage. Seismic acquisition, processing and presentation techniques have advanced considerably in recent years,

*From* Ziegler, K., Turner, P. & Daines, S. R. (eds), 1997, *Petroleum Geology of the Southern North Sea: Future Potential*, Geological Society Special Publication No. 123, pp. 123–141.

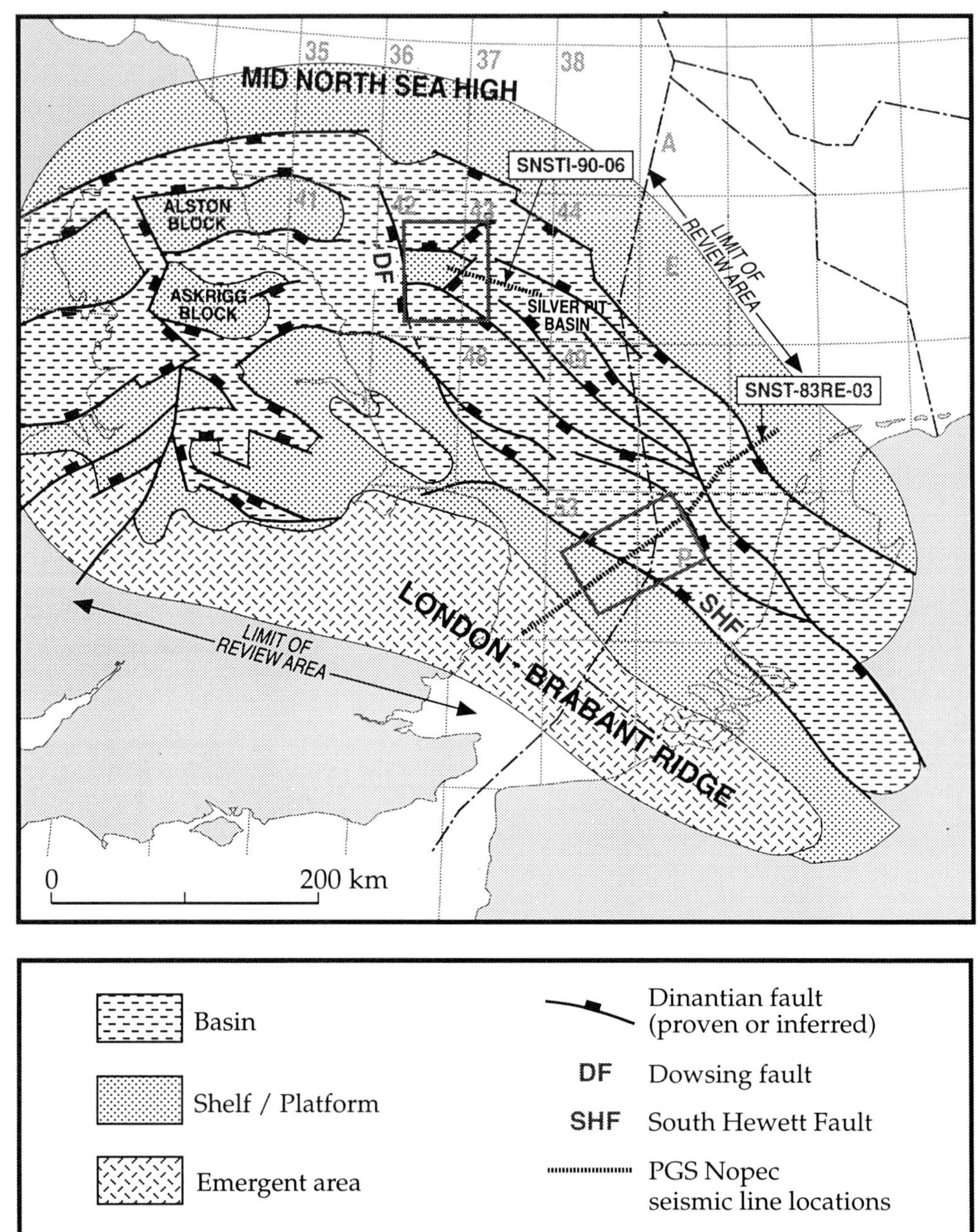


**Fig. 1.** Schematic Dinantian Palaeography of the Southern North Sea. Figures 2–6 and 8 reproduce the proposed Dinantian fault framework shown on this figure. The outlined boxes show the location of the opportunity areas presented in this review. The two seismic lines used to illustrate these opportunities are also shown.

and it is timely to return to the literature base, using high-quality regional seismic lines for control, to re-examine the pre-Permian section. It is also pertinent to reappraise the pre-Permian section as replacement reserves are required to supplement the first generation of discoveries. Whilst the reservoir quality is likely to be more problematic than in the Bunter and Rotliegend, the existing infrastructure may assist the economics.

Despite the number of wells in the area, few penetrate far below the top Namurian, and the deep-geology of much of the southern North Sea remains very poorly constrained. Thus, whilst the middle tier (Bunter/Rotliegend reservoirs) is increasingly well understood, many aspects of the pre-Permian geology can still be treated only in frontier basin terms.

## New source-rock possibilities

As previously noted, the stratigraphic control for the deep-geology of the SNS is limited. As a result, the illustrations (Figs 1–6 and 8) and the text supporting this section of the paper are necessarily generalized and in some cases speculative. The foundation for the illustrations is the Dinantian structural framework presented by Besly (in press) and, for the offshore region, our interpretation of a 10 by 15 km grid acquired by PGS Nopec. The offshore structural elements are expressed at the Base Zechstein level as late Carboniferous (Variscan) inversion frequently complicates mapping at lower horizons. The assumption is made that the Zechstein extensional faulting is co-incident with Carboniferous extensional trends.

Emphasis is placed on new potential rather than the re-examination of the documented geology. Besly (in press) provides a comprehensive account of the Westphalian coal basin and the shallow-water Namurian and Dinantian coaly horizons present on the southern margin of the Mid North Sea High.

### Pointers provided by
### palaeogeographic reconstructions

*New Westphalian source potential.*   Coals within the Westoe Coal and Caister Coal Formations (Cameron 1993) are widely considered to be the source for the middle-tier fields and also many of the lower-tier fields (Cornford 1990). However, other sources are possible; these are commented on in the next section. The most encouraging location for new source-rocks of Westphalian age is the lacustrine basin located to the northeast of the London–Brabant Ridge and to the northwest of the Variscan Front of mainland Europe (Quirk 1993a). The position of this basin in relation to the regional framework is shown on Fig. 2. Oil-prone source rocks are identified from this basin by De Jager *et al.* (1993) and Cole *et al.* (1995). Their existence is also suspected in the Roer Valley Graben region

of The Netherlands (Winstanley 1993). In the United Kingdom, lacustrine source rocks are most likely to be developed in Quadrants 48, 49, 53 and 54. In The Netherlands, the most favourable quadrants are P Q and S. In view of their lacustrine origin, cannel coals in the North Staffordshire Coalfield and at Point of Air in North East Wales (Armstrong *et al.* 1995), may define the continuation of the lacustrine basin westwards across mainland England.

*Namurian source potential.*   Namurian oil-prone source rocks are present onshore and in the East Irish Sea (Fig. 3). They are associated with the initial infill of former rift basins (Fraser & Gawthorpe 1990). Geochemical fingerprinting indicates that the East Midlands oils are derived from the Namurian (Fraser *et al.* 1990). The recent discoveries in the East Irish Sea Basin have also been typed to a Namurian source, the Holywell Shale Formation (Armstrong *et al.* 1995). Both oil analyses (Pering 1973) and modelling (Walkden & Williams 1991) suggest that the commonly occurring residual oil shows in the Castleton area of Derbyshire are derived from the adjacent Edale Shale Formation. The source-rocks of the western Pennines, the Sabden and Bowland Shale Formations, are discussed by Lawrence *et al.* (1987) and Hardman *et al.* (1993). To the southeast in The Netherlands, Namurian source rocks are considered to be present in the Roer Valley Graben region (Winstanley 1993).

In the SNS, though a major deepwater, shale basin existed for much of the Namurian (Cameron 1993), there are no descriptions of oil-prone source-rocks. Two reasons appear to account for the deficiency of oil-prone source-rocks. Firstly, the palaeogeographic reconstructions presented by Cameron (1993) suggest that terrigenous input from the Mid North Sea High will have produced an initially gas-prone source in the extreme northern region of the study area. Secondly, the succession in areas other than the London–Brabant Ridge and the Mid North Sea High is in the gas window and has, therefore, lost its initial oil potential. Vitrinite reflectance ($R_o\%$) values ranging between 1.5% and 1.8% characterize the otherwise standard Pennine deepwater shale succession described from the 48/3-3 well by Leeder *et al.* (1990). Total organic carbon (TOC) peaks of 4–5% remain in this interval of the well (Leeder & Hardman 1990). It is, therefore, considered that Namurian shales of the type present at the base of the 48/3-3 well will have generated substantial

                              N. CAMERON & T. ZIEGLER

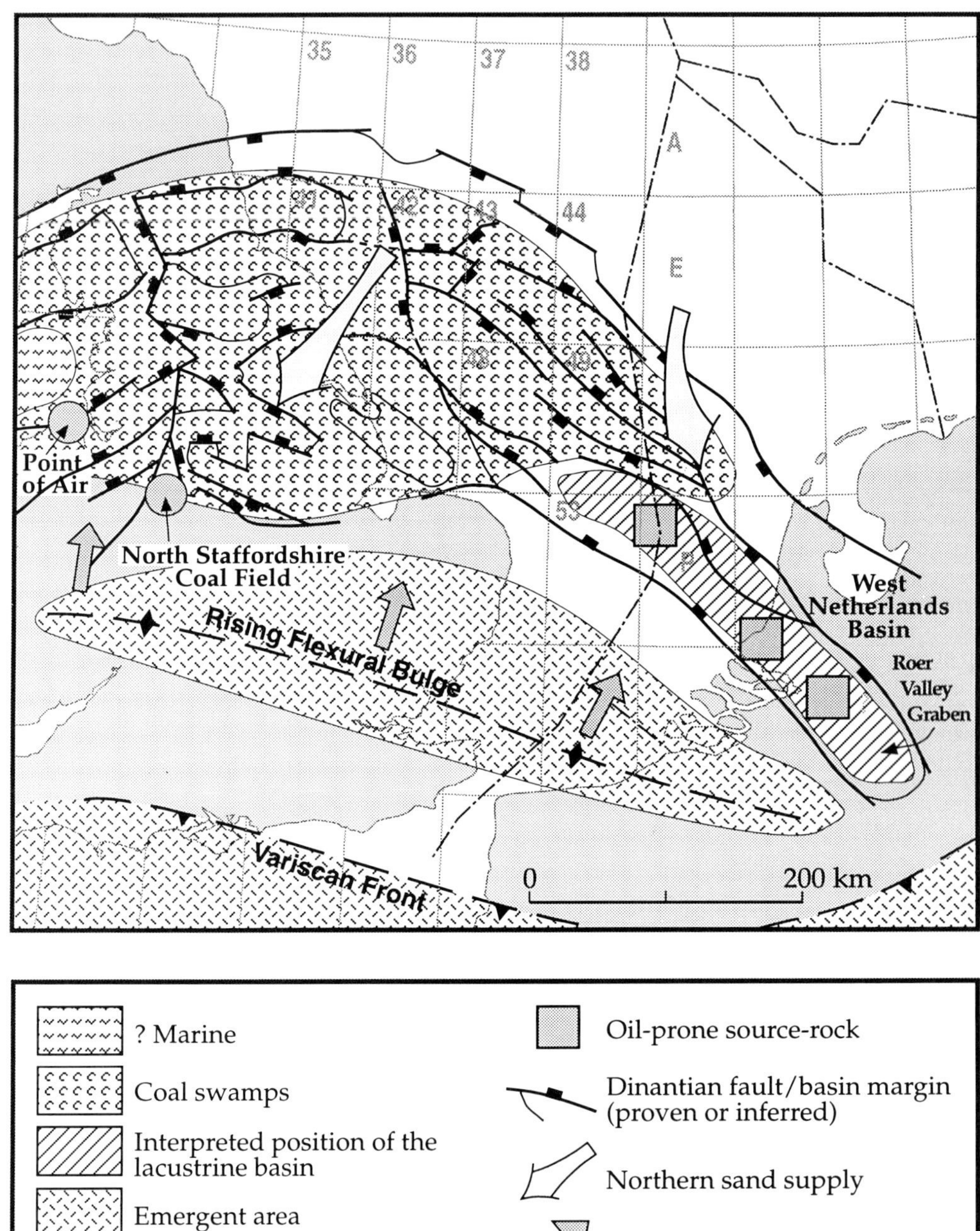



**Fig. 2.** Westphalian hydrocarbon play elements.

volumes of hydrocarbons during their earlier burial history.

It is possible that the Bowland Shale Formation (Cameron 1993) of the SNS may be a major contributor to the lower-tier. Bénard & Bouché (1991) argue, on the basis of maturation modelling, that the Sole Pit area gas fields are sourced from the Namurian and Dinantian succession rather than Westphalian coals. The effectiveness of the Namurian as a thermogenic gas source is

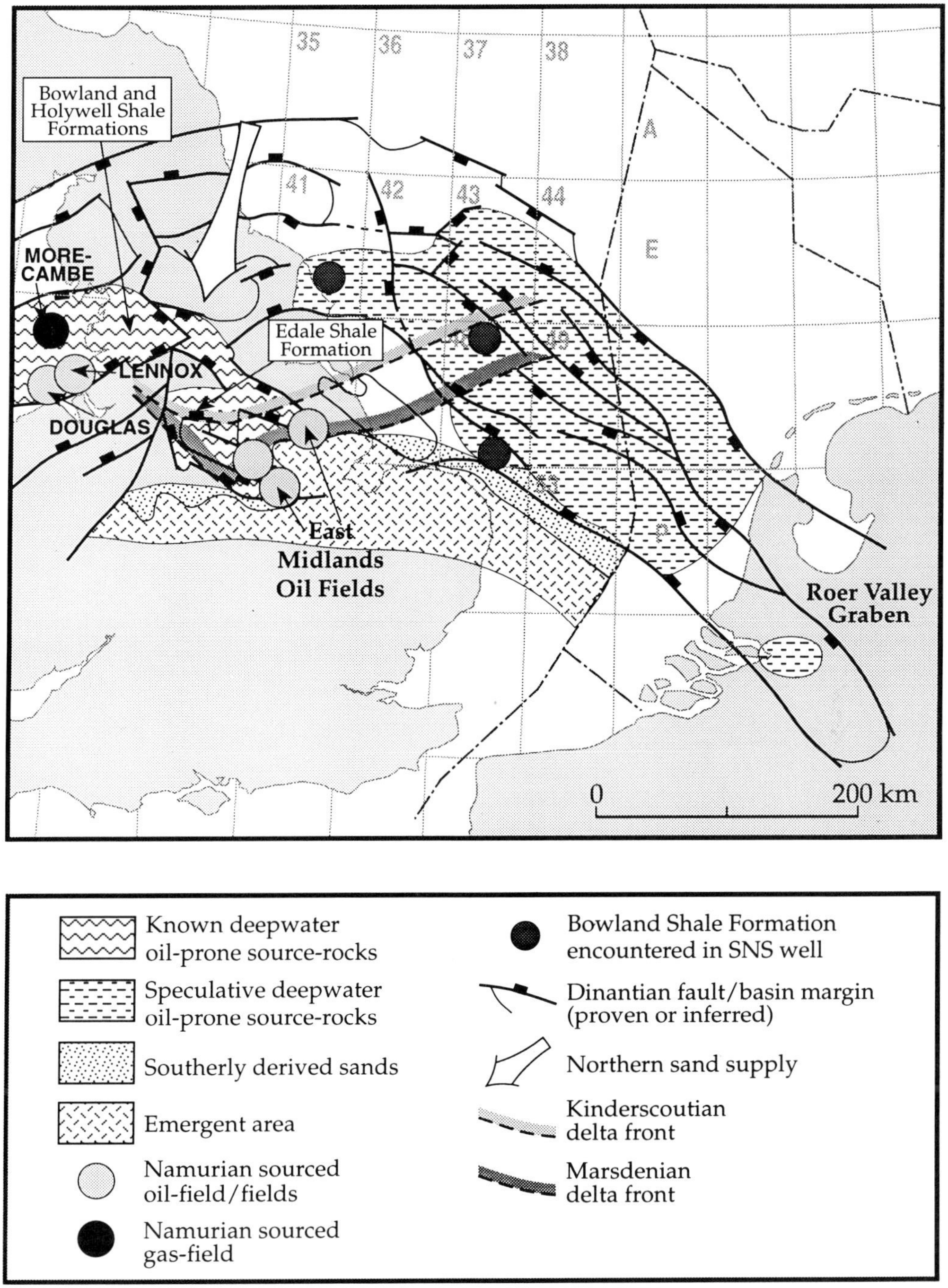

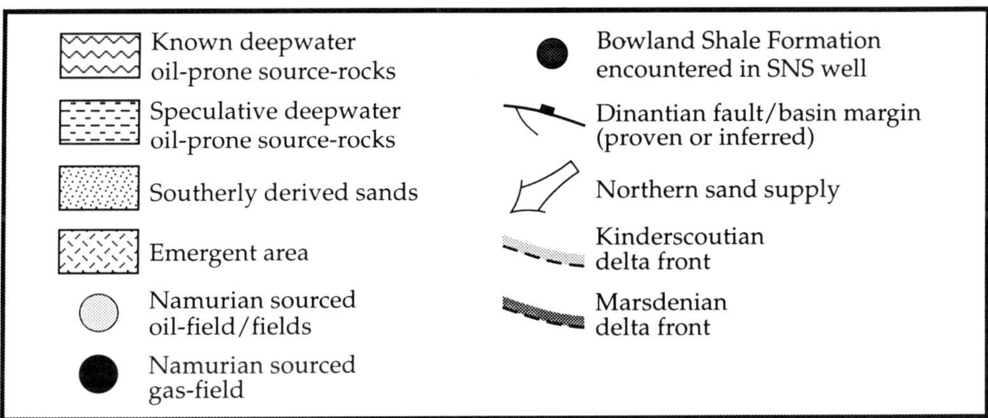


**Fig. 3.** Namurian hydrocarbon play elements.

demonstrated by the size of the Morecambe Gas Fields in the East Irish Sea (Hardman *et al.* 1993). The initial oil-generation phase from the Namurian section of the SNS may be recorded by the oil staining in the Rotliegend reservoirs of the Jupiter Fields (Quadrant 49) (Leveille *et al.* 1996). Alternatively, the source could be the lacustrine horizons in the Westphalian.

As noted by Cameron *et al.* (1992) and Cameron (1993), the published well control available to define the extent of deepwater Bowland Shale Formation is very limited (Fig. 3). Unpublished PGS Nopec 2D seismic lines were used to improve the delineation of the northern limit of the Formation by identifying those areas where the lower Namurian is thick. The southwards progradation of the Millstone Grit Formation deltas (Cameron 1993) is illustrated on Fig. 3 using the results of Fraser & Gawthorpe (1990), Leeder & Hardman (1990) and Collinson *et al.* (1993). There are no accounts of the uppermost Namurian infill history of the Quadrant 48/49 region or the offshore Netherlands. However, a drop in base level at the end of the Namurian may have removed the topmost part of this succession in some of these areas (Quirk pers. comm. 1995). The PGS seismic interpreted for this review indicates that the Namurian in depocentres such as the Silver Pit and Sole Pit Basins is between 500 and 1500 m thick. Thick basin floor fans are present. However, since they do not occupy more than 20% of the succession their presence will not significantly impact the volume of potential source-rocks.

*Dinantian (Viséan–Tournaisian) source potential.* Little is known about the Dinantian geology of the southern North Sea away from the immediate extensions of the onshore highs and platforms. The PGS seismic suggests that a broad, rift-induced basin is present. It also suggests, as shown on Fig. 13, that this basin contains a thick sedimentary infill. Pro-delta to basinal marine environments, as suggested by Cameron *et al.* (1992), are envisaged. Figure 4 shows the anticipated position of the basin in relation to the surrounding carbonate platforms and the southwards prograding Yoredale delta. Since the bulk of the basin was geographically remote from the Yoredale delta, anoxic bottom conditions are expected to have existed in this area, at least for part of the time. Onshore, Dinantian source rocks are known from the possibly structurally and sedimentologically analogous Bowland Basin (Gawthorpe 1987a; Lawrence *et al.* 1987). Source rocks are also known in northeast Wales from the southern margin of the East Irish Sea Basin (Armstrong *et al.* 1995). Black-shale infill of the shelf-margin Boulder Bed at Castleton (Derbyshire) may be of Viséan age (discussion in Simpson & Broadhurst 1969). Offshore, the PGS seismic indicates that the Dinantian succession in the depocentres is more than 1000 m thick and, if pro-delta to basinal

facies are present, substantial volumes of source-rocks are anticipated to exist in this area.

*Mid- and late Devonian source potential.* Even less is known about the Devonian of the SNS. It is uncertain when and where sedimentation was initiated following the final Caledonian (Acadian) deformation phase in the early and middle Devonian (Woodcock 1991). Because of these uncertainties, one illustration (Fig. 5) is used to present the elements of the ensuing Devonian palaeogeography.

Sufficient vertical section exists on the PGS seismic lines, for example in the Silver Pit portion of Fig. 13, to permit a substantial Devonian succession beneath the Carboniferous thicks of the Southern North Sea. However, it is not possible to pick the position of the Devonian–Carboniferous boundary in these areas. Evolving rift settings, similar to those described from onshore by Fraser & Gawthorpe (1990), are indicated.

Belgian and Dutch well control indicate that marine environments are present to the southeast of the United Kingdom median line (Ziegler 1990). The envisaged mid- and late Devonian tectonic setting of the SNS would permit marine facies to extend north-westwards from The Netherlands along the centre of the developing basin. Support for this possibility is provided by Corfield & Gawthorpe (1995) who found that the marine, middle Devonian limestones in the Mid North Sea High region (Quadrants 37 and 38) are more widespread than indicated by Ziegler (1990). At present it is considered that the Mid North Sea High limestones are related entirely to a transgression that progressed northwards from The Netherlands along the Proto-Central Graben (Ziegler 1990).

If marine strata are present in the United Kingdom sector of the SNS, the palaeogeography depicted by Ziegler (1990) for The Netherlands suggests that source- rocks should occur. The most favourable area for source-rocks is the axial region of the subsequent Carboniferous rift system. Quadrants 48, 49, 53 and 54 offer the best possibilities. This area is shown on Fig. 5.

*Silurian source potential.* As a result of a recent British–Belgian research programme, there is a modern and comprehensive literature base for the Silurian succession in the London–Brabant Ridge region (Pharaoh *et al.* 1993). Beneath East Anglia, Belgium and NW France, the dark colour of the rocks and the common lack of bioturbation point to initially high TOC values. However, slaty cleavage related to the Acadian

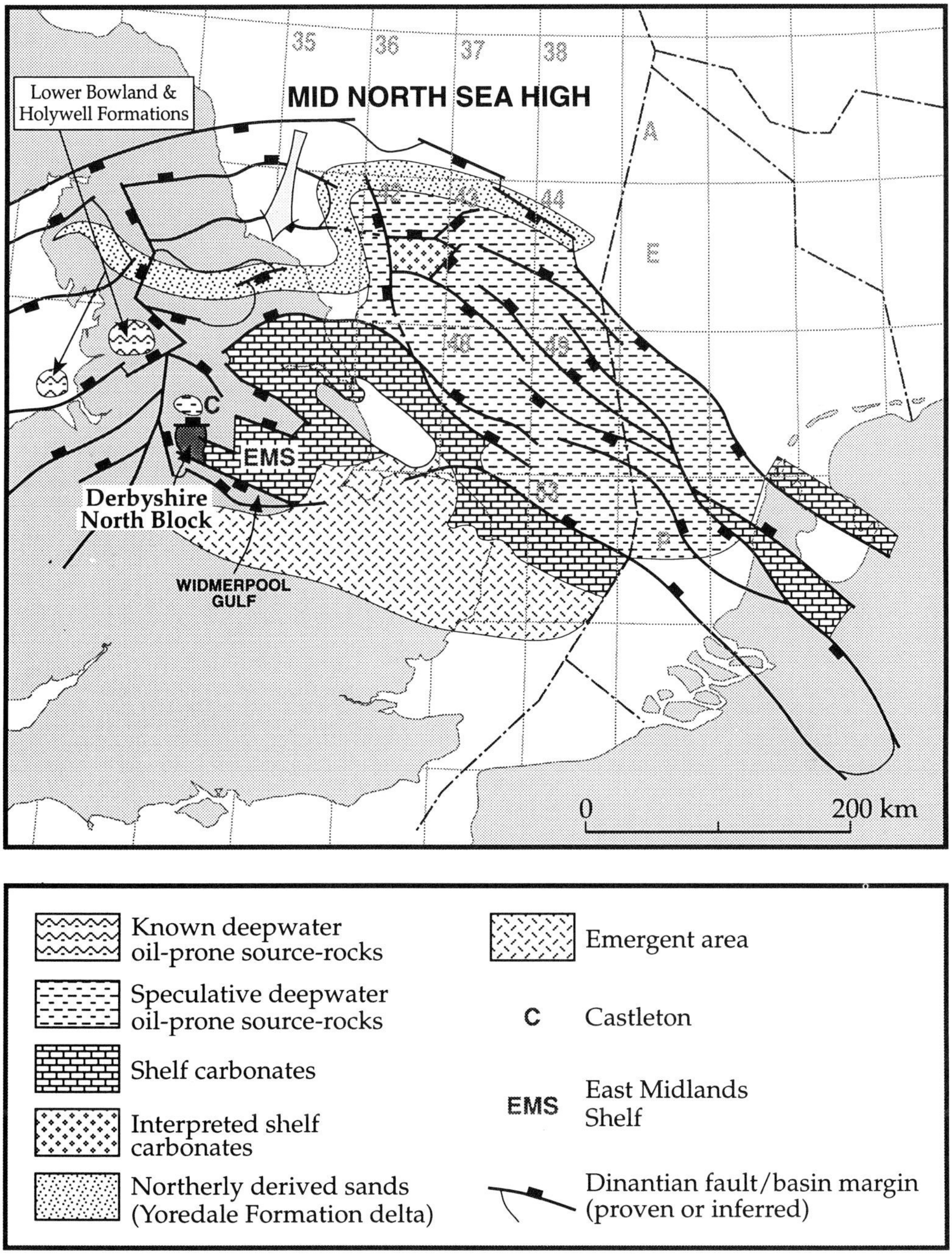

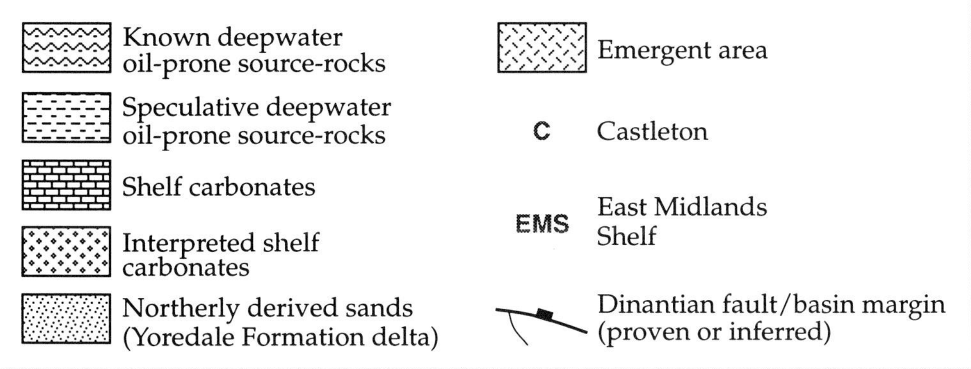


**Fig. 4.** Dinantian hydrocarbon play elements.

deformation phase is widespread. It is described from East Anglia borehole cores (Woodcock & Pharaoh 1993), and from outcrop and wells in Belgium (Verniers & van Grootel 1993) (Fig. 6). The Silurian section in the Boulonnais (northern France) is cleaved (Morton & Bristow 1989). Since cleavage is initiated at a vitrinite reflectance ($R_o$%) equivalent of 3.5–4% (Teichmüller 1987; Smith 1993), no source potential can remain in these areas. High maturity levels are

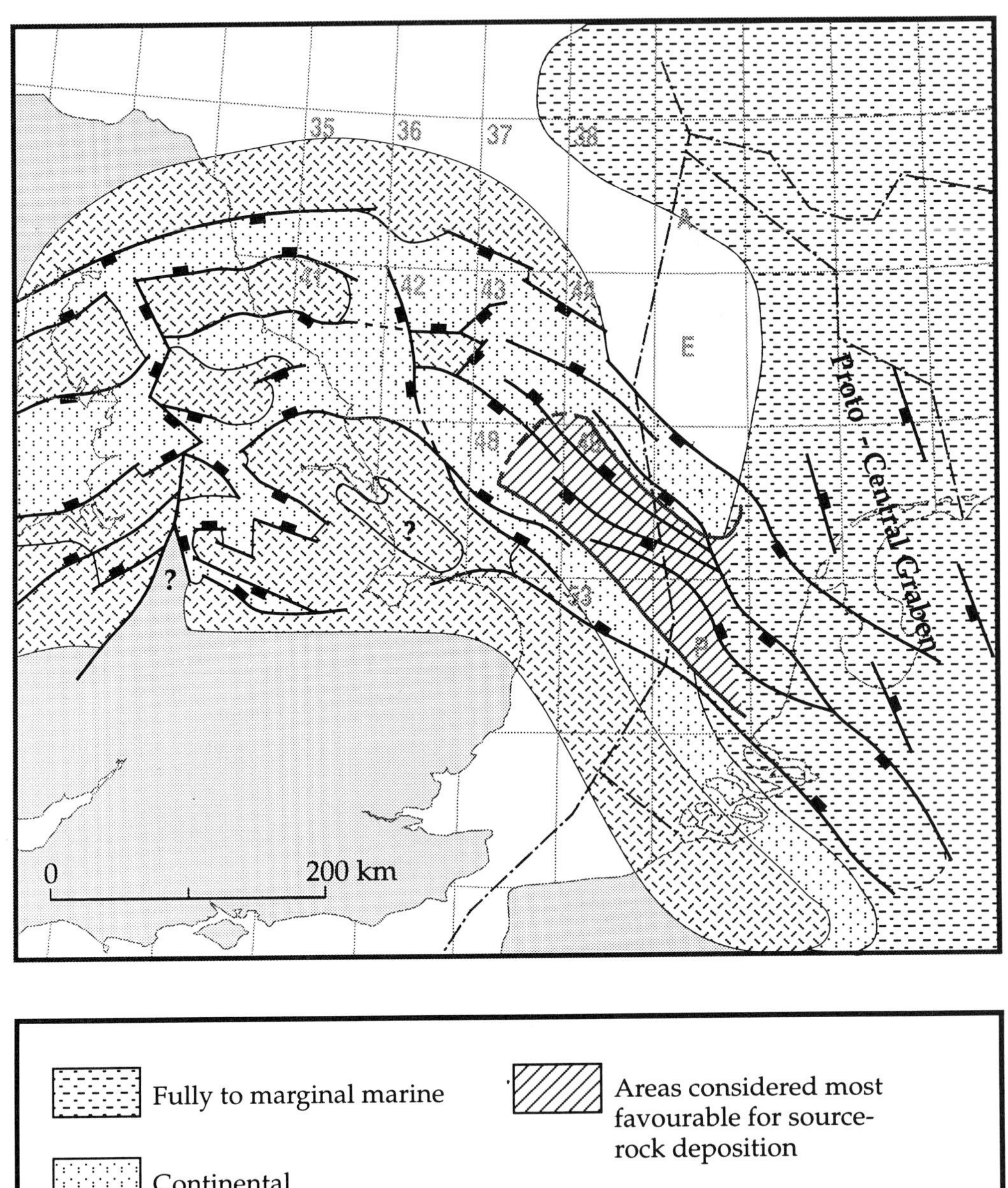



**Fig. 5.** Mid- and later Devonian hydrocarbon play elements.

also revealed by the white mica (illite) crystallinity study of Merriman *et al.* (1993). They
show that the Pre-Acadian section throughout
East Anglia has reached anchizone and epizone
(greenschist metamorphism) metapelite grades.

The anchizone equates to vitrinite reflectance
equivalent values of between 2.6% and 3.4%, the
epizone begins at a vitrinite reflectance equivalent of 3.4% (see Meere 1995 table 1). Devonian
schists are present in the Brabant region of

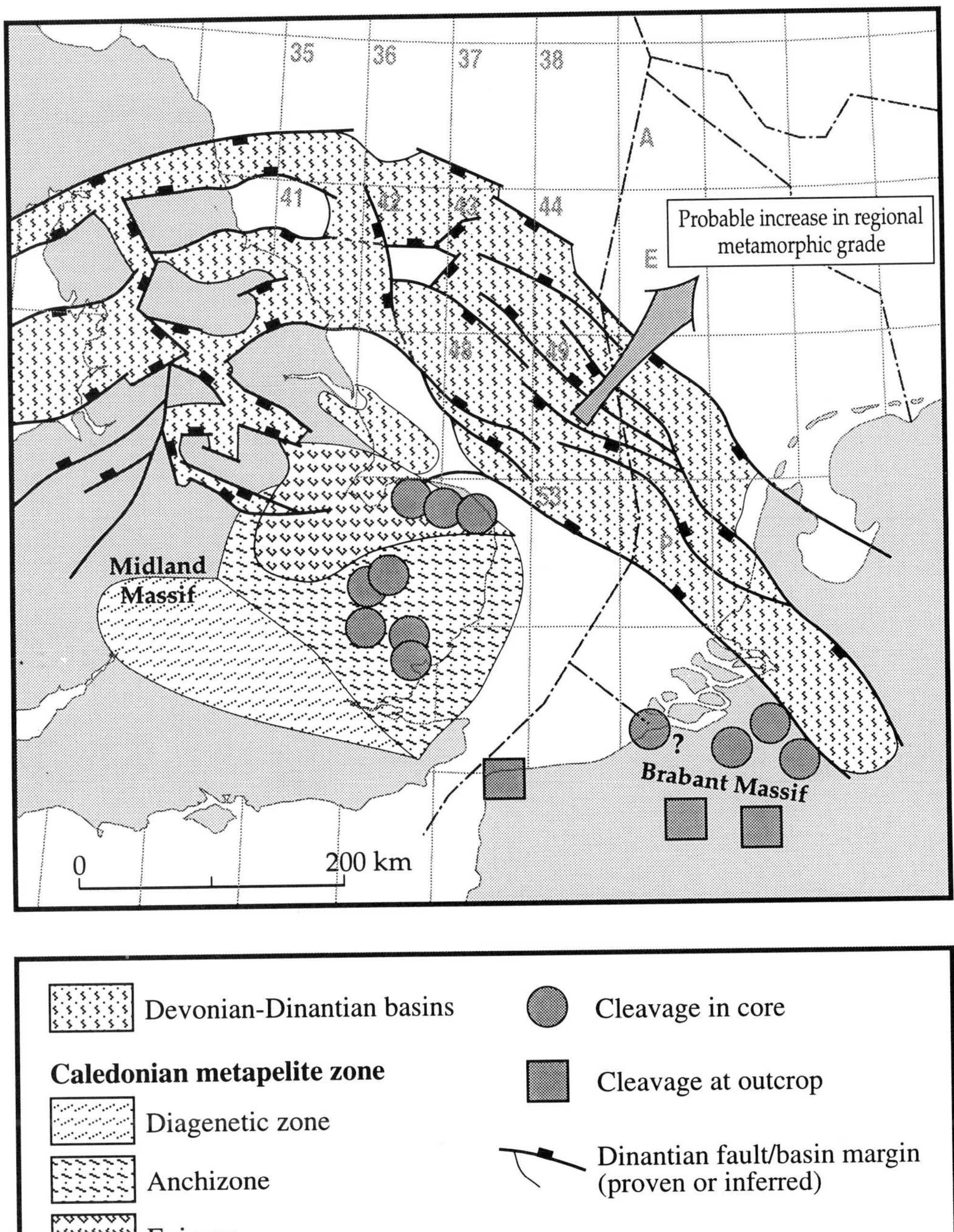


**Fig. 6.** Map of Silurian metamorphic grade.

Belgium (Ziegler 1990). Offshore, palaeogeographic reconstructions for the Caledonian orogeny suggest that regional metamorphic grade increases northeastwards towards the Tornquist Line (Ziegler 1990). Diagenetic zone grade sediments are only present onshore (Merriman *et al.* 1993) in association with the stable Midland Massif. Tremadocian beds in this area have vitrinite reflectance equivalent values of 1.3% and 2.6% (Smith 1993).

*Pointers provided by the contribution of pre-Westphalian source-rocks to the global hydrocarbon budget*

Ulmishek & Klemme (1990) found that major source-horizons are restricted to a few intervals in the stratigraphic column. It was also evident that major source horizons tend to occur globally. Their results mean that a second, albeit crude method exists of predicting the source-rock potential of underexplored regions.

In relation to the Palaeozoic, Ulmishek & Klemme (1990) show that approximately 20% of the world's hydrocarbon reserves originated from pre-Westphalian source rocks. As illustrated in Fig. 7 of this paper, 9% of these reserves were supplied from the Silurian and 8% from the late Devonian to Tournaisian succession. In contrast to the global picture, 1.6% of reserves in the combined area of the United Kingdom, Ireland, The Netherlands, Denmark and the North Sea region of Norway are derived from pre-Westphalian source-rocks. The United Kingdom area figure is based on Leckie *et al.* (1995). An arbitrary Namurian/Westphalian source split of 50/50 was used for the lower-tier discoveries in the SNS. The middle tier of the SNS fairway was assumed to be entirely sourced from the Westphalian.

The mismatch between the global and the United Kingdom region figures in some cases reflects geological history. The Acadian induced overmaturity of the Silurian section in the SNS is an example of a geological factor. In other areas the mismatch may reflect the lack of exploration. The preceding regional review suggests that the latter explanation is probable for the basinal, late Devonian to Tournaisian succession of the SNS. The preservation potential of hydrocarbons associated with basinal settings is examined in the next section.

*Restrictions imposed by Post-Variscan tectonics*

The Mesozoic depocentres of the Southern North Sea were affected by late Jurassic–early Cretaceous, late Cretaceous–early Tertiary and mid-Tertiary phases of inversion (van Hoorn 1987). These areas present considerable risks for deeper-tier exploration as the integrity of the pre-inversion traps may have been disrupted. Although fracture-induced permeability and porosity might have been increased by inversion, complete recharging of structures is unlikely to have recurred in those areas where source rocks had previously passed through the peak gas-generation window. The areas of maximum inversion are illustrated on Fig. 8. They were defined by seismic mapping by the authors and by merging the published vitrinite reflectance ($R_o$%) contours for the near top-Carboniferous (Leeder & Hardman 1990; Bailey *et al.* 1993). The position of the resulting middle-tier fairway is indicated, as is the lower-tier fairway of Quadrants 43 and 44. Variscan inversion is also an important element in controlling the position of the lower-tier fairway (Besly in press). The two examples of new opportunities discussed below are positioned outside the regions of strong inversion.

## Examples of new opportunities

A wide variety of opportunities was identified from the PGS seismic grid. These are related to both the existing lower-tier of the fairway and the anticipated basal-tier. Two of these opportunities are detailed below.

The first example, whose location is defined by the southern of the two dashed boxes on Fig. 1 has adequate well control. The traps are structural and the target horizons are sandstones in the Westphalian to Devonian succession. The second example, whose location is indicated by the northern dashed box on Fig. 1 is very different and illustrates the region's upside potential. The target is located in a region with few deep Carboniferous penetrations. The trap is stratigraphic and the target is a progradational wedge of Dinantian aged carbonates.

The seismic data were selected from PGS Nopec's regional grid of 2D surveys. Our results, although not based on detailed mapping, have been checked by cross-tying and are accurate on the gross scale. Seismic sections exhibiting the targets regions are shown in Figs 9, 10 and 13. The locations of these lines are indicated on Fig. 1. Visits were made to the Castleton area of Derbyshire (Fig. 4) to examine in the field the outcrop equivalents of the Dinantian reservoir section considered to be present in the SNS.

*Quad 53: Westphalian to Devonian structural traps set up by Variscan compression along the northeast margin of the London–Brabant Ridge*

This area was first investigated ten years ago by Tubb *et al.* (1986) who identified a suite of Palaeozoic targets. Our findings confirm many

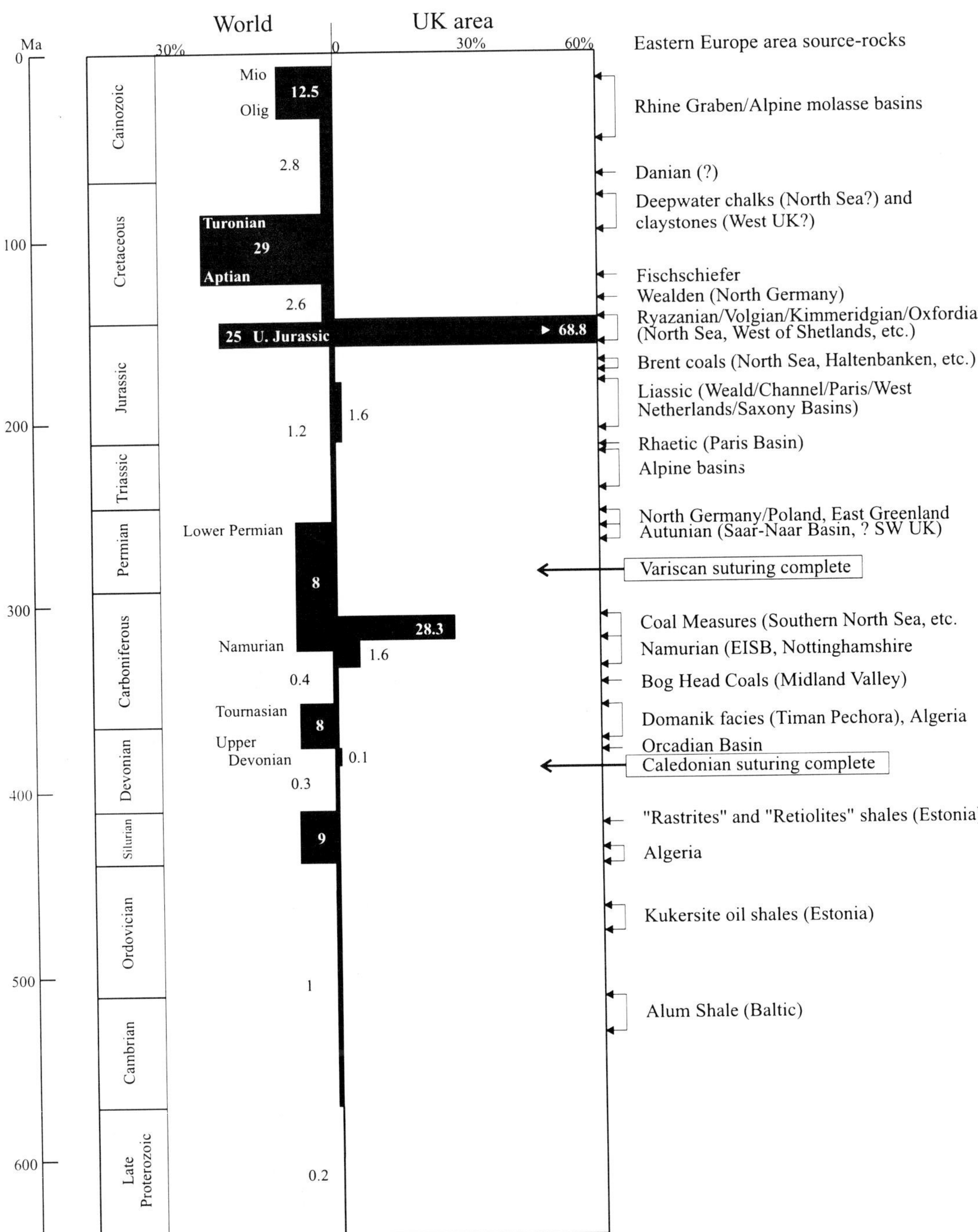



**Fig. 7.** World and UK area hydrocarbon reserves compared by source rock age. The world column is derived from Ulmishek & Klemme (1990), the United Kingdom area column is based on Leckie *et al.* (1995). The right hand column shows, for comparison purposes, the ages of the western Europe area source-rocks. The completion dates of the Caledonian and Variscan orogenies are also indicated. The Harland *et al.* (1990) time scale is used.

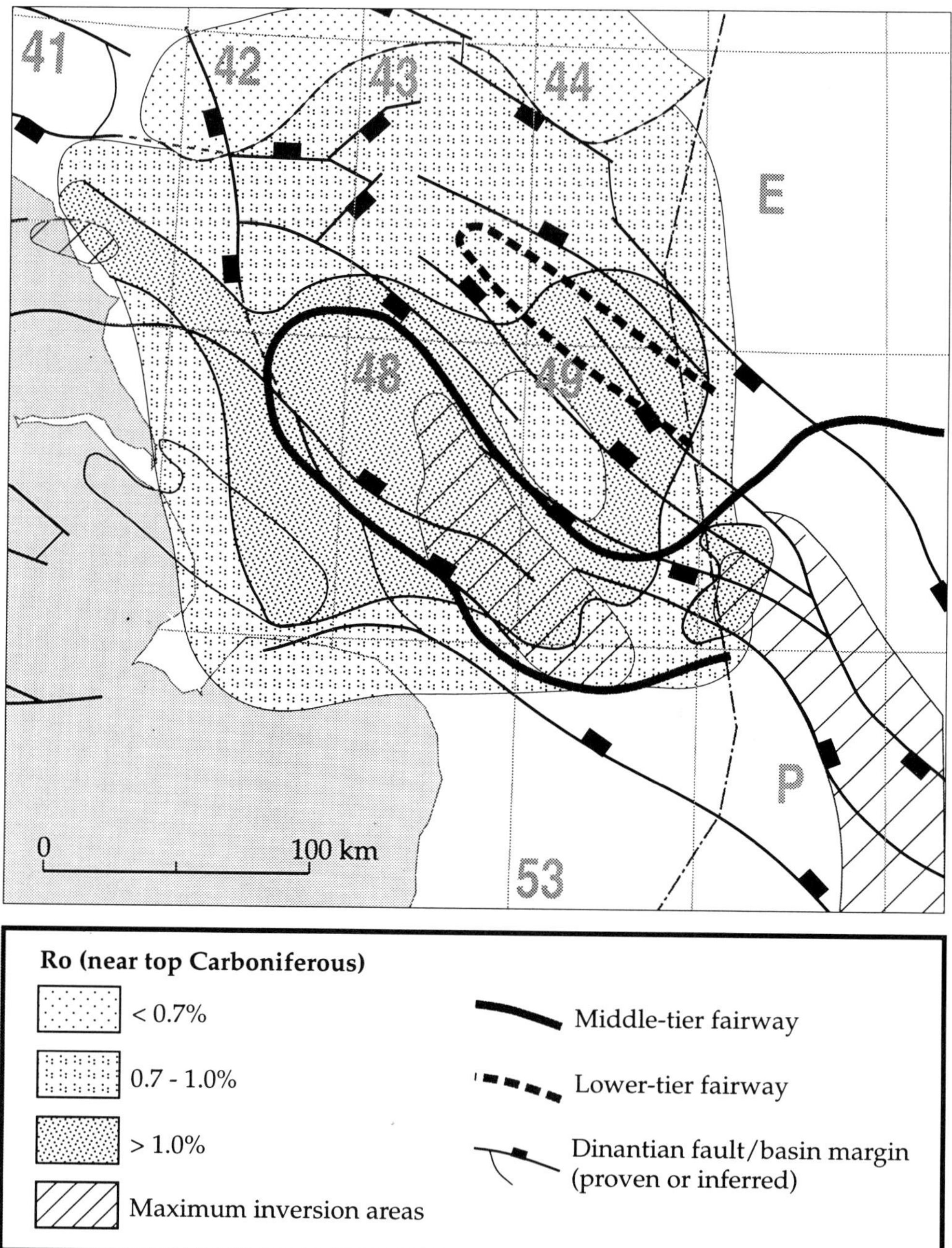



**Fig. 8.** Map of near top Carboniferous maturity superimposed on area of maximum post-Variscan inversion.

of the predictions of that earlier study. However, the regional approach used this time study has allowed the dynamics of the play concepts to be more clearly demonstrated. In addition, we have also considered the implications of an oil-prone succession in the Westphalian.

The area of interest lies towards the southwestern end of line SNST-83RE-03 (Figs 9 and 10). The location of the line is shown on Fig. 1. This line runs in a southwest to northeast direction from the edge of the London–Brabant Ridge to the Broad Fourteens Basin in the

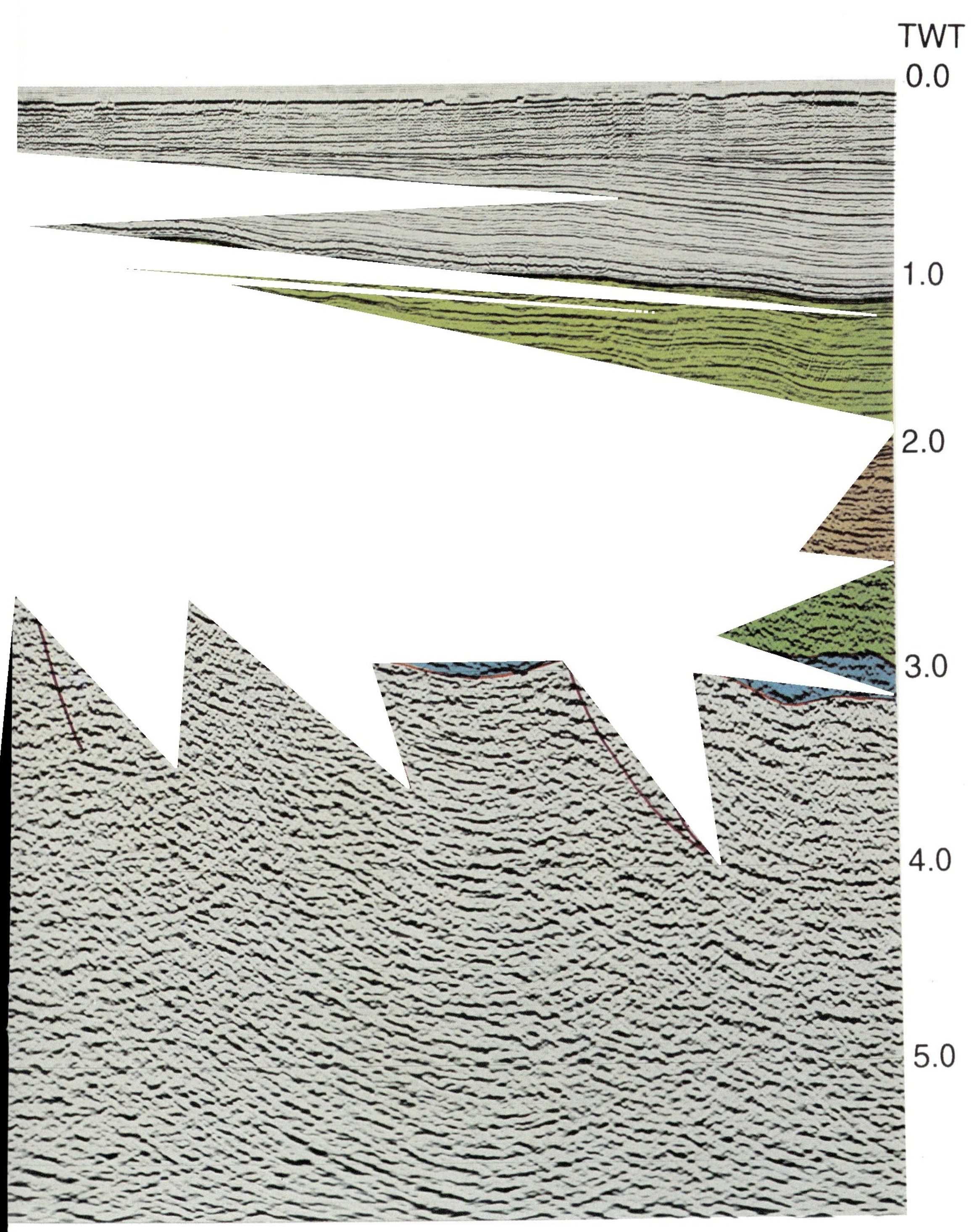

NE
TWT
0.0
1.0
2.0
3.0
4.0
5.0

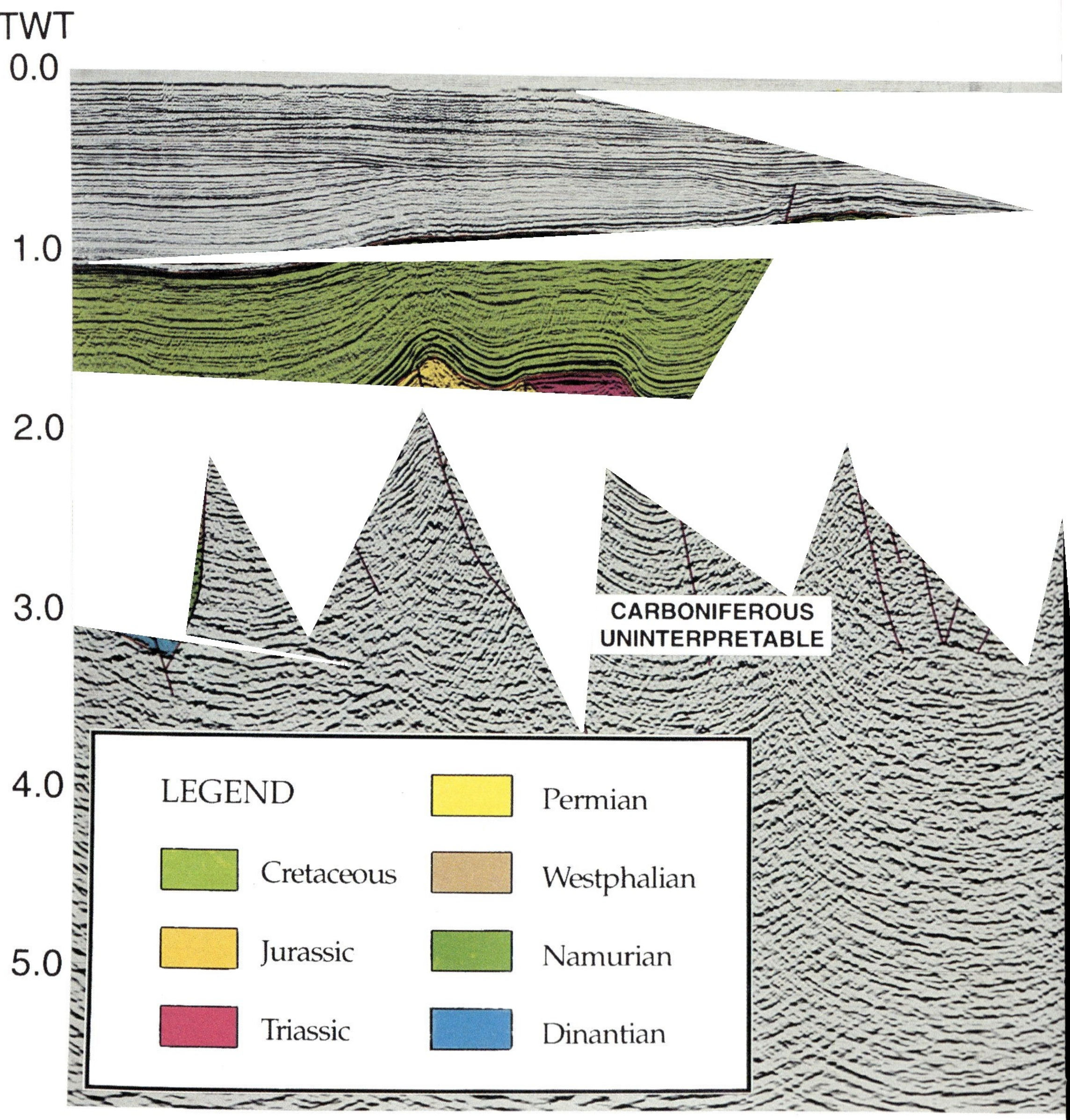


**Fig. 10.** Broad Fourteens Basin: northeastern part of interpreted PGS Nopec seismic line SNST-83RE-03. This exhibits a thick inverted Mesozoic section. Carboniferous events are difficult to resolve beneath the Broad Fourteens Basin.

# SNST-83RE-03

BRO

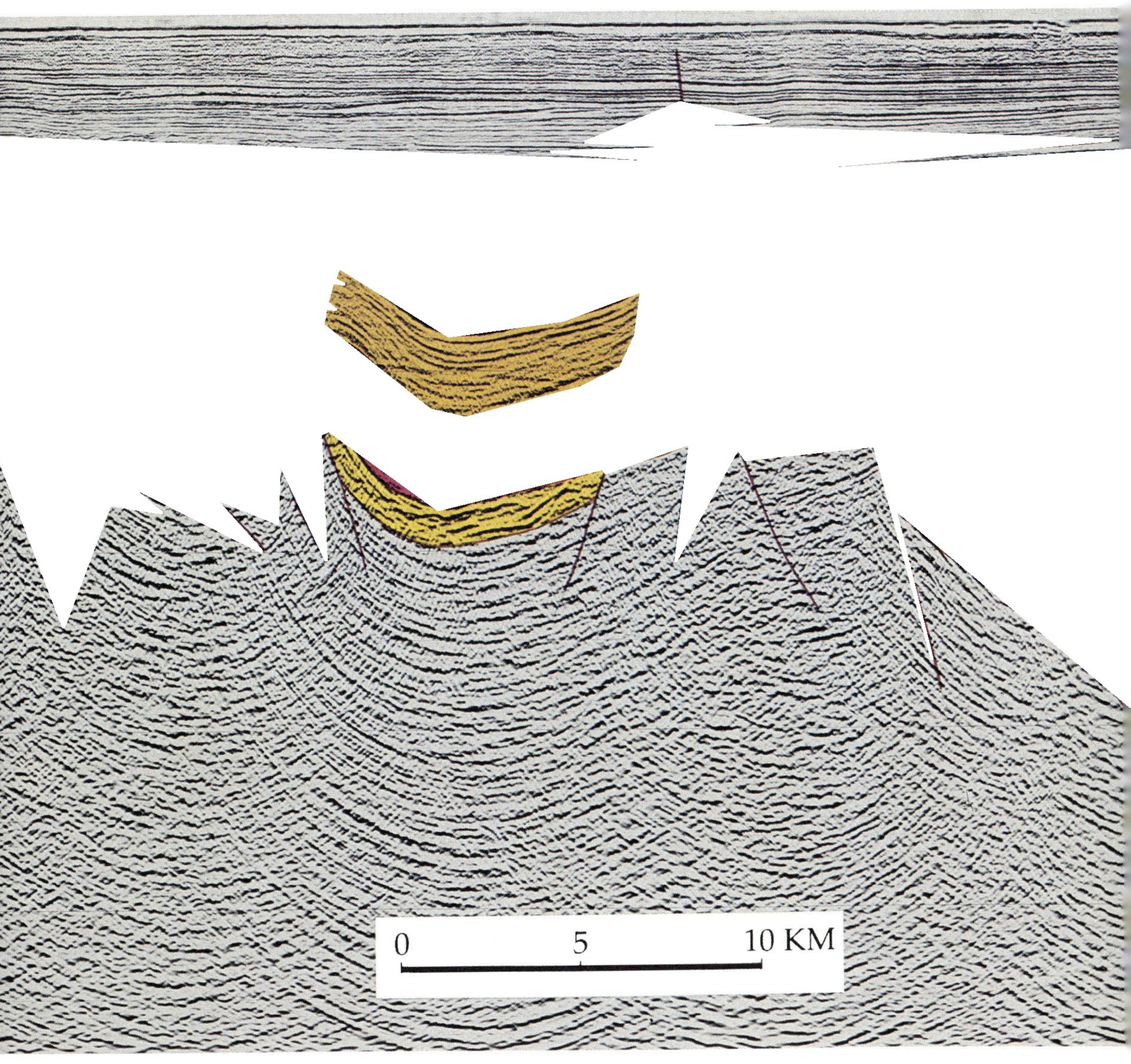

# SNST-83RE-03

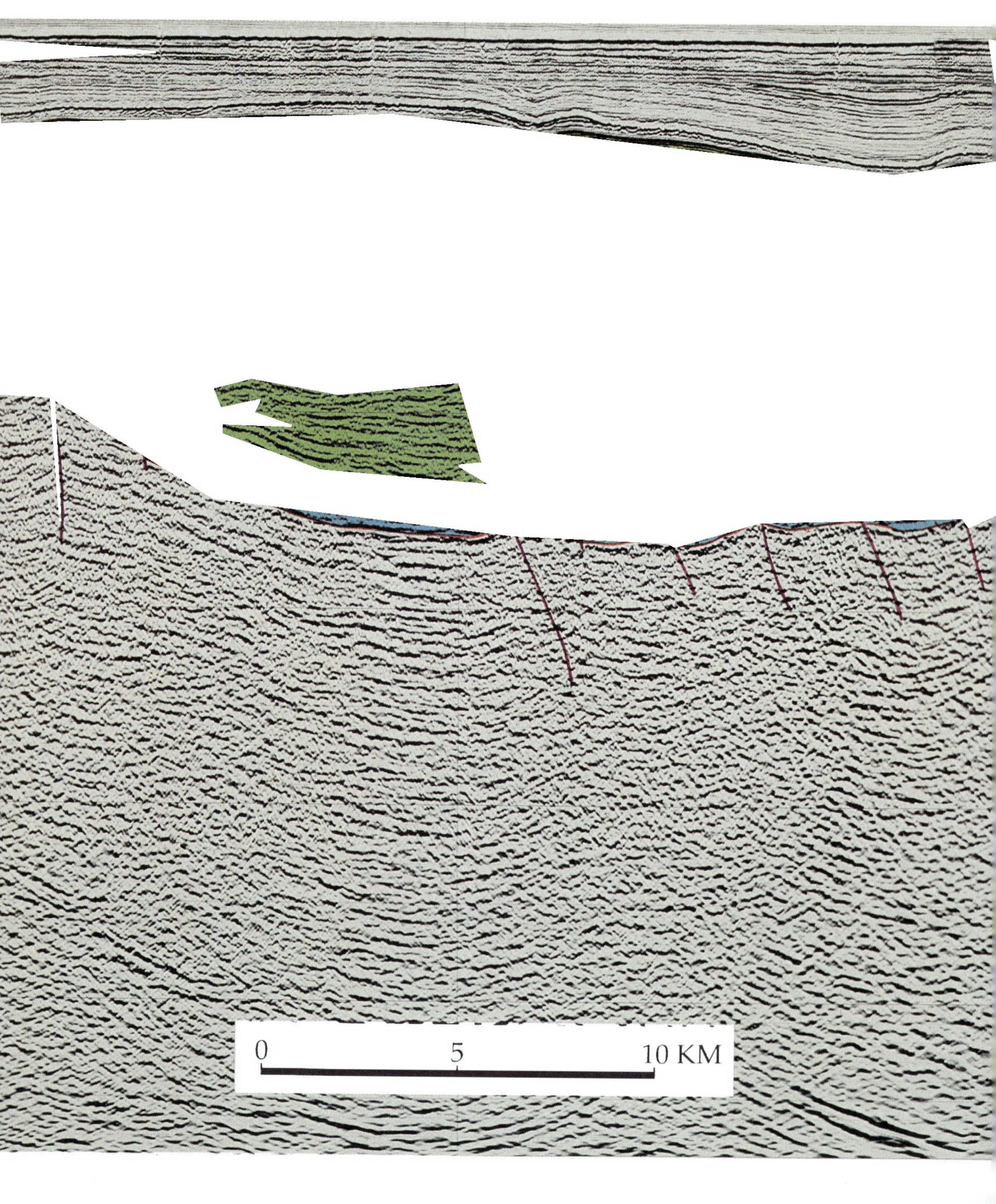

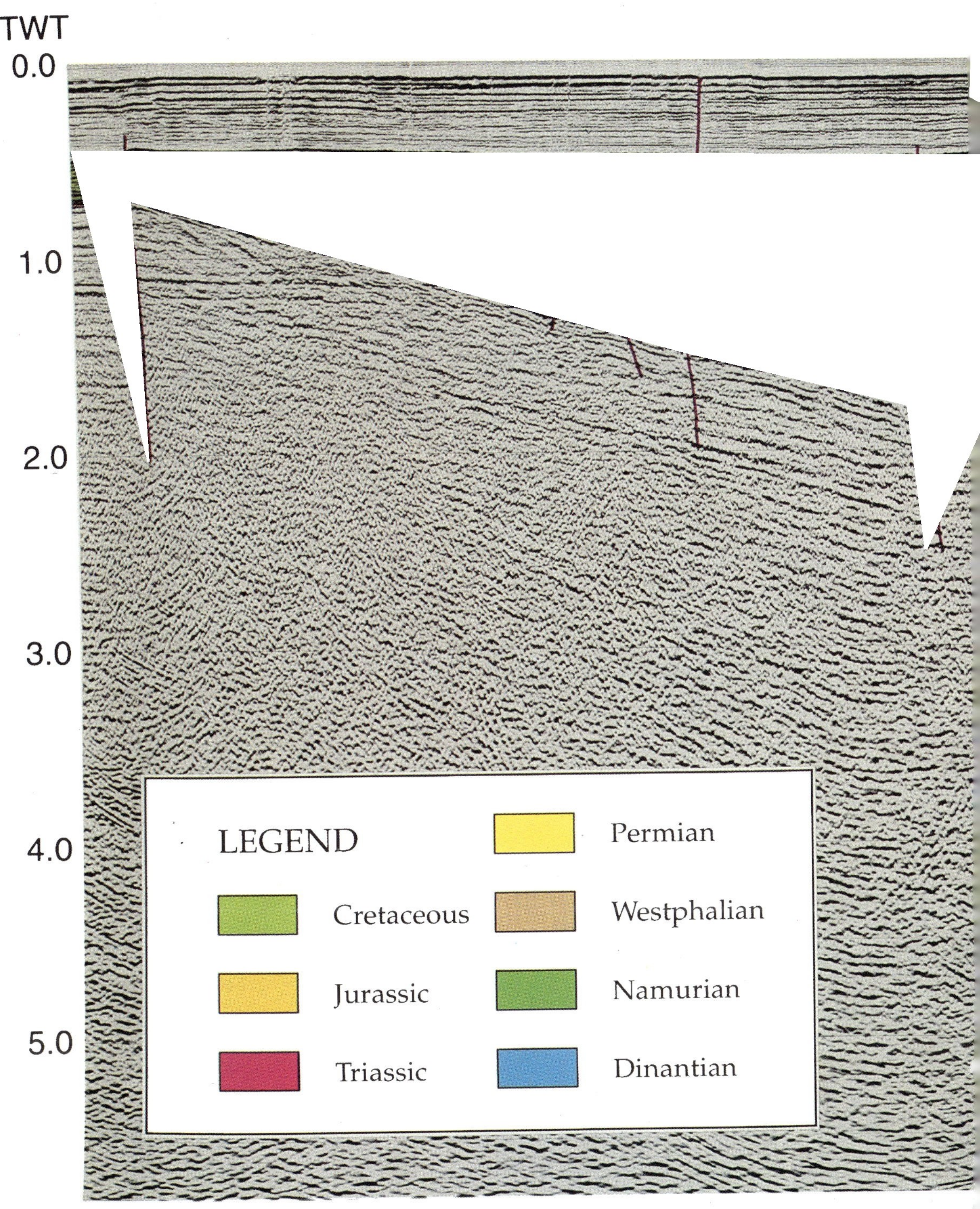

**Fig. 9.** London–Brabant Ridge flank: the southwestern part of interpreted PGS Nopec seismic line SNST-83RE-03. This shows a Carboniferous section rapidly thickening to the northeast. A wide variety of structural traps is evident (Sy, Symon Unconformity).

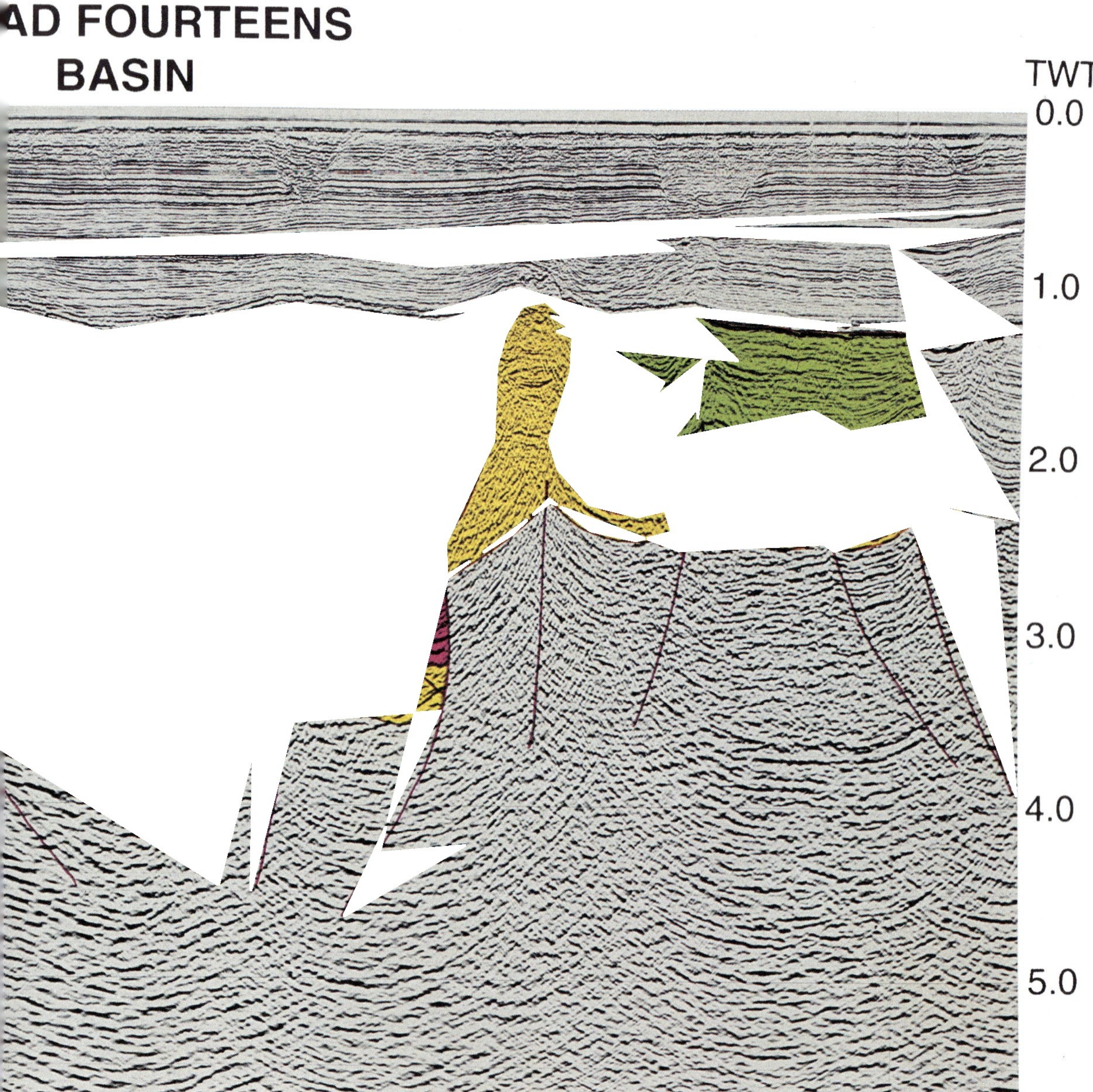

AD FOURTEENS
BASIN
NE
TWT
0.0
1.0
2.0
3.0
4.0
5.0

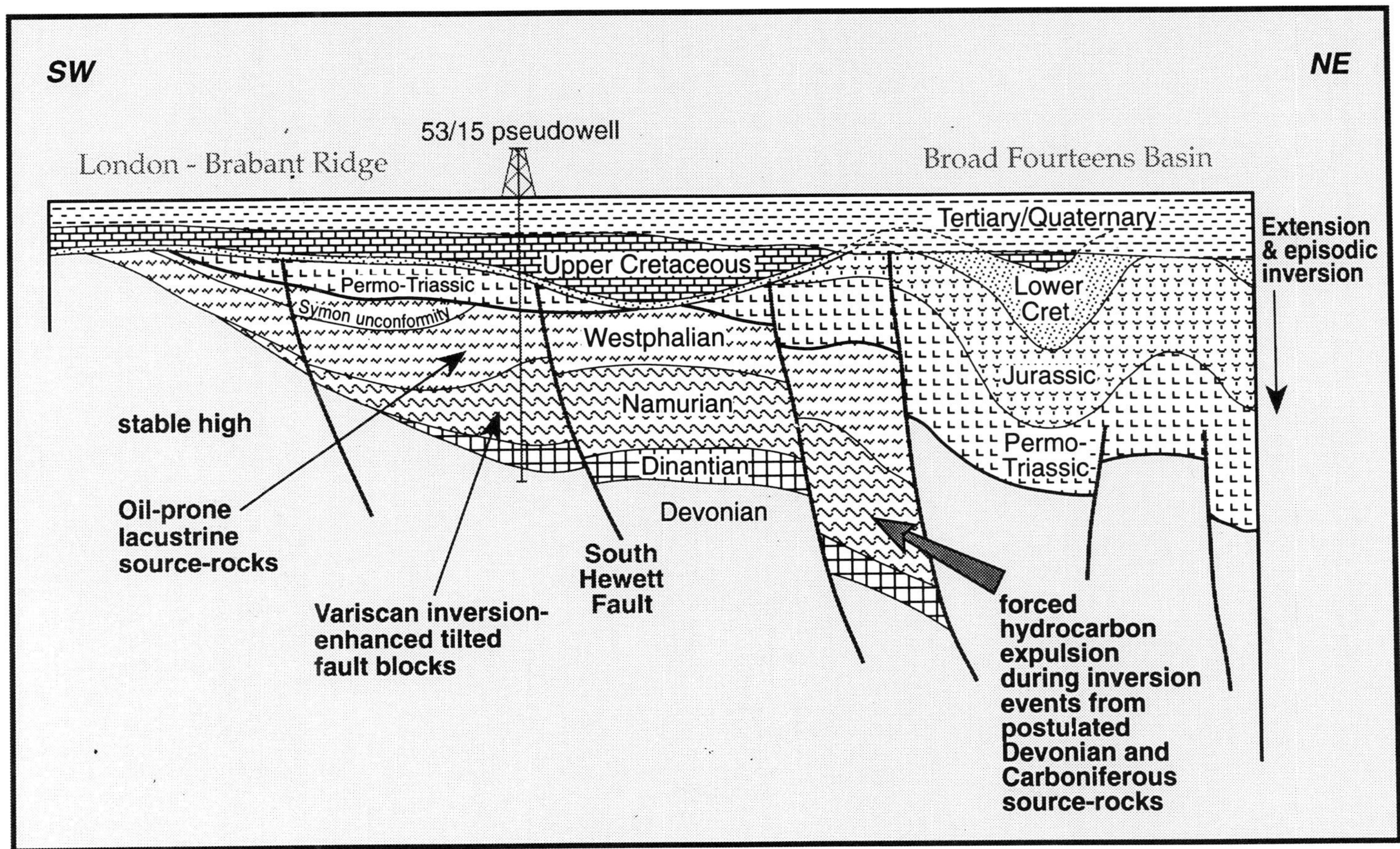


**Fig. 11.** London–Brabant Ridge flank play concept. Palaeozoic-generated hydrocarbons are forced up-dip into the flank of the London–Brabant Ridge during inversion events.

Netherlands. The line is tied to well 53/19a-1 which has a total depth (TD) of 2317 m below mean sea level in the Westphalian 'A'. As illustrated in the figures, the entire section thickens to the northeast across northeasterly dipping faults. Strong inversion tectonics dominate the region of the Broad Fourteens Basin. Initially, readily traceable Carboniferous reflections progressively weaken to the northeast towards the Broad Fourteens Basin. Two factors are responsible for the loss of reflection strength. The first is the increase in thickness and tectonic complexity of the post-Carboniferous cover. The second is the change in facies within the older Carboniferous section from a shelf setting in the southwest to a basinal setting in the northeast. The intra-Westphalian 'C' Symon Unconformity is expressed on Fig. 9 at the southwest end of the line. Reflections below the Dinantian are attributed to Devonian clastics.

The area is examined in terms of the maturation history of the succession present in a pseudowell located immediately to the southwest of the South Hewett Fault (Figs 1 and 11). The target, which is positioned on the seismic line, is a Variscan inversion enhanced, tilted fault block in Block 53/15. The pseudowell is located on the crest of the structure. The structural setting permits charging from both local Westphalian oil-prone source rocks and laterally from the Broad Fourteens Basin region to the northeast. Though Toarcian source-rocks are in the oil window in the Broad Fourteens Basin (Roelofsen & de Boer 1991), the barrier presented to lateral migration by the Permo-Triassic evaporites makes it unlikely that oils from this source have entered the Palaeozoic succession. Even if there are gaps in the evaporite barrier, hydrocarbons would preferentially migrate up-dip through the Rotliegend sandstones. Instead, the forced expulsion of pre-Permian derived hydrocarbons during inversion events towards the London–Brabant Ridge is envisaged as the charge mechanism from the Broad Fourteens Basin region.

Figure 12 is a BMOD-maturation history plot for the pseudowell. The crestal location was selected to determine the relationship between the timing of structural growth, local Westphalian maturation, and regional Namurian, Dinantian and Devonian maturation. Modelling in areas without thermal calibration with wells requires, as noted by Tubb *et al.* (1986), assumptions on basin evolution and heat flow. We used a constant heat flow 50 mW/m$^2$, the present-day value for the region (Lee *et al.* 1987, fig. 1). The model was run using an uniform seabed temperature of 10°C and a constant 50 m

of water cover. Lithologies and the burial history curve were derived from the local wells and the seismic line. An equal mix of Type I and Type II kerogens was used. Times are related to the Harland *et al.* (1990) timescale.

The Dinantian section entered the early oil window ($R_o$ 0.5–0.7%) at the end of the Namurian (320 Ma). The Westphalian section entered the early oil window ($R_o$ 0.5–0.7%) in the mid-Eocene (45 Ma). At these maturity levels significant on-site generation has not occurred. To the southwest of the structure, there is an additional cover of some 500 m of Carboniferous sediments below the Permian unconformity (Fig. 11). The model, when run for this location, showed that the base of the Westphalian section entered the early oil window ($R_o$ 0.5–0.7%) in the late Triassic (210 Ma) and the mid-mature window ($R_o$ 0.7–1.0%) in the Oligocene (25 Ma). This means significant amounts of oil could have been generated off-structure from the Westphalian section between the late Oligocene and the onset of uplift in the mid-Miocene. The burial history of the pseudowell demonstrates that closure existed before generation commenced. Top and side seal are provided by intraformational Palaeozoic shales.

The model results also imply that oil generation from the more deeply buried Westphalian section to the northeast of the South Hewett Fault occurred after the formation of the end-Carboniferous inversion structures (Fig. 11). Therefore, unlike Tubb *et al.* (1987), who modelled the 53/12-1 well, gas is not considered to be the primary target in the immediate vicinity of the pseudowell. However, gas and condensate prospectivity will increase to the northeast of the South Hewett Fault because of the progressively deeper burial of the Palaeozoic succession and the predicted, increasingly older base of the source section.

*Quads 42 and 43: potential Dinantian stratigraphic traps associated with the hanging-wall margins of a newly defined, fault-bounded high*

This example, which involves a basal-tier opportunity, has been selected to illustrate the potential for entirely new plays. It is located to the south of the Mid North Sea High in Quadrants 42 and 43 (Fig. 1). Whilst we have few concerns about the structural interpretation, the ages and facies attributed to the Carboniferous succession are open to debate because of the lack of well ties. However, our interpretation does not

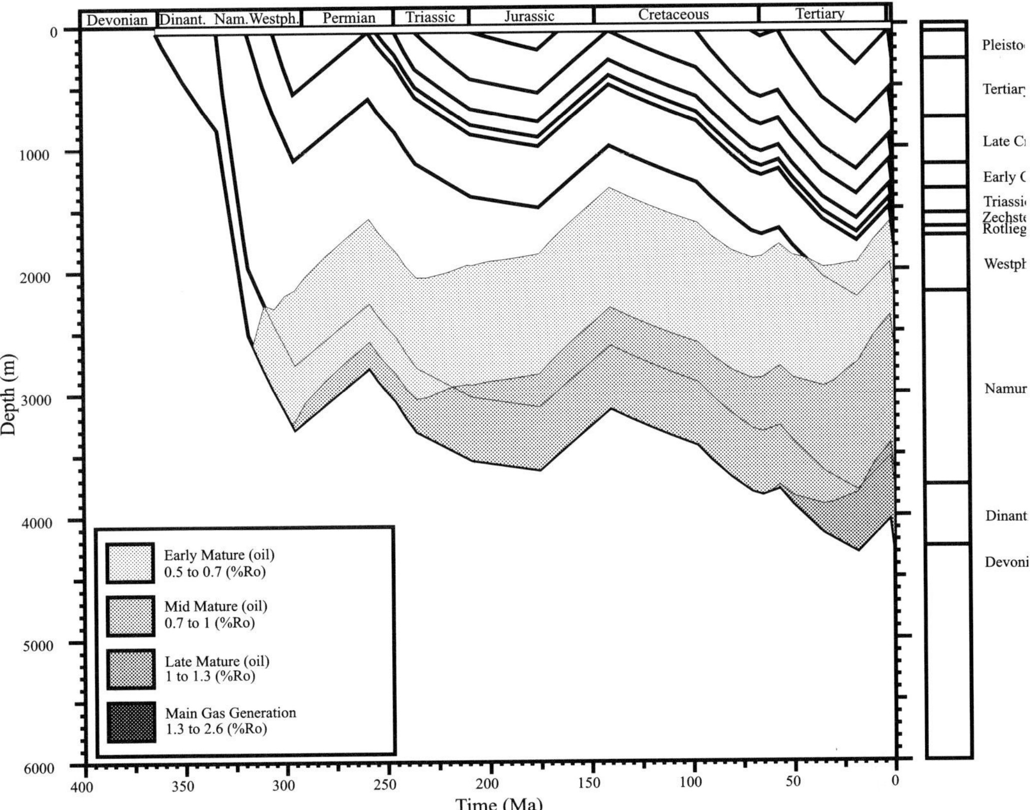


**Fig. 12.** London–Brabant Ridge flank. Maturation model for the pseudowell shown on Fig. 11. The pseudowell is located in Block 53/15 immediately to the southwest of the South Hewett Fault.

stretch the limits of the published regional geological framework. Also, since there are seismic and outcrop analogues supporting our interpretation, we consider this example to be an excellent demonstration of the capacity of the SNS to support significant new opportunities.

The target is a high, defined for the first time by our mapping. As indicated on Fig. 1 most of this high is located in Quadrant 42 to the east of the Dowsing Fault. Regionally the high represents a recurrence to the east of the Dowsing Fault of the east–west Alston Block and Askrigg Block trends present onshore. We failed to identify the large Silverwell High considered by Collinson *et al.* (1993), on the basis of gravity mapping, to extend across the southern portion of Quadrant 43. Instead, NW–SE-trending fault blocks were found to be present in this area.

Line SNSTI-90-06 illustrates the structural features of the target. Its location is shown on Fig. 1 and the interpretation is presented in Fig. 13. The line runs ESE from the high into the Silver Pit Basin of Quadrant 43. The high is

considered to have been formed by block faulting in the middle to late Devonian. Thicker Carboniferous section is present in the hanging wall of the fault defining the eastern limit of the high than in the footwall. To the WNW across the crest of the high the Carboniferous section progressively pinches out beneath the Permian Silver Pit Formation shales (Bailey *et al.* 1993). A broad Variscan inversion arch is present in the Silver Pit Basin. This very large structure is aligned with the Quadrants 43 and 44 gas discoveries to the southeast.

Of the multiple exploration targets in the area, we found the progradational wedge developed on the south-east margin of the high to be the most interesting. This feature is marked PW on Fig. 13. We consider that much of the wedge consists of Dinantian detrital carbonates derived from a platform succession analogous to that capping the onshore Dinantian blocks of northern England. The high frequency/strong amplitude seismic reflections associated with the rhythmic beds of the southwards prograding, Yoredale Formation (Cameron 1993) overlie the progradational wedge.

The bed forms of the progradational wedge resemble the carbonate wedges contained in the EC4 and EC6 seismic sequences of the Widmerpool Gulf (Ebdon *et al.* 1990). The location of the Widmerpool Gulf is shown on Fig. 4. The play potential of United Kingdom carbonate wedges is illustrated by Fraser and Gawthorpe (1990). Bailey *et al.* (1993) include a diagram suggesting that Dinantian shelf and deeper water carbonates are associated with highs in the Sole Pit and Silver Pit regions. Outcrop and subcrop analogues are present in the Bowland region (Gawthorpe 1987*a,b*) and at Castleton in the Peak District of Derbyshire. Residual oil shows are present in both these areas. The Castleton district is of particular interest as the depositional morphology of upper portion of the northwest margin of the Dinantian, North Derbyshire Block (Fig. 4) is exhumed by erosion. The subcrop geology of the immediate area to the north and south of Castleton is illustrated by Gutteridge (1991). The regional subcrop setting is portrayed by Quirk (1993*b*). Figure 8 of Quirk (1993*b*) shows that our model for the Quadrants 42/43 high has tectonic features in common with the North Derbyshire Block.

The abundance and variety of residual oil shows in the Castleton area suggest that, before the removal of the Namurian cover, the shelf margin crest and the northwards inclined shelf margin succession could have hosted a significant oil accumulation. The elaterite of Windy Knoll to the west of Castleton provides the best exposure of a residual oil body (Peacock & Taylor 1966, Pering 1973). Windy Knoll forms a small culmination on the ridge crest that was previously completely top and side sealed by the Edale Shale Formation. Our field work suggests that the base of the elaterite defines a residual oil–water contact (OWC). Bitumen impregnation in limestone breccias below the OWC may indicate the charge pathway. These breccias were found by Peacock & Taylor (1966) to be radioactive. If the bitumen and uranium are genetically associated, an indication of the area formerly open to oil migration is provided by Peacock & Taylor's map of surface uranium anomalies. Radioactive anomalies occupy the entire 6000 m long, 500 m wide and 150 m high exposed portion of the slope to the south of Castleton. Peacock & Taylor (1966) note that the anomalies persist in depth. They illustrate samples of bituminous calcite and uraniferous phosphatic limestone collected from caves in the Castleton area.

The hydrocarbons of the Castleton and Bowland region are, as previously indicated, considered to have been predominantly sourced from the onlapping shales of the Namurian Edale Shale Formation. In the Quadrants 42 and 43 region, Namurian and Dinantian shales are the required source.

The Dinantian carbonate outcrops in the Pennines and North Wales commonly exhibit very poor reservoir properties due to pervasive calcite cementation. Walkden & Williams (1991) show for the North Derbyshire Block that the loss of reservoir quality was most probably related to meteoric water flow during the Namurian and early Westphalian. At this time westwards inclined tilting had exposed the East Midlands Shelf carbonates (Fig. 4) and east to west water flow was possible. The resulting cements form Walkden & Williams' diagenetic Z3 zone. Walkden & Williams (1991) consider that Z3-type cements will not occur where there was protection from late meteoric water influx. We consider the Quadrant 42/43 high was protected. Though the high is approximately the same size as the North Derbyshire Block, it is not connected to a larger shelf system. Also, our interpretation of seismic line SNSTI-90-06 suggests that the eastwards dip of the crest was not induced by differential subsidence during the Carboniferous.

Walkden & Williams (1991) describe a post-Z3, porosity and permeability enhancing event. This event, designated Z4, is considered to be related to the release of aggressive, mineral and hydrocarbon-bearing fluids from the flanking basins during maximum Carboniferous burial (Walkden & Williams 1991). However, the amount and areal extent of the porosity and permeability enhancement associated with the Z4 event is uncertain. For example, there are two possible origins for the cavernous 'Blue John' fluorite mineralized voids within the Boulder Bed at Treak Cliff Cavern near Castleton (Ford 1969). They may represent primary depositional porosity that was protected by partial clay infill of pores from Z3 fluids (Walkden pers. comm. 1995). Alternatively, they may represent the recreation of porosity and permeability in Z4 times before the 'Blue John' fluorite was deposited. If the former origin is correct protection from cementation is possible. If the latter explanation is correct, then there is evidence that the Z4 event can open cavernous voids. Either mechanism is beneficial to the Quadrant 43 opportunity. Quirk (pers. comm. 1995) considers some of the cavernous porosity in Treak Cliff Cavern was enhanced by aggressive hydrothermal fluids.

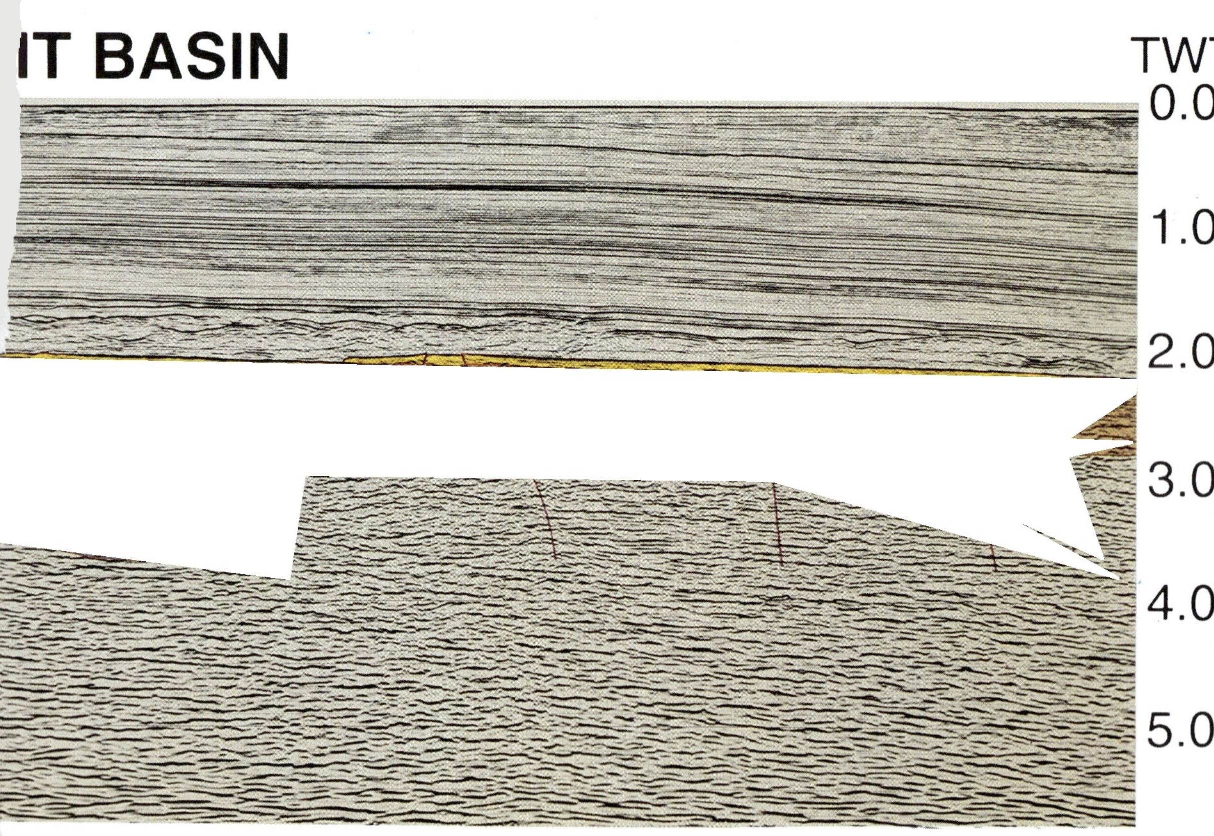

ESE
IT BASIN
TWT
0.0
1.0
2.0
3.0
4.0
5.0

# SNSTI-90-06

SILVER P

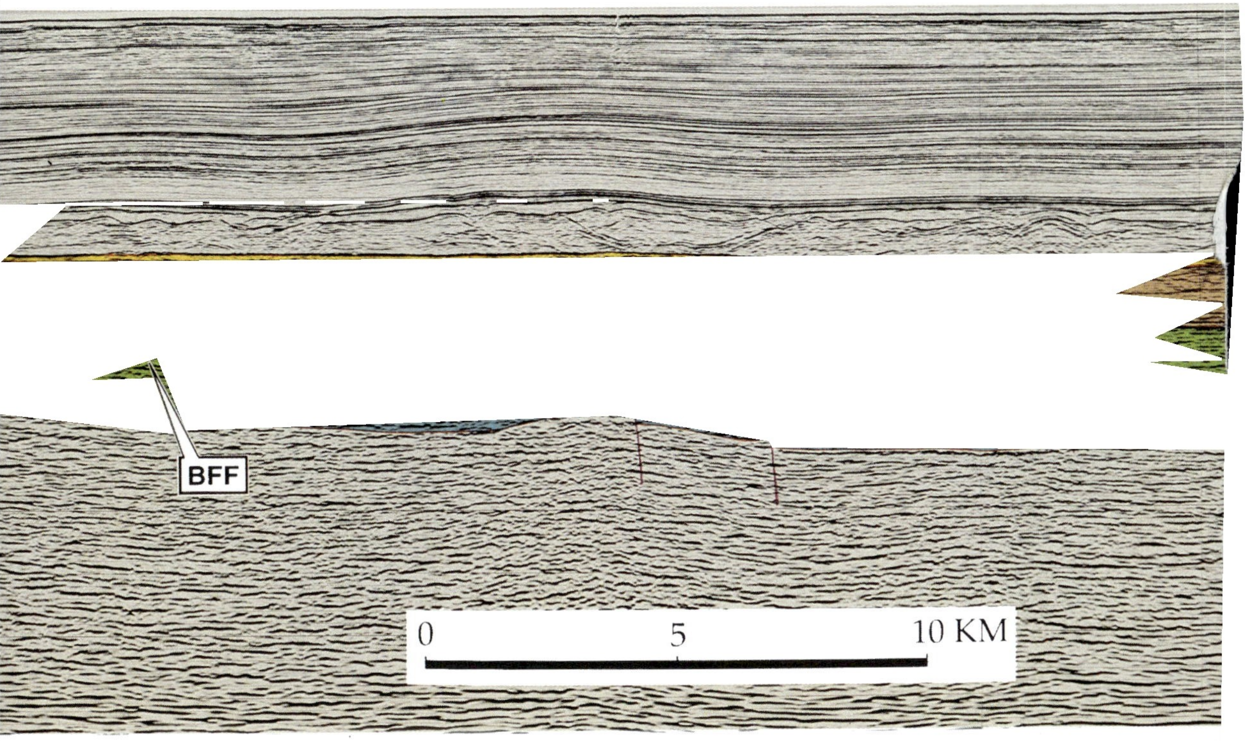

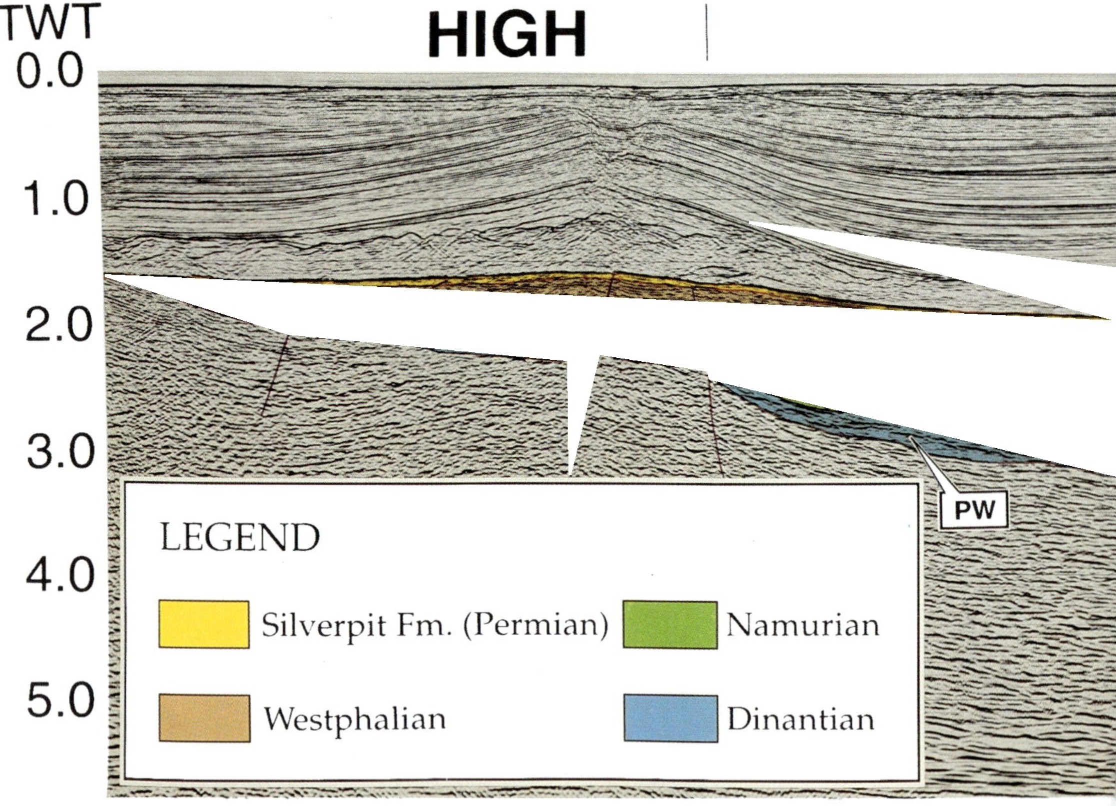



**Fig. 13.** Quad 42/43 High: interpreted PGS Nopec seismic line SNSTI-90-06. In addition to the Dinantian progradational wedge (PW) discussed in the text, the section exhibits a variety of other trap types including a large Namurian basin-floor fan (BFF).

Depth conversion, based on the few surrounding wells and the seismic stacking velocities, shows that a Dinantian test of the hanging-wall succession will require a 5500 m well. Gas is the objective. However, the initial hydrocarbon charge may have been of oil, some of which could still be in place. If, as predicted, Z3 cements are missing the Dinantian shelf carbonates prognosed to cap the high will form an up-dip continuation of the projected progradational wedge reservoirs. Top-seal is provided by the Namurian cover sequence and, at the western pinchout of the Carboniferous, by the Silver Pit Formation. In addition to the Dinantian opportunities, Fig. 13 indicates that the Namurian succession over the high is of interest. Up-dip, pinchout and drape traps are possible. Down-dip in the Silver Pit Basin, structural and stratigraphic traps related to thick basin-floor fans within the Bowland Shale Formation are present. One of these fans is marked BFF on Fig. 13.

## Discussion and conclusions

The exploration and development history of basins containing stacked fairways shows that the pattern of discoveries evolves in discrete evolutionary bursts: the rapid emergence of the lower tier of the SNS fairway is a current case in point. Typically, either the shallowest (the upper tier of the SNS fairway) or the largest structures (the middle tier of the SNS fairway) are found first. Other portions of the fairway stack are subsequently added, often initially by serendipity. We have presented evidence, derived both from the literature and PGS seismic, that the regional geology can accommodate further significant, and in many cases entirely new, opportunities. We consider that we have prepared the conceptual template needed to begin the systematic and methodical exploration of these opportunities and, in the process, determine the lower limit of the SNS fairway.

Both the regional literature and the PGS seismic indicate that significant amounts of Namurian, and probably also Dinantian, source rocks are probable. In addition, Westphalian oil-prone source rocks are present in a lacustrine basin located to the north-east of the London–Brabant Ridge. However, their importance remains to be determined. Mid- to late Devonian source rocks are possible in Quadrants 48, 49, 53 and 54. Quadrants 53 and 54 lie on the migration pathway for all hydrocarbons moving northwestwards out of depocentres located in The Netherlands. Only the Silurian succession proved to have no potential as it passed through the gas window during the early and mid-Devonian.

Our demonstration of new structural features and new play concepts within a basin, considered by many to be mature, provides an excellent illustration of the risks associated with using the still incomplete understanding of the SNS to predict ultimate reserves. It is salutary to remember that ten years ago few would have envisaged the existence of the lower tier of the SNS fairway.

We gratefully acknowledge technical assistance and guidance provided by J. Brooks (DTI) and G. Walkden (Aberdeen University). H. Illich of GeoMark Inc. in Houston ran the maturation models for the 53/15 pseudowell area. PGS Nopec supplied the seismic. PGS Nopec and S. King (King Computer Graphics) prepared the illustrations. We also wish to thank the three reviewers for their improvements to the manuscript. Our particular thanks go to D. Quirk who shared with us his expertise on the region.

## References

ARMSTRONG, J., D'ELIA, V. & TRUEBLOOD, S. 1995. Oil to source correlation in the Liverpool Bay area. *In: The Petroleum Geology of the Irish Sea and Adjacent Areas.* The Geological Society, London, 7–9. February (abstracts).

BAILEY, J. B., ARBIN, P., DAFFINOTI, O., GIBSON, P. & RITCHIE, R. S. 1993. Permo-Carboniferous plays in the Silver Pit Basin. *In:* PARKER, J. R. (ed.) *Petroleum Geology of North West Europe: Proceedings of the 4th Conference.* The Geological Society, London, 707–715.

BÉNARD, F. & BOUCHÉ, P. 1991. Aspects of the petroleum geology of the Variscan foreland of western Europe. *In:* SPENCER, A. M. (ed.) *Generation, Accumulation and Production of Europe's Hydrocarbons.* Special Publications of the European Association of Petroleum Geoscientists, **1**. Oxford University Press, Oxford, 119–138.

BESLY, B. M. in press. Carboniferous. *In:* GLENNIE, K. W. (ed.) *Petroleum Geology of the North Sea: Basic Concepts and Recent Advances* (fourth edn). Blackwell Scientific Publications.

BODENHAUSEN, J. W. A. & OTT, W. F. 1981. Habitat of the Rijswijk Oil Province. *In:* ILLING, L. V. & HOBSON, G. D. (eds) *Petroleum Geology of the Continental Shelf of North-West Europe.* Institute of Petroleum, London, 301–309

CAMERON, T. D. J. 1993. Carboniferous and Devonian of the Southern North Sea. *In:* KNOX, R. W. O'B. & CORDEY, W. G. (eds) *Lithostratigraphic Nomenclature of the UK North Sea.* British Geological Survey and UK Offshore Operators Association, 91.

——, CROSBY, A., BALSON, P. S., JEFFERY, D. H., LOTT, G. K., BULAT, J. & HARRISON, D. J. 1992. *United Kingdom Offshore Regional Report: The Geology of the Southern North Sea,* London, HMSO for the British Geological Survey.

COLE, J. M., KIRK, M. & CRITTENDEN, S. 1995. The recognition and hydrocarbon potential of depositional sequence stratigraphic cycles in the Westphalian (Conybeare Group) of the east Midlands of England and the Anglo-Dutch Southern North Sea gas basin. *In: Stratigraphic Advances in the Offshore Devonian and Carboniferous Rocks, UKCS and Adjacent Areas.* The Geological Society, London 19th. January 1995 (abstracts).

COLLINSON, J. D., JONES, C. M., BLACKBOURN, G. A., BESLY, B. M., ARCHARD, G. M. & MCMAHON, A. H. 1993. Carboniferous depositional systems of the southern North Sea. *In:* PARKER, J. R. (ed.) *Petroleum Geology of North West Europe: Proceedings of the 4th Conference.* The Geological Society, London, 677–687.

CORFIELD, S. M. & GAWTHORPE, R. L. 1995. Tectonostratigraphic evolution of the Upper Devonian and Dinantian seismic sequences of the Mid North Sea High, Southern North Sea. *Stratigraphic Advances in the Offshore Devonian and Carboniferous Rocks, UKCS and Adjacent Areas.* The Geological Society, London 19 January 1995 (abstracts).

CORNFORD, C. 1990. Source rocks and hydrocarbons of the North Sea. *In:* GLENNIE, K. W. (ed.) *Introduction to the Petroleum Geology of the North Sea* (third edn). Blackwell Scientific Publication, 294–361.

DE JAGER, J., DOYLE, M., GRANTHAM, P. & MABILLARD, J. 1993. Hydrocarbon habitat of the West Netherlands Basin. *American Association of Petroleum Geologists Bulletin,* **77**, 1618 (abstract).

EBDON, C. C., FRASER, A. J., HIGGINS, A. J., MITCHENER, B. C. & STRANK, A. R. E. 1990. The Dinantian stratigraphy of the East Midlands: a seismostratigraphic approach. *Journal of the Geological Society, London,* **147**, 519–536.

FORD, T. D. 1969. The Blue John fluorspar deposits of Treak Cliff, Derbyshire, in relation to the Boulder Bed. *Proceedings of the Yorkshire Geological Society,* **37**, 153–157.

FRASER, A. J. & GAWTHORPE, R. L. 1990. Tectonostratigraphic development and hydrocarbon habitat of the Carboniferous in northern England. *In:* HARDMAN, R. F. P. & BROOKS, J. (eds) *Tectonic Events Responsible for Britain's Oil and Gas Reserves.* The Geological Society, London, Special Publications, **55**, 49–86.

——, NASH, D. F., STEELE, R. P. & EBDON, C. C. 1990. A regional assessment of the intra-Carboniferous play of Northern England. *In:* BROOKS, J. (ed.), *Classic Petroleum Provinces.* The Geological Society, London, Special Publications, **50**, 417–440.

GAWTHORPE, R. L. 1987a. Tectono-sedimentary evolution of the Bowland Basin, N. England. *Journal of the Geological Society, London,* **144**, 59–72.

——1987b. Burial dolomitization and porosity development in a mixed carbonate-clastic sequence: an example from the Bowland Basin, northern England. *Sedimentology,* **34**, 533–558.

GLENNIE, K. W. 1997. The history of exploration in the southern North Sea. *This volume.*

GUTTERIDGE, P. 1991. Aspects of Dinantian sedimentation in the Edale Basin, North Derbyshire. *Geological Journal,* **26**, 245–269.

HARDMAN, M., BUCHANAN, J., HERRINGTON, P. & CARR, P. 1993. Geochemical modelling of the East Irish Sea Basin: its influence on predicting hydrocarbon type and quality. *In:* PARKER, J. R. (ed.) *Petroleum Geology of North West Europe: Proceedings of the 4th Conference.* The Geological Society, London, 809–821.

HARLAND, W. B., ARMSTRONG, R. L., COX, C. V., CRAIG, L. E., SMITH, A. G. & SMITH, D. G. 1990. *A Geologic Time Scale 1989.* Cambridge University Press, Cambridge.

LAWRENCE, S. R., COSTER, P. W. & IRELAND, R. J. 1987. Structural development and petroleum potential of the northern flanks of the Bowland Basin (Carboniferous), North-west England, *In:* BROOKS, J. & GLENNIE, K. (eds) *Petroleum Geology of North West Europe.* Graham and Trotman, London, 217–224.

LECKIE, G. G., SPENCER, A. M. & CHEW, K. J. 1995. North Sea hydrocarbon plays and their resources. *57th EAGE Conference and Technical Exhibition, Glasgow, Scotland, 29 May–2 June 1995,* Oral and Poster presentations, EAPG Division, extended abstracts, **2**, F018, 2.

LEE, M. K., BROWN, G. C., WEBB, P. C., WHIELDON, J. & ROLLIN, K. E. 1987. Heat flow, heat production and thermo-tectonic setting in mainland UK. *Jour-nal of the Geological Society, London,* **144**, 35–42.

LEEDER, M. R. & HARDMAN, M. 1990. Carboniferous of the Southern North Sea Basin and controls on hydrocarbon prospectivity. *In:* HARDMAN, R. F. P. & BROOKS, J. (eds) *Tectonic Events Responsible for Britain's Oil and Gas Reserves.* The Geological Society, London, Special Publications, **55**, 87–105.

——, RAISWELL, R., AL-BIATTY, H., MCMAHON, A. & HARDMAN, M. 1990. Carboniferous stratigraphy, sedimentation and correlation of well 48/3–3 in the southern North Sea Basin: integrated use of palynology, natural gamma/sonic logs and carbon/sulphur geochemistry. *Journal of the Geological Society, London,* **147**, 287–300.

LEVEILLE, G. P., PRIMMER, T. J., DUDLEY, G., ELLIS, D. & ALLINSON, G. J. 1997. Diagenetic controls on reservoir quality in Rotliegendes Sandstones, Jupiter Fields area, southern North Sea. *This volume.*

MEERE, P. A. 1995. Sub-greenschist facies metamorphism from the Variscides of SW Ireland: an early syn-extensional peak thermal event. *Journal of the Geological Society, London,* **152**, 511–521.

MERRIMAN, R. J., PHAROAH, T. C., WOODCOCK, N. H. & DALY, P. 1993. The metamorphic history of the concealed Caledonides of eastern England and their foreland. *Geological Magazine,* **130**, 613–620.

MORTON, N. & BRISTOW, C. 1989. *Geology of the Boulonnais, Northern France.* Petroleum Exploration Society of Great Britain, Field Trip Guide No. 24.

PEACOCK, J. D. & TAYLOR, K. 1966. Uraniferous collophane in the Carboniferous Limestone of Derbyshire and Yorkshire. *Bulletin of the Geological Survey of Great Britain*, **25**, 19–32.

PERING, K. L. 1973. Bitumens associated with lead, zinc and fluorite ore minerals in North Derbyshire, England. *Geochimica et Cosmochimica Acta*, **37**, 410–417.

PHARAOH, T. C., MOLYNEUX, S. G., MERRIMAN, R. J., LEE, M. K. & VERNIERS, J. 1993. The Caledonides of the Anglo-Brabant Massif reviewed. *Geological Magazine*, **130**, 561–562.

QUIRK, D. G. 1993*a*. Interpreting the Upper Carboniferous of the Dutch Cleaver Bank High. *In*: PARKER, J. R. (ed.) *Petroleum Geology of North West Europe: Proceedings of the 4th Conference*, The Geological Society, London, 697–706.

——1993*b*. Origin of the Peak District Orefield. *UK Journal of Mines and Minerals*, **12**, 4–15.

ROELOFSEN, J. W. & DE BOER, W. D. 1991. Geology of the Lower Cretaceous Q/1 oil-fields, Broad Fourteens basin, The Netherlands. *In*: SPENCER, A. M. (ed.) *Generation, Accumulation and Production of Europe's Hydrocarbons*. Special Publications of the European Association of Petroleum Geoscientists, **1**. Oxford University Press, Oxford, 201–216.

SIMPSON, I. M. & BROADHURST, F. M. 1969. A Boulder Bed at Treak Cliff, North Derbyshire. *Proceedings of the Yorkshire Geological Society*, **37**, 141–151.

SMITH, N. J. P. 1993. The case for exploration of deep plays in the Variscan fold belt and its foreland. *In*: PARKER, J. R. (ed.) *Petroleum Geology of North West Europe: Proceedings of the 4th Conference*. Geological Society, London, 667–675.

TEICHMÜLLER, M. 1987. Recent advances in coalification studies and their application to geology. *In*: SCOTT, A. S. (ed.) *Coal and Coal-bearing Strata: Recent Advances*. The Geological Society, London, Special Publications, **32**, 127–169.

TUBB, S. R., SOULSBY, A. & LAWRENCE, S. R. 1986. Palaeozoic prospects on the northern flanks of the London-Brabant Massif, *In*: BROOKS, J., GOFF, J. C. & VAN HOORN, B. (eds) *Habitat of Palaeozoic Gas in N.W. Europe*. Geological Society, London, Special Publications, **23**, 55–72.

ULMISHEK, G. F. & KLEMME, H. D. 1990. *Depositional Controls, Distribution, and Effectiveness of World's Petroleum Source Rocks*. United States Geological Survey Bulletin 1931.

VAN HOORN, B. 1987. Structural evolution, timing and tectonic style of the Sole Pit inversion. *Tectonophysics*, **137**, 239–284.

VERNIERS, J. & VAN GROOTEL, G. 1991. Review of the Silurian in the Brabant Massif, Belgium. *Annales de la Société Géologique de Belgique*, **114**, 163–193.

WALKDEN, G. M. & WILLIAMS, D. O. 1991. The diagenesis of the late Dinantian Derbyshire-east Midland carbonate shelf, central England. *Sedimentology*, **38**, 643–670.

WINSTANLEY, A. M. 1993. A review of the Triassic play in the Roer Valley Graben, SE onshore Netherlands. *In*: PARKER, J. R. (ed.) *Petroleum Geology of North West Europe: Proceedings of the 4th Conference*. The Geological Society, London, 595–607.

WOODCOCK, N. H. 1991. The Welsh, Anglian and Belgian Caledonides compared. *Annales de la Société Géologique de Belgique*, **114**, 5–17.

WOODCOCK, N. H. & PHARAOH, T. C. 1993. Silurian facies beneath East Anglia. *Geological Magazine*, **130**, 681–690

ZIEGLER, P. A. 1990. *Geological Atlas of Western and Central Europe*. Shell Internationale Petroleum Maatschappij BV, Amsterdam, second edition.

# The structure of the Westphalian in the northern part of the southern North Sea

DAVID G. QUIRK & JOHN F. AITKEN

*Geology & Cartography Division, Oxford Brookes University,
Gipsy Lane Campus, Oxford, OX3 0BP, UK*

**Abstract:** A clear understanding of the tectonic framework of the Westphalian in the northern part of the southern North Sea has improved confidence during seismic interpretation and offers the possibility of finding new exploration targets within structures that have already been drilled. Two aspects of the tectonic framework are crucial. Firstly, Westphalian strata were deposited within a major foreland basin; consequently, thickness variations on a local scale are minimal. Secondly, the area was affected by a phase of rifting, known as the Saalian event, during the early Permian. This rifting event led to the formation of major E–W and NW–SE normal faults that tilt and offset Westphalian strata. The Upper Rotliegend, which lies unconformably on top of the Westphalian, was deposited during thermal sag after rifting ceased. Identifying where Westphalian reservoirs are present within structure requires careful mapping of both intra-Westphalian and base Rotliegend levels.

This paper addresses the pre-Rotliegend structure of the Westphalian–early Stephanian in the Silverpit Basin and Dutch Cleaver Bank High in the northern part of the Southern North Sea between longitudes 2°E and 4°E and latitudes 54°N and 54° 30′ N (Fig. 1a & b). The work is based mostly on proprietary 2D and 3D seismic data provided across most of the study area by Fina, Goal Petroleum (now Talisman Energy), Statoil and their partners. Released and unreleased data were also made available from approximately 40 wells.

In order to correlate and map reservoir-prone intervals in the northern part of the southern North Sea, particularly where the density or quality of seismic data is less than optimal, an understanding of the structure is essential. Stratigraphic details of the Westphalian–early Stephanian are discussed separately in Quirk (this volume). In view of the difficulties involved in accurately dating the succession, particularly the 'barren red' Westphalian C–early Stephanian, the entire interval above the top of the Namurian and below the base of the Rotliegend will be referred to simply as Westphalian throughout the rest of this paper.

## Structural setting

A representative geological cross-section summarizing the structure and stratigraphy of the northern part of the southern North Sea (SNS) is shown in Fig. 2. The area is characterized by thick Carboniferous, Permo-Triassic, Cretaceous and Tertiary sections. The basin was initiated during the late Devonian by a rifting episode which affected much of NW Europe (Leeder & Hardman 1990; Quirk 1993*a*). Rifting continued into the Dinantian associated with the deposition of shallow water carbonates on up-faulted areas and deep water carbonates and mudstones in down-faulted areas. The rift climax was probably reached during the early Namurian associated with transgression. This transgression led to the widespread deposition of basinal shales which form source rocks for oil and gas in the East Midlands onshore UK and in the Irish Sea. After rifting ceased the basin continued to subside due to thermal sag (Leeder 1982, 1987; Leeder & Hardman 1990; Quirk 1993*a*). Gradually the basin was infilled by deltas prograding from the north (Collinson *et al.* 1993). By the end of the Namurian the basin had been filled approximately back to base level and the area was dominated by fresh water delta top sediments.

A thick succession of alluvial and lacustrine strata was deposited in the Westphalian (Quirk this volume). This contains the main hydrocarbon targets in the study area consisting of coarse-grained fluvial sandstone reservoirs and gas source rocks in the form of coal and organic-rich shale (Ritchie & Pratsides 1993; Quirk 1993*a*). A major erosional surface, the Saalian unconformity, is present at the top of the Carboniferous (Quirk 1993*a*). On the basis of seismic interpretation and well log correlations across, for example, UK Block 44/29, it can be shown that, in places, at least 900 m of section above the Westphalian A has been cut out by the Saalian unconformity.

*From* Ziegler, K., Turner, P. & Daines, S. R. (eds), 1997, *Petroleum Geology of the Southern North Sea: Future Potential*, Geological Society Special Publication No. 123, pp. 143–152.

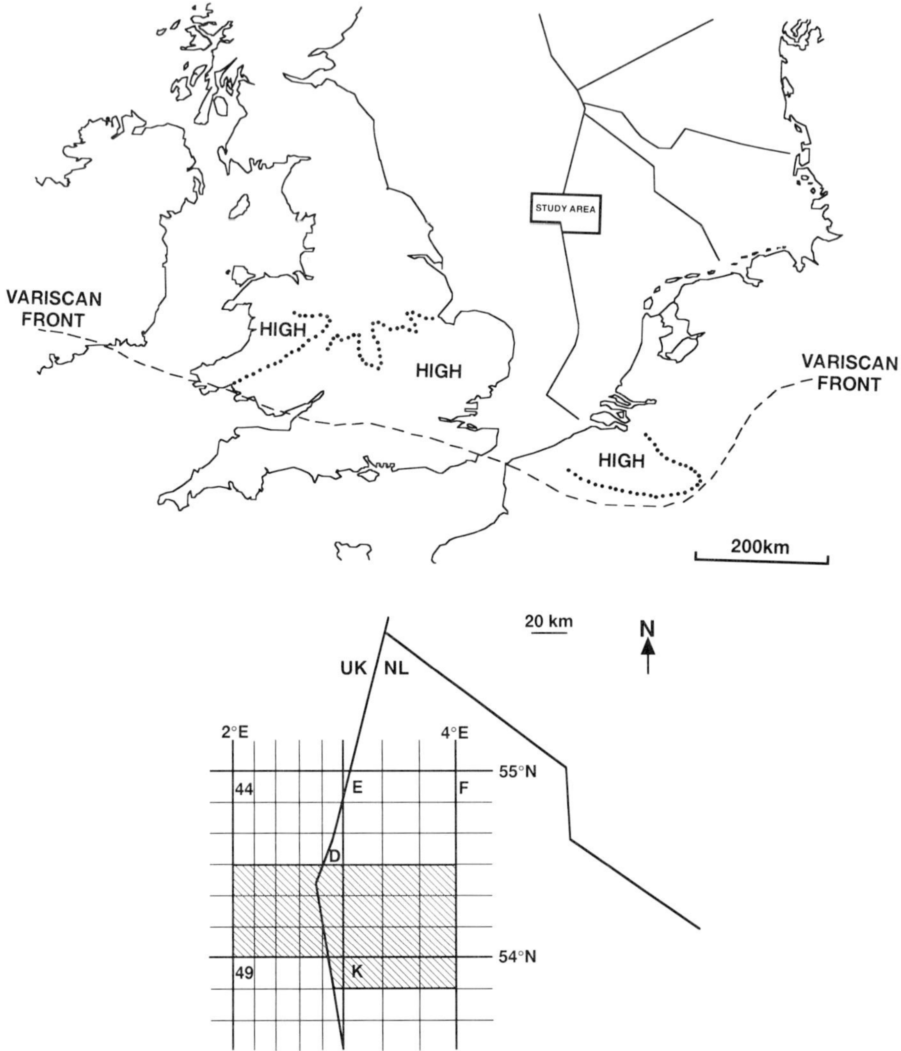


**Fig. 1.** Location of study area in the northern part of the Southern North Sea.

Red claystones and evaporites of Upper Rotliegend (late Permian) age (Silverpit facies) overlie the Saalian unconformity and these beds form the seal to underlying Westphalian targets. Rapid marine transgression occurred at the end of the Rotliegend associated with the deposition of a thin, copper-rich, black shale known as the Kupferschiefer. The Kupferschiefer marks the base of the Zechstein. The rest of the Zechstein consists mostly of thick evaporites, approximately 90% of which is halite.

Continental Triassic sediments lie conformably on top of the Zechstein and contain a secondary reservoir target known as the Bunter Sandstone (Cameron *et al.* 1992; Ritchie & Pratsides 1993). In a few places marine Liassic rocks are preserved but a major surface of erosion called the Kimmerian unconformity has removed most of the Lower Jurassic and some or all of the Triassic over large parts of the study area.

The Kimmerian unconformity corresponds with a period of rifting in the late Jurassic which has affected large parts of the North Sea (Underhill & Partington 1993). Gentle halokinesis occurred in the SNS at this time, probably in association with the development of NW–SE-oriented horsts and grabens in the underlying Carboniferous–Rotliegend section. However, evidence for the presence of syn-rift sediments

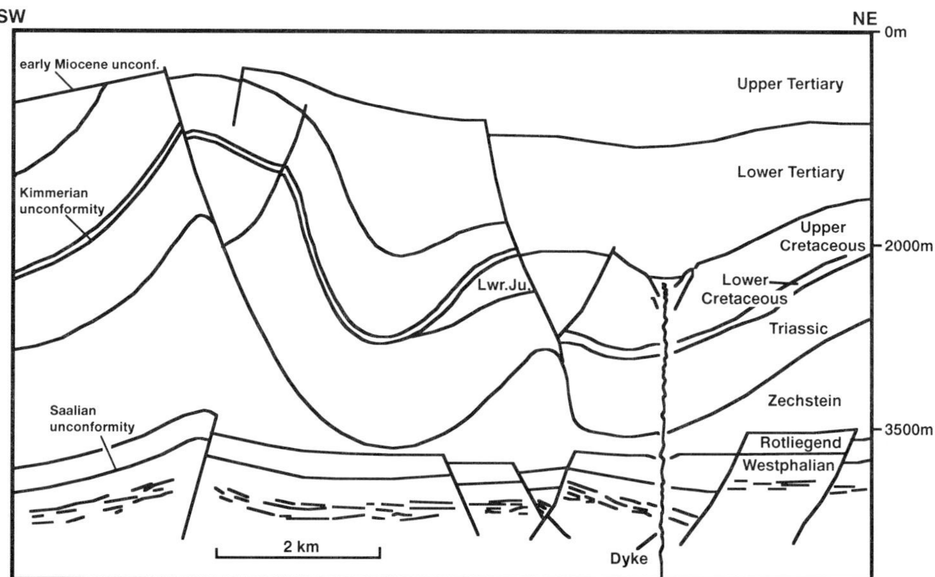



**Fig. 2.** Schematic cross-section illustrating the general structure observed on seismic in the study area.

of late Jurassic age is rarely seen in the SNS probably because the effects of thermal doming were dominant in this area.

A thick Cretaceous interval occurs above the Kimmerian unconformity. The lowermost part of the Cretaceous consists of thin marine siliciclastics which are overlain by a monotonous succession of carbonates comprising mostly chalk. Towards the end of the Cretaceous a period of gentle compression affected the SNS leading to a limited amount of folding, sediment slumping and erosion (Quirk 1993*a*). Some older normal faults in the Westphalian and Rotliegend were probably reactivated as reverse faults at this time.

Subsidence continued during the early Tertiary when a thick succession of fine-grained marine sediments was deposited. However, a significant period of inversion occurred during the early Miocene and led to the formation of a major unconformity which intersects the present UK land surface farther west. This inversion was associated with the main episode of salt movement in the SNS and the development of normal faults above zones of rising salt. NW–SE-oriented dolerite dykes were also intruded during the early Tertiary in the SNS, the walls of which may provide important migration routes for gas from the Carboniferous into the Bunter Sandstone (Brown *et al.* 1994). Subsidence was resumed during the late Miocene and is still occurring.

## Pre-Rotliegend structure of the Westphalian

Two observations have the greatest significance in understanding the tectonic evolution of the Westphalian in the study area:

(1) the thickness of Westphalian strata is remarkably uniform suggesting that they were not affected by syn-depositional faulting or folding;
(2) Westphalian strata have been affected by a post-depositional phase of extension and uplift prior to the Rotliegend.

Figure 3 shows a typical 3D seismic line from UK Quadrant 44 which illustrates the above observations. On Fig. 3a the pre-Rotliegend structure of the Westphalian is at first difficult to visualise due to the presence of an E–W-oriented reverse fault that was probably active during compression at the end of the Cretaceous. In order to remove the effects of post-Rotliegend faulting, the interpreted line (Fig. 3a) has been horizon flattened using GeoQuest 'Charisma' software at two separate levels in order to show (i) the intra-Westphalian structural configuration (Fig. 3b) and (ii) the base Rotliegend structural configuration (Fig. 3c).

### Intra-Westphalian structuration

The flattened section shown in Fig. 3b helps to illustrate that the Westphalian has a tram-line-like appearance. Faults which affect the

Westphalian on the original seismic section (Fig. 3a) are apparently post-Westphalian in age because no thickening of Westphalian strata is visible across the faults (Fig. 3b).

Well correlations across the study area also support the view that syn-depositional faults are absent. Thus, Quirk (this volume) shows that over a north–south distance of over 50 km only minimal thickness variations can be detected. Nonetheless, to the south of latitude 54°N the Westphalian begins to thicken noticeably (Fig. 4). It continues to thicken towards The Netherlands and NW Germany where more than 3500 m of Westphalian strata were deposited in the vicinity of the Variscan front (David 1990; Drozdzewski 1993). The basin is therefore strongly asymmetric in cross-sectional shape. Fluvial sandstones derived from north of the study area form an important part of the Westphalian section in the northern part of the SNS (Collinson *et al.* 1993; Quirk 1993*a*). They die out south of latitude 54° 45′ N where the succession is dominated by fine-grained lacustrine sediments. Fluvial sandstones re-appear close to the Variscan front in The Netherlands and NW Germany. Onshore UK, Upper Carboniferous sediments in the Central Pennine Basin, to the north of the Midlands Microcraton (Fig. 4), are interpreted by most authors to have been deposited during a period of thermal sag following rifting during the Lower Carboniferous (Leeder 1982, 1987; Leeder & Hardman 1990). A similar rift-sag morphology has been proposed for the Dinantian–Silesian section in the SNS by several authors (Cameron *et al.* 1992; Bailey *et al.* 1993; Hollywood & Whorlow 1993). However, the shape and lithological trends described for the Westphalian across the SNS, The Netherlands and NW Germany (Fig. 4) is more reminiscent of a foreland basin (Quirk 1993*a*). It seems likely, therefore, that subsidence during the Westphalian in this area was, at least in part, caused by downward bending of the plate on the footwall side of the Variscan front during the final stages of subduction. This model helps to explain why no evidence is seen for the existence of syn-depositional extensional faults within the Westphalian.

Both the shape and scale of the Westphalian basin in NW Europe is reminiscent of the present day Ganga Basin which has developed in the foreland of the Himalayas (Lyon-Caen & Molnar 1985). Most of the sediment in the Ganga basin has been derived by rivers from the Himalayas. The Westphalian basin differs in that there was also a major source of sediment to the north on the footwall plate, namely the Laurentian–Fennoscandian Shield. The Westphalian basin extended east into Poland and west towards Newfoundland (Ziegler 1990), although the SNS was partially separated from the western extension of the basin by the Midlands Microcraton/London–Brabant Massif (Fig. 4). Any connection to open ocean was restricted and for this reason sediment in the centre of the basin is dominated by lacustrine rather than marine strata.

## Early Permian (Saalian) structuration

In Fig. 3c the seismic section shown in Fig. 3a has been flattened on a horizon interpreted to represent the base of the Rotliegend. The resultant flattened section helps to reveal the structure of the Westphalian at or soon after the onset of sedimentation during the Upper Rotliegend. Figure 3c clearly shows the existence of a set of normal faults which offset the Westphalian below the Saalian unconformity at the base of the Rotliegend. These faults appear to have formed sometime after deposition of the Westphalian (–early Stephanian) and prior to the Upper Rotliegend. This period of time corresponds with the stratigraphic section that is missing at the unconformity and is hence assigned an early Permian (Autunian) age. Similar early Permian faults are common in the study area and the largest of these have a general E–W orientation (e.g. Fig. 5). Some of these faults have throws of as much as 500 m. In plan view they are often very straight and may be up to a few tens of kilometres in length (e.g. Fig. 5). Often it is difficult to recognize the nature of these faults on 2D seismic, particularly those oriented E–W, because they occur oblique to the major NW–SE-trending faults which define the main Rotliegend structures in the SNS.

It can be seen on Fig. 3c that the Upper Rotliegend is slightly thicker in the hanging wall of one of these E–W normal faults. However, the very flat, layer-cake appearance of the Upper Rotliegend suggests that this thickening was not due to active growth on the normal fault as syn-sedimentary faulting would normally cause reflections to diverge. Instead the thickening is thought to indicate that, rather than complete peneplanation having occurred during the early Permian, a relict escarpment is preserved which became subsequently buried beneath Upper Rotliegend strata.

NW–SE faults were also active prior to the Rotliegend apparently at the same time as the E–W normal faults. The pre-Rotliegend offset on the NW–SE faults is also almost entirely normal

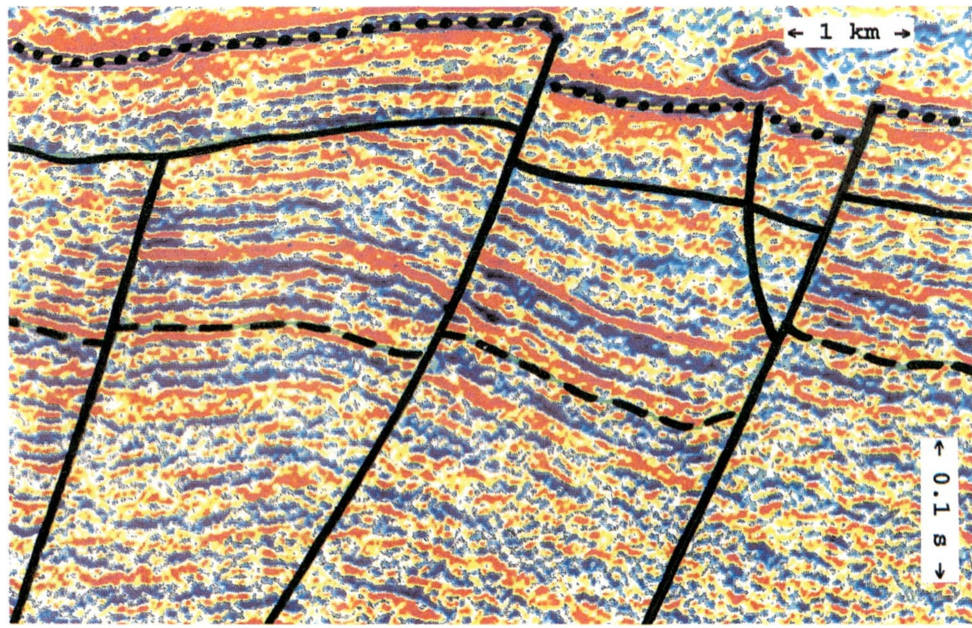



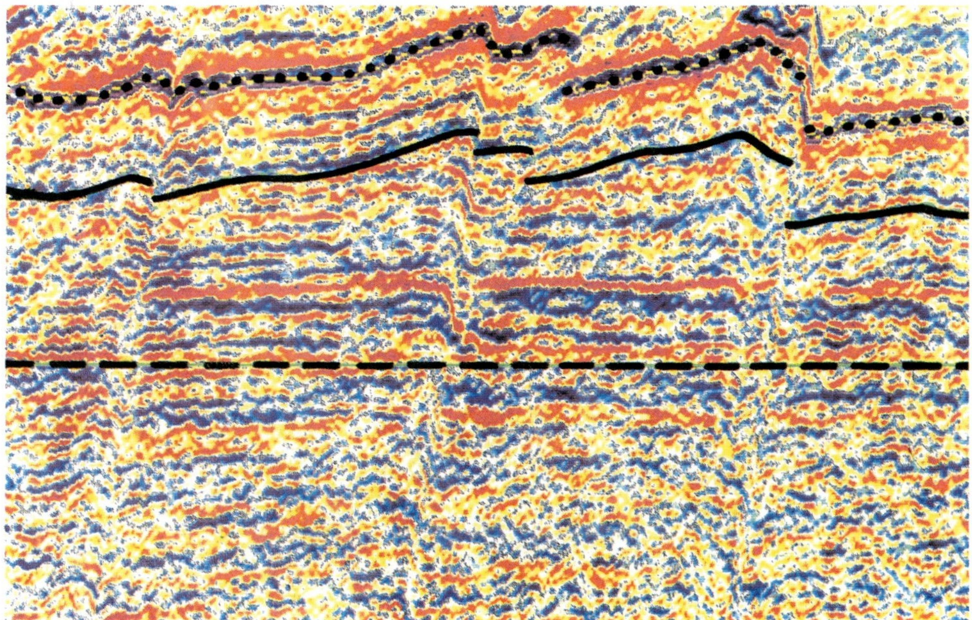

**Fig. 3.** Typical seismic section from the study area showing: (**a**) original data; (**b**) data flattened on an intra-Westphalian horizon (c. base Westphalian B); (**c**) flattened on the base Rotliegend (Saalian unconformity). Reflection interpreted with dotted line is base Zechstein; reflection interpreted with solid line is base Rotliegend; reflection interpreted with dashed line is intra-Westphalian horizon. Horizon flattening was achieved using GeoQuest 'Charisma' software.

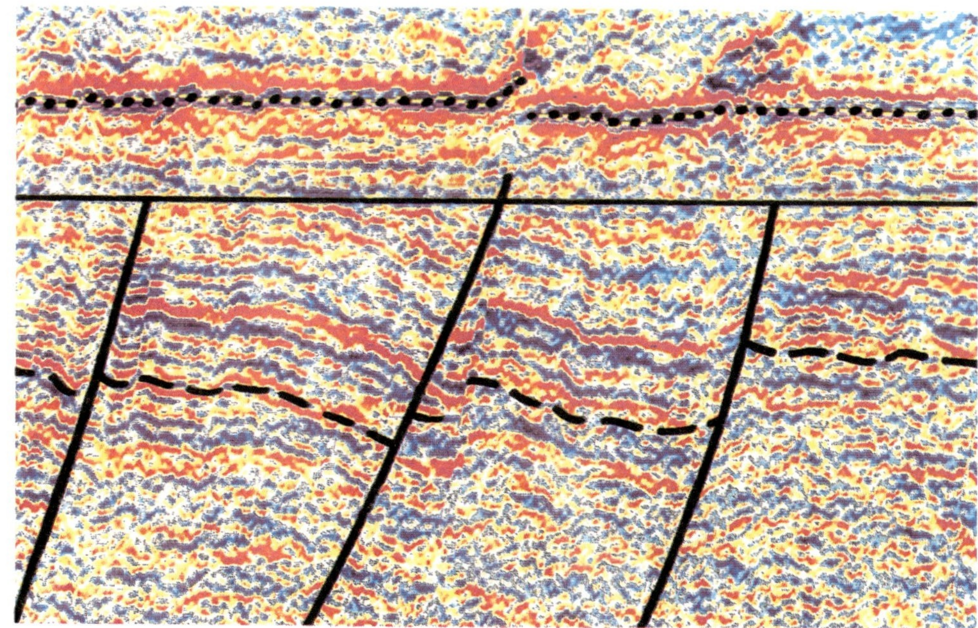

**Figure 3.** (*Continued*)

(Quirk 1993*a*) despite the fact that monoclines and anticlines are often observed in the hanging-walls of these faults (e.g. Fig. 3a). These folds have sometimes mistakenly been interpreted as indicating that the NW–SE faults are compressional in origin. However, on good-quality seismic data most of the monoclines and anticlines can be shown to represent either rollover structures associated with extensional movement along one fault or else tulip structures

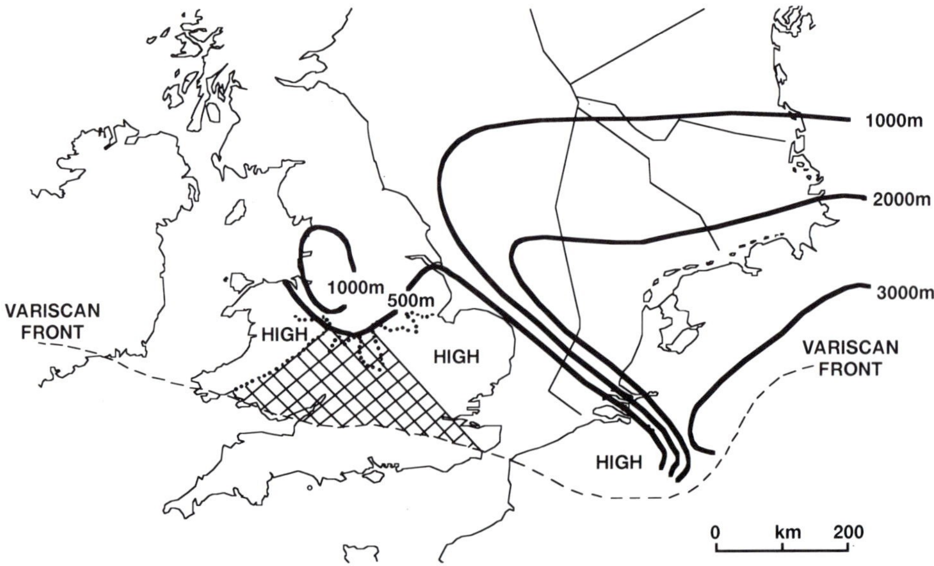


**Fig. 4.** Restored isopachyte map of the Westphalian-early Stephanian (based on Calver 1969; Drozdzewski 1993; and proprietary oil company data). Cross-hatching indicates position of Midlands Microcraton.

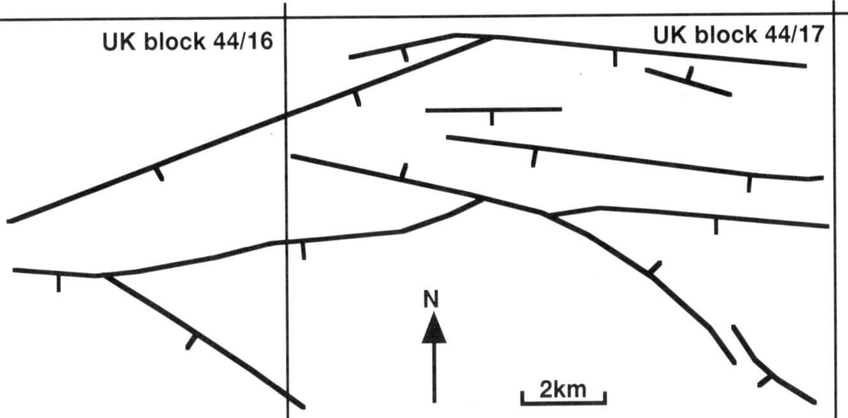



**Fig. 5.** Map showing major E–W faults of early Permian (Saalian) age that offset Westphalian strata in the northern part of UK block 44/17 (based on 1988 ECL 2D seismic data).

associated with oblique slip along two opposing faults.

From the evidence above, it is clear that the study area was affected by a period of dominant extension and uplift during the early Permian. A major episode of rifting is known to have occurred in NW Germany and surrounding areas at this time where thick volcanics and associated syn-rift sediments belonging to the Lower Rotliegend were deposited (Fig. 6; Ziegler 1990; Glennie this volume). A period of early Permian rifting has also affected the Irish Sea (Quirk & Kimbell 1997). In Scotland, a major suite of E–W-oriented dolerite dykes were intruded during the late Stephanian–Autunian (Francis 1978). At approximately the same time, the East Midlands of England were affected by N–S extension and a period of hydrothermal mineralization (Quirk 1993b). It appears, therefore, that during the early Permian large parts of

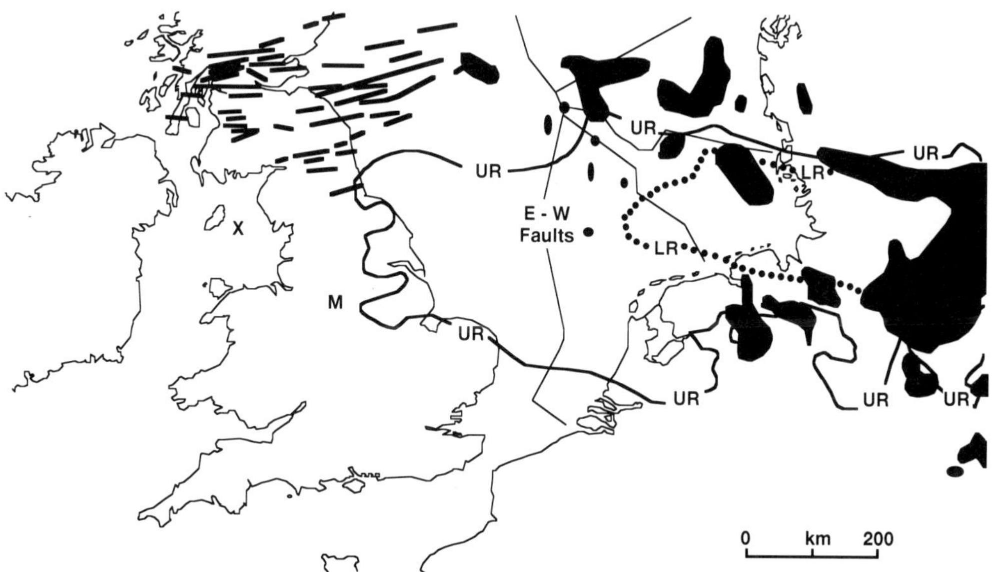



**Fig. 6.** Map showing Permian basin features (based on Francis 1978; Ziegler 1990; Quirk 1993b; Quirk & Kimbell 1997). Black shading and thick black lines indicate position of early Permian volcanics and dolerite dykes; X, minor rifting in Irish Sea; M, an area of hydrothermal mineralization in the East Midlands; LR, outer limit of Lower Rotliegend sediments; UR, outer limit of Upper Rotliegend sediments.

NW Europe were affected by rift tectonics and related activity associated with N–S extension and thermal doming. This tectonic phase is known as the Saalian event after the name given to the unconformity at the base of the Rotliegend in NW Germany and The Netherlands (e.g. Quirk 1993a; Glennie this volume). The Saalian event is frequently overlooked in the UK because of the importance given to older Variscan compressional tectonics (e.g. Fraser & Gawthorpe 1990). Reverse faults and folds that were formed by the Variscan orogeny are usually oriented N–S to NE–SW in the English Pennines and Irish Sea (e.g. Quirk & Kimbell 1997), i.e. orthogonal to Saalian extensional structures. These compressional structures are most prominent to the north and west of the Midlands Microcraton, and are sub-parallel to its NW margin; they are virtually absent from the study area in the northern part of the SNS which lies approximately 300 km NE of the Midlands Microcraton. In contrast, current work in progress by the authors in the East Midlands, onshore UK, shows that, close to the NE margin of the Midlands Microcraton, Variscan structures are predominantly strike slip in origin.

## Post-Saalian basin evolution

After rifting ceased in the early Permian the lithosphere in the SNS region began to cool and subside to form the Southern Permian Basin. This basin, which is therefore interpreted as a thermal sag basin, developed above the area of the lithosphere thinned by extension (Fig. 6). It is oriented E–W similar to the main faults formed during rifting. Syn-rift sediments are absent from the study area and the earliest sediments deposited on top of the Saalian unconformity are late Permian in age (Glennie this volume). These sediments, which belong to the Upper Rotliegend, onlap the Saalian unconformity to the west and north where the succession becomes thinner. Typical of a thermal sag basin, the size of the Southern Permian Basin expanded with time from the Upper Rotliegend through to the Triassic (see Fig. 6; Ziegler 1990).

The climate was very arid throughout most of the Permian and yet evidence is now coming to light that the Upper Rotliegend and Zechstein were deposited within a relatively short period of time (Glennie this volume). An explanation for this might be that the Southern Permian Basin subsided well below sea level before sediment began to accumulate. Thereafter, possibly associated with a wetter climate, Rotliegend and Zechstein strata were deposited rapidly in the accommodation space (between basin floor and sea level) that had already been created.

## Discussion

Two important observations have been made that are relevant to interpreting the Westphalian in the northern part of the SNS. These are that

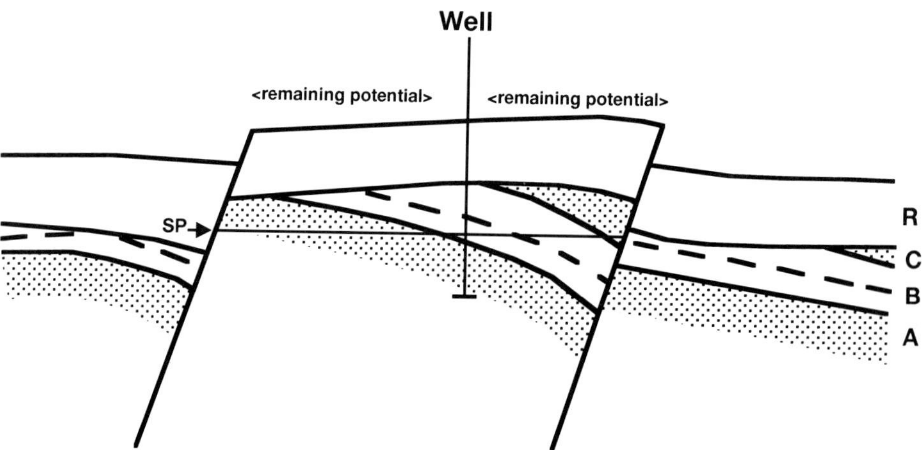



**Fig. 7.** Schematic section through a 'dry' well that has unfortunately failed to test two Westphalian reservoir-prone units (A & C) within structure. The well has instead encountered a mud-dominated unit (B) below the base Rotliegend (Saalian unconformity) and above spillpoint (SP). The dotted line indicates the position of an intra-Westphalian reflection, below and above which the reservoir-prone units occur at relatively constant interval separations. The seal to the structure is provided by the Upper Rotliegend (R) which consists of red claystones and evaporites (Silverpit facies).

(i) intra-Westphalian thickness variations are minimal; and (ii) the sub-crop of the Westphalian at the base Rotliegend has been greatly affected by an early Permian (Saalian) rifting event. As a consequence of (i), it is possible to predict the occurrence of known reservoir intervals on seismic by adding or subtracting constant thicknesses from interpreted intra-Westphalian reflections. However, due to (ii), the mapping of intra-Westphalian reflections is complicated by the presence of E–W and NW–SE faults that often have a different structural expression or no expression at Upper Rotliegend level. It is, therefore, worth noting that many Rotliegend structures in the northern part of the SNS which have been drilled for Westphalian targets were tested 'dry' due to the absence of Westphalian reservoir intervals within base Rotliegend closure. As illustrated in Fig. 7 there is still significant potential for additional gas-bearing Carboniferous reservoirs present in the undrilled parts of some of these structures due to the angular nature of the Saalian unconformity. Hence, in order to predict where Westphalian reservoirs occur within structure, it is necessary to carefully map both intra-Westphalian intervals and the base of the Rotliegend (top seal) on seismic.

## Conclusions

An understanding of the structural setting of the Westphalian is essential to exploration success in the northern part of the SNS. Two observations are of fundamental importance:

(1) thickness variations within the Westphalian are minimal;
(2) a set of post-Westphalian E–W and NW–SE normal faults greatly affect sub-crop at the base Rotliegend.

The explanation for (1) is that the Westphalian in the SNS was deposited within a major foreland basin in which syn-depositional faults were not active. The basin was uplifted and extended during the early Permian when a tectonic episode, known as the Saalian event, led to the formation of the faults mentioned in (2). The Southern Permian Basin developed due to thermal sag after rifting ceased.

To summarize, reservoirs and source rocks are confined to pre-rift (Westphalian) strata and the top seal is formed by post-rift strata (Upper Rotliegend Silverpit facies) which is separated from the Westphalian by an angular (Saalian) rift unconformity.

The authors are grateful to Fina and block 44/29 partners, Goal Petroleum, Statoil, Schlumberger Geo-Quest, Exploration Consultants and Oxford Brookes University who have provided support and sponsorship for several research projects in the study area. Thanks are also due to the Nederlandse Aardolie Maatschappij (NAM) where, as an employee of Shell, D.G.Q. acquired many of his original ideas whilst working on the Cleaver Bank High. The authors also wish to acknowledge the assistance and advice provided by P. Mason, D. Evans, D. Lawton, J. Kunka, B. Evans, N. Cameron, K. Ziegler, J. Maynard and L. Hill.

## References

BAILEY, J. B., ARBIN, P., DAFFINOTI, O., GIBSON, P. & RITCHIE, J. S. 1993. Permo-Carboniferous plays of the Silver Pit Basin. *In:* PARKER, J. R. (ed.) *Petroleum Geology of Northwest Europe: Proceedings of the 4th Conference.* Geological Society, London, 707–715.

BROWN, G., PLATT, N. H. & MCGRANDLE, A. 1994. The geophysical expression of Tertiary dykes in the southern North Sea. *First Break*, **12**, 137–146.

CALVER, M. A. 1969. Westphalian of Britain. *Compte Rendu du sixiéme Congrès International de Stratigraphie et de Géologie du Carbonifère*, Sheffield 1967, **1**, 233–254.

CAMERON, T. D. J., CROSBY, A., BALSON, P. S., JEFFERY, D. H., LOTT, G. K., BULAT, J. & HARRISON, D. J. 1992. *United Kingdom offshore regional report: the geology of the Southern North Sea.* HMSO for the BGS, London.

COLLINSON, J. D., JONES, C. M., BLACKBOURN, G. A., BESLY, B. M., ARCHARD, G. M. & MCMAHON, A. H. 1993. Carboniferous depositional systems in the Southern North Sea. *In:* PARKER, J. R. (ed.) *Petroleum Geology of Northwest Europe: Proceedings of the 4th Conference.* Geological Society, London, 677–687.

DAVID, R. 1990. Sedimentologie und Beckenanalyse im Westfal C und D des nordwestdeutschen Oberkarbons. *Deutsche Wissenschaftliche Gesellschaft für Erdöl, Erdgas und Kohle E.V.* Hamburg, Forschungsbericht, **384–3**.

DROZDZEWSKI, G. 1993. The Ruhr Coal Basin (Germany): structural evolution of an autochthonous foreland basin. *International Journal of Coal Geology*, **22**, 231–250.

FRANCIS, E. H. 1978. The Midland Valley as a rift, seen in connection with the late Palaeozoic European rift system. *In:* RAMBERG, I. B. & NEUMAN, R. E. (eds) *Tectonics and Geophysics of Continental Rifts.* Riedel, Dordrecht, 133–147.

FRASER, A. J. & GAWTHORPE, R. L. 1990. Tectonostratigraphic development and hydrocarbon habitat of the Carboniferous in northern England. *In:* HARDMAN, R. F. P. & BROOKS, J. (eds) *Tectonic events responsible for Britain's Oil and Gas Reserves.* Geological Society, London, Special Publications, **55**, 49–86.

GLENNIE, K. W. 1997. *The history of exploration in the Southern North Sea. This volume.*

HOLLYWOOD, J. M. & WHORLOW, C. V. 1993. Structural development and hydrocarbon occurrence of the Carboniferous in the UK Southern North Sea. *In:* PARKER, J. R. (ed.) *Petroleum Geology of Northwest Europe: Proceedings of the 4th Conference.* Geological Society, London, 689–696.

LEEDER, M. R. 1982. Upper Palaeozoic basins of the British Isles – Caledonide inheritance versus Hercynian plate margin processes. *Journal of the Geological Society, London,* **139**, 479–491.

LEEDER, M. R. 1987. Tectonic and palaeogeographic models for Lower Carboniferous Europe. *In:* MILLER, J., ADAMS, A. E., & WRIGHT, V. P. (eds) *European Dinantian Environments.* Wiley, Chichester, 1–20.

—— & HARDMAN, M. 1990. Carboniferous geology of the Southern North Sea Basin and controls on hydrocarbon prospectivity. *In:* HARDMAN, R. F. P. & BROOKS, J. (eds) *Tectonic events responsible for Britain's oil and gas reserves.* Geological Society, London, Special Publications, **55**, 87–105.

LYON-CAEN, H. & MOLNAR, P. 1985. Gravity anomalies, flexure of the Indian plate, and the structure, support and evolution of the Himalaya and Ganga Basin. *Tectonics,* **4**, 513–538.

QUIRK, D. G. 1993a. Interpreting the Upper Carboniferous of the Dutch Cleaver Bank High. *In:* PARKER, J. R. (ed.) *Petroleum Geology of Northwest Europe: Proceedings of the 4th Conference.* Geological Society, London, 697–706.

——1993b. Origin of the Peak District Orefield. *UK Journal of Mines and Minerals,* **12**, 4–15.

——1997. The sequence stratigraphy of the Westphalian of the Southern North Sea. *This volume.*

—— & KIMBELL, G. S. 1997. Structural evolution of the Isle of Man and central part of the Irish Sea. *In:* MEADOWS, N., TRUEBLOOD, S. & COWAN, G. (eds) *The Petroleum Geology of the Irish Sea and Adjacent Areas.* Geological Society, London, Special Publications, in press.

RITCHIE, J. S. & PRATSIDES, P. 1993. The Caister Fields, Block 44/23a, UK North Sea. *In:* PARKER, J. R. (ed.) *Petroleum Geology of Northwest Europe: Proceedings of the 4th Conference.* Geological Society, London, 759–769.

UNDERHILL, J. R. & PARTINGTON, M. A. 1993. Jurassic thermal doming and deflation in the North Sea: implications of the sequence stratigraphic evidence. *In:* PARKER, J. R. (ed.) *Petroleum Geology of Northwest Europe: Proceedings of the 4th Conference.* Geological Society, London, 337–347.

ZIEGLER, P. A. 1990. *Geological Atlas of Western and Central Europe* (2nd edn). Shell Internationale Petroleum Maatschappij B.V., The Hague.

# Sequence stratigraphy of the Westphalian in the northern part of the Southern North Sea

D. G. QUIRK

*Geology & Cartography Division, Oxford Brookes University,
Gipsy Lane, Oxford OX3 0BP, UK*

**Abstract:** A series of stacked fluvio-lacustrine sequences in the Westphalian of the northern part of the Southern North Sea can be correlated over areas of greater than 2500 km$^2$. Each sequence is similar in terms of thickness (100–300 m) and the internal arrangement of lithofacies but they differ in colour and in the amount of sandstone and coal that they contain. A typical sequence consists of a sand-rich lower unit, a mud-dominated middle unit which fines upwards towards its top and a mud-dominated upper unit which generally coarsens upwards. The main reservoir-prone intervals are confined to the lower–middle parts of sand-rich units and consist of coarse-grained braided or low-sinuousity channel deposits. These fluvial sandstones are laterally extensive and are neither typical of incised valley fills (in terms of stratal geometries) nor of progradational systems tracts (in terms of facies distribution). Instead they appear to have been deposited during periods of widespread fluvial aggradation caused by transgression as a result of relative lake-level rise. Each lower sand-rich unit is overlain by a dark mudstone with a high gamma ray expression which is interpreted as an initial (lake) flooding surface marking the onset of predominantly lacustrine and mire sedimentation associated with the development of the fining-upwards mud-dominated unit. The top of each fining-upwards unit is marked by a maximum gamma ray peak interpreted as a maximum flooding surface which defines the end of lake transgression. Above the maximum flooding surface, each coarsening-upwards mud-dominated unit shows increasing deltaic influence upwards associated with a slowing in the rate of relative lake level rise. The tops of some of these units are eroded suggesting that in some cases deltaic progradation was followed by a relative fall in lake level; in other cases the boundary with the overlying transgressive sand-rich unit in the next sequence is wholly conformable. Coal mires developed in coastal areas, particularly during transgression when coarse-grained sediment was trapped up-dip on the coeval alluvial plain. Relative changes in lake level had a periodicity of 1.5 Ma or less indicating that they may be related to third order eustatic cycles.

The Westphalian of the Southern North Sea is becoming an increasingly important target for gas exploration (e.g. Leeder & Hardman 1990; Ritchie & Pratsides 1993) and an extensive seismic and well data base relating to the Westphalian now exists. However, current lithostratigraphic schemes (e.g. Cameron 1993) lack the detail required to map and correlate known reservoir-prone intervals (see Ritchie & Pratsides 1993) and are unable to predict the occurrence of new reservoir targets (e.g. Quirk & Aitken this volume). This paper attempts to partly redress this shortfall by providing a relatively low resolution (exploration-scale) sequence stratigraphic interpretation for the Westphalian in the Silverpit Basin and Dutch Cleaver Bank High, specifically between longitudes 2°E and 4°E and latitudes 54°N and 54° 40′ N (Fig.1), and complements a separate paper discussing the geological setting and structure of the Westphalian (Quirk & Aitken this volume). Although the early Stephanian is not specifically addressed in this paper it has been found that many of the observations

reported here apply equally well where younger Carboniferous strata are preserved in the study area (Quirk 1993).

## General stratigraphy and depositional setting

Cameron (1993) assigns the Westphalian of the Southern North Sea to four formations, namely the Caister Coal, Westoe Coal, Schooner and Brig Formations comprising the Conybeare Group, which in total is at least 1200 m thick (Quirk 1993). The Westphalian A–lower Westphalian C section (Caister Coal, Westoe Coal and lower part of the Schooner Formations) consists of coal-bearing mudstones, siltstones and sandstones previously described by Collinson *et al.* (1993) and Ritchie & Pratsides (1993). The upper Westphalian C–early Stephanian section (the upper part of the Schooner Formation and the Brig Formation) comprise 'barren' red beds which are discussed by Tubb *et al.* (1986) and Besly *et al.* (1993).

Collinson *et al.* (1993) suggest that facies and depositional styles in the Westphalian in the

*From* Ziegler, K., Turner, P. & Daines, S. R. (eds), 1997, *Petroleum Geology of the Southern North Sea: Future Potential*, Geological Society Special Publication No. 123, pp. 153–168.

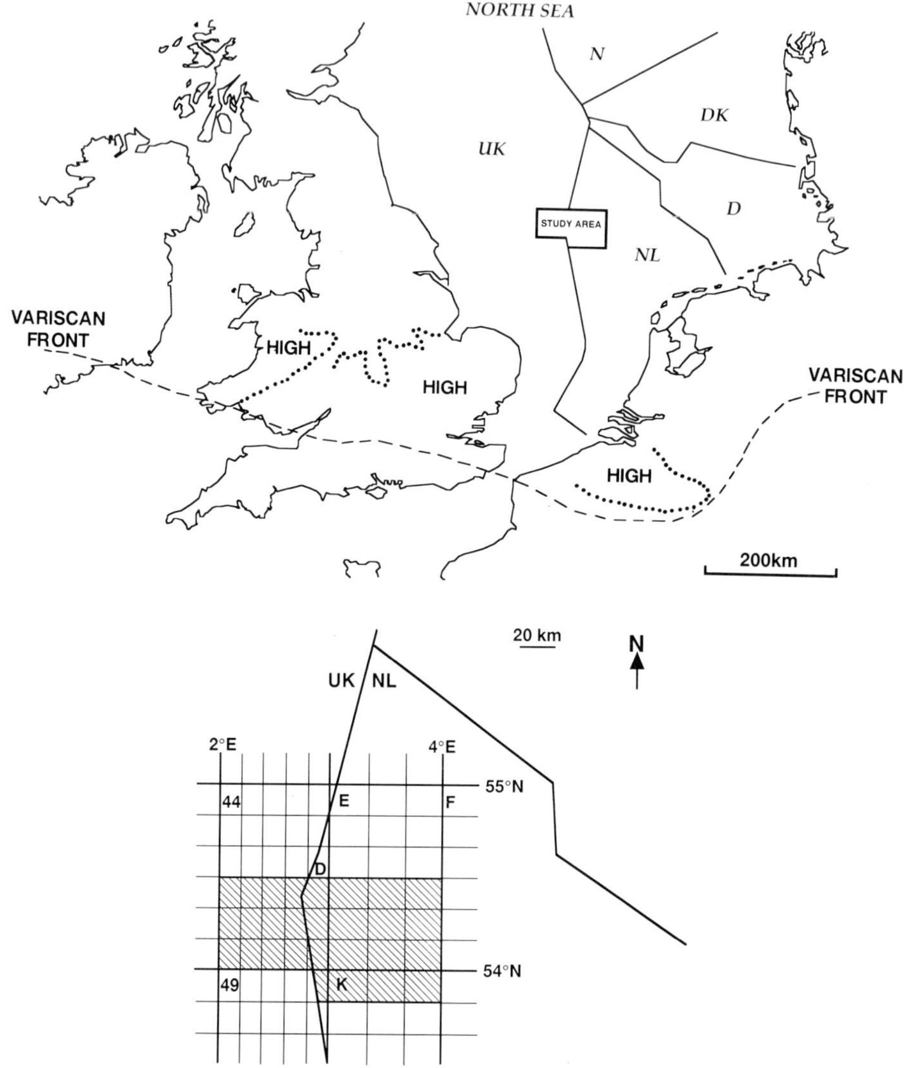



**Fig. 1.** Location of study area in the northern part of the Southern North Sea.

offshore are similar to those encountered on-shore in the UK. They record that the offshore Westphalian A and early Westphalian B is characterized by stacked fluvial sandstones analogous to the deposits of the Durham coalfield (e.g. Haszeldine & Anderton 1980; Fielding 1984, 1986; Haszeldine 1984). However, the middle and upper parts of the Westphalian B are notably sand free in the offshore and are more similar to the successions in the East Midlands (e.g. Rothwell & Quinn 1987; Fraser & Gawthorpe 1990; Fraser *et al.* 1990; Cope *et al.* 1992; Storey & Nash 1993). Similarly, Besly *et al.* (1993) report that the red bed succession within the Southern North

Sea is analogous to that onshore (e.g. Besly 1988) and that two distinct sets of alluvial red beds are present, one of Westphalian C and the other of Westphalian D age. Reddening is present locally throughout the Westphalian C as a result of syndepositional oxidation of overbank sediments in well-drained soils and by Westphalian D times probably as a consequence of climatic change (e.g. Besly 1987; Quirk 1993). Locally the Westphalian B is also reddened as a result of deep weathering on erosional highs during the Permian (Besly *et al.* 1993). In contrast to the onshore UK, Quirk (1993) indicated that the northern part of the Southern North Sea was dominated by the effects of a major river system coming from the north. Hence, the area was influenced mostly by large scale basin processes rather than local autocyclic effects (cf. Guion *et al.* 1995). Discharge and sediment input rates were high in the Southern North Sea but of an order similar to subsidence rates (Quirk 1993).

Quirk & Aitken (this volume) demonstrate that the Westphalian of the Southern North Sea was deposited within a major foreland basin which is asymmetric in cross-section and elongate in plan lying to the north of the Variscan front (Coward 1990; Ziegler 1990; Drozdzewski 1993; Gayer *et al.* 1993; Quirk 1993; Rijkers & Duin 1994). Coarse-grained sediment was derived dominantly from the Laurentian–Fennoscandian Caledonides farther to the north (Leeder *et al.* 1990; Besly *et al.* 1993; Collinson *et al.* 1993; Quirk 1993), with a limited amount of sediment coming from the Variscan mountains in the south and the Anglo-Brabant Massif to the SE (Collinson *et al.* 1993). A thick, shale-dominated succession occurs in the southern part of the Southern North Sea away from the sources of coarse-grained sediment (Quirk 1993). The basin was essentially non-marine, probably because river discharge was high and because the elongate nature of the foreland restricted the inflow of ocean water.

## Correlating the Westphalian

This study was based on the correlation of 40 wells (see appendix) tied into 2500 km of 2D seismic and approximately 1200 km$^2$ of 3D seismic data. Much of the data is unreleased or proprietary and was used with the permission of various companies, notably Fina Petroleum, Statoil and partners. Gas has been produced from the Westphalian in half of the study wells; the Rotliegend, which unconformably overlies the Westphalian forms the seal. Most of the unsuccessful wells have failed because a non-reservoir interval was encountered within base Rotliegend closure.

Although seismic data clearly show that *overall* thickness variations in the Westphalian of the study area are minimal (Quirk & Aitken this volume; Fig. 2), there are a number of problems associated with correlation:

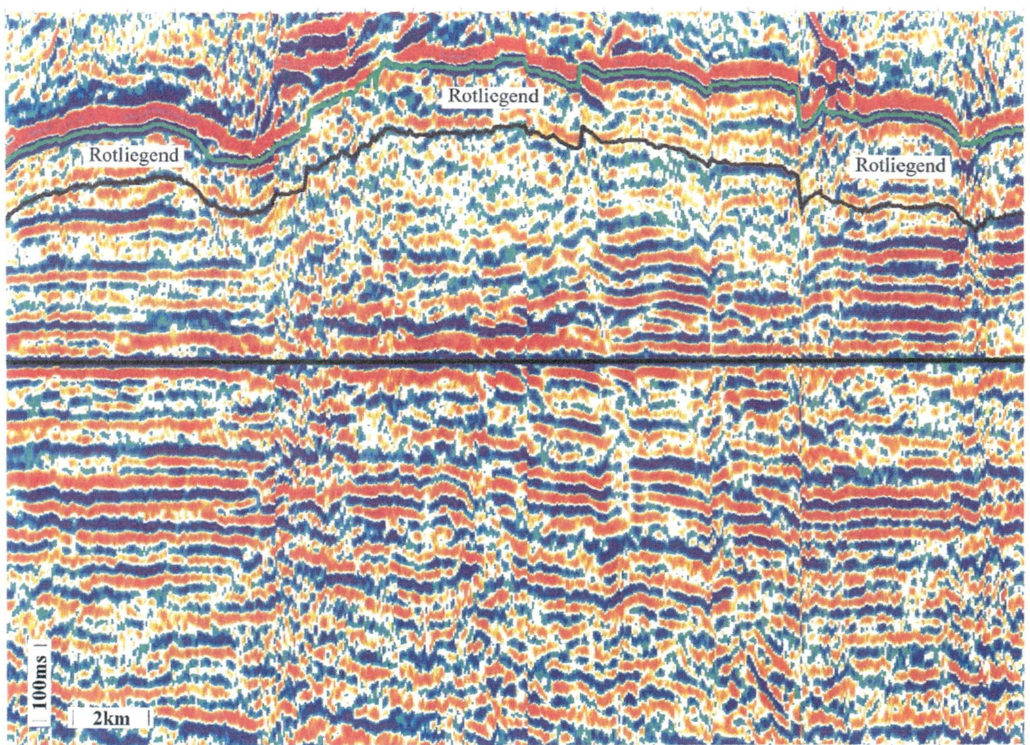


**Fig. 2.** Seismic section flattened on an interpreted intra-Westphalian horizon using GeoQuest Charisma software in order to illustrate the uniform thickness of Westphalian strata in the study area.

(1)  the sub-crop of the Westphalian at the base Rotliegend (Saalian) unconformity is complex due to the affects of normal faulting and uplift during the early Permian (see Quirk 1993);

(2)  the stratigraphic resolution of palynological data from different sources across the study area is no better than ±150 m (see Ritchie & Pratsides 1993);

(3)  it is difficult to positively identify specific coal beds, marine bands or other marker beds in wells due to a lack of comprehensive core material (see Whittaker *et al.* 1985).

Despite these problems, accurate correlation is possible when well log data and biostratigraphic information is properly integrated with seismic data. The results of such an integrated approach are reported here.

*Lower Westphalian seismic-well tie*
One of the key aspects in correlating the Westphalian is the recognition of a highly reflective intra-Westphalian package on good quality 2D seismic data (Fig. 3). This reflective package is of chronostratigraphic significance corresponding to the Westphalian A–early Westphalian B based on an average age determined from palynological data (Quirk 1993). The upper part of the reflective package, which is early Westphalian B in age, is the most strongly reflective and can be mapped with confidence over most of the study area. It consists of two or three seismic wavelets accounting for a two way travel time of approximately 40–60 ms, equivalent to a thickness of 90–130 m based on an average seismic interval velocity of 4400 m s$^{-1}$, derived from well data. Despite the fact that the reflective package has a relatively uniform seismic appearance on 2D data (Quirk 1993), well logs show that it has great lithological variation, ranging from more than 75% sandstone with no coal in the north of the study area to less than 25% sandstone with a significant number of coal seams in the south. It is strongly reflective due to seismic tuning (constructive interference of wavelets) within high order, coarse-grained/fine-grained sediment cycles or parasequences (Fig. 4; Quirk 1993), even though the amounts of sandstone, mudstone and coal vary within the parasequences.

A coal-bearing shaly interval occurs at the top of the reflective package (Fig. 4) which is roughly equivalent to the Westoe Coal Formation of Cameron (1993). This shaley interval is approximately middle–upper Westphalian B in age and is 120–160 m thick. It has a similar well

log expression over the entire study area (Fig. 5). Apart from its basal part, it is not highly reflective on 2D seismic data; although it contains numerous coals, these are generally less than 50 cm thick and are not resolved on seismic. The lower boundary of the shaley interval is usually marked by a high gamma ray peak corresponding with a dark mudstone which approximately ties with the upper part of the reflective package on seismic (Fig. 4). By combining all of the available palynological data from the 40 wells in the study, it appears that this mudstone roughly corresponds with an upper Westphalian B interval which includes the Maltby and Clown marine bands onshore UK (see Calver 1969; Ramsbottom *et al.* 1978).

It has been noted on recent high resolution 3D data that the seismic character differs significantly from 2D data such that the reflectivity of the middle part of the Westphalian A–lower Westphalian B interval decreases and the reflectivity of the upper Westphalian B and Westphalian C increases. The reason for this is thought to be because the 3D seismic data contains higher frequency data, which has the effect of decreasing the thickness at which tuning occurs. Nonetheless, the seismic generally has a consistent appearance within the same survey, such that different reflective intervals can be mapped with confidence, particularly if variable amplitude displays are used. By tying the seismic to wells, it can be shown that these intervals are of stratigraphic significance similar to observations already made on 2D data (Quirk 1993). However, even with high resolution 3D data, it has so far proved impossible to predict lithofacies from seismic character.

The Westphalian has a tramline-like appearance on seismic with no evidence for syndepositional fault activity. However, the succession is broken up by post-Carboniferous faults (see Quirk & Aitken this volume). The ability to artificially flatten interpreted intra-Westphalian reflections on a seismic workstation greatly aids interpretation by removing the effects of this later tectonism (e.g. Fig. 2).

*Lithostratigraphic sub-division*
The base of the middle–upper Westphalian B shaly interval (top of the reflective package) is used as a datum for more detailed well log correlations and lithostratigraphic interpretations because it can be confidently picked both on seismic and well logs. The basic process involves dividing the succession above and below the base of the shaley interval into 50–150 m thick 'sand-rich units' and 'mud-dominated units' on the basis of gamma ray, sonic log, density

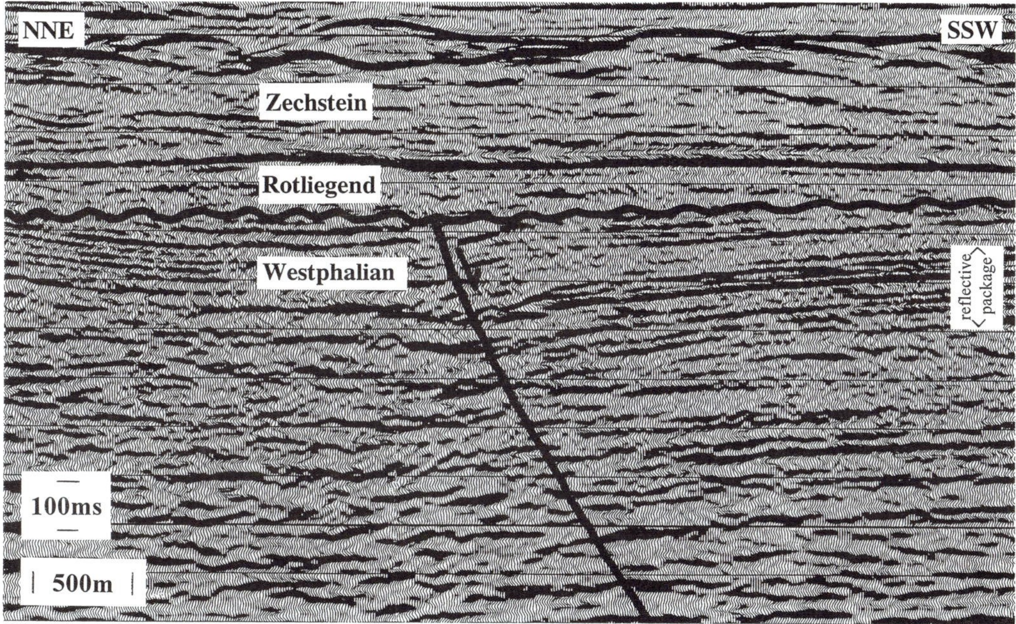


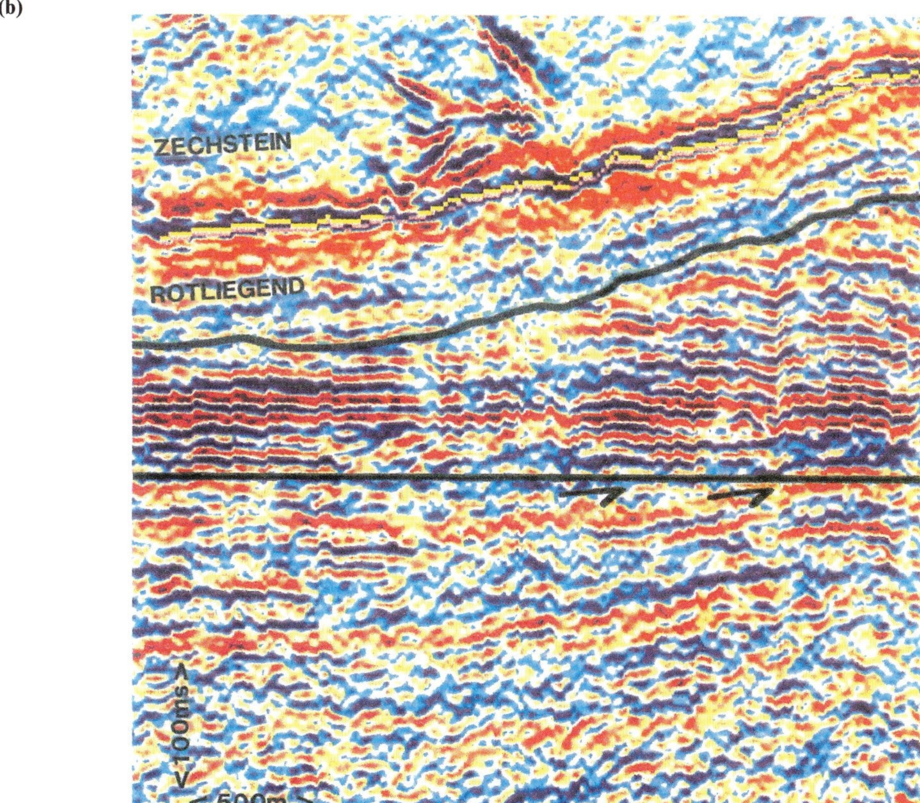


**Fig. 3. (a)** Typical 2D seismic section in the study area showing the intra-Westphalian reflective package (Westphalian A-early Westphalian B). Data courtesy of Schlumberger GeoQuest. **(b)** Typical 3D seismic section in the study area showing the base of a sand-rich unit of early Westphalian B age flattened using GeoQuest Charisma software in order to illustrate rare erosional truncation of underlying strata (indicated by arrows).

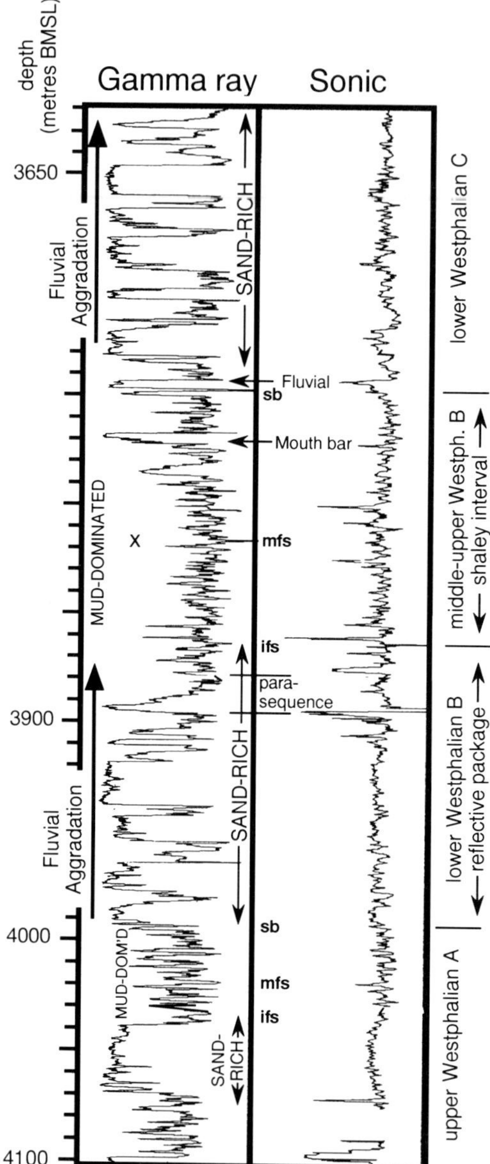


**Fig. 4.** Lithostratigraphic and sequence stratigraphic well log interpretation of UK well 44/19-3. Deflections to the left on the gamma ray log are sandstones; sharp peaks to the left on the sonic log are coals; sb, sequence boundary; mfs (X), maximum flooding surface; ifs, initial flooding surface.

and neutron porosity data. Sand-rich units contain more than 30% sandstone, an example of which is the lower Westphalian B reflective package (Fig. 4). In contrast, mud-dominated units contain less than 30% sandstone such as the middle-upper Westphalian B shaley interval

(Fig. 4). Six to seven sand-rich units and six to seven mud-dominated units have been identified within the Westphalian of the study area. However, several stages of iteration involving interpretation of gamma ray and sonic logs, seismic data and palynological information are required before a consistent correlation can be produced.

On the basis of limited (confidential) core data and an electrofacies scheme similar to that proposed by Collinson *et al.* (1993), it has been observed that sand-rich units are dominated by thick, medium- to very coarse-grained fluvial sandstones with a blocky or fining-upwards gamma ray expression, interbedded with thin, fine-grained interfluvial sediments. Mud-dominated units consist largely of lacustrine facies with occasional coarsening-upwards mouth bar deposits in their upper parts (Fig. 4). On gamma ray and sonic logs mud-dominated units display sharp upper contacts with overlying sand-rich units. The base of any mud-dominated unit is more gradational in character although a high gamma ray peak corresponding to a dark mudstone is normally observed within the transition from the underlying sand-rich unit and provides a convenient pick for the lower contact (Fig. 4). In Westphalian A–B mud-dominated units coal, indicative of mire conditions, is common except in the extreme north and the lacustrine facies are grey to dark grey in colour. In Westphalian C–D units coal is absent except in the extreme south of the study area. Westphalian C–D mud-dominated intervals have rarely been cored, but what little material is available suggests that they consist predominantly of red to light grey lacustrine strata. However, the presence of floodplain fines cannot be ruled out.

Sand-rich units generally contain discrete sandstone bodies with high reservoir potential. For example, porosities of greater than 15% and permeabilities of more than 500 mD are not uncommon. The best reservoirs have a blocky gamma ray expression and are more common in the middle and lower parts of sand-rich units, particularly in the north of the study area where they are more often coarse grained. They consist predominantly of quartz; lithic fragments are fairly rare. In core, individual sandstone bodies tend to exhibit different scales of bounding surfaces including small scale planar and trough cross-beds and larger scale erosional surfaces which are often overlain by conglomeratic sandstones. Internally there is little evidence of any overbank sediment. They are therefore interpreted as representing stacked, braided or low sinuousity channel deposits. At present, well spacing within the study area is insufficient to determine over what distance individual sandstone bodies extend but sand-rich units, which are generally composed of between two and nine individual sandstone bodies, depending on the total thickness of the sand-rich units, can themselves be correlated over areas of more than 2500 km$^2$ (e.g. Fig. 6a). It is therefore probable that the sandstones are amalgamated and they are thought to have a sheet-like geometry.

Mud-dominated units are correlatable over the entire study area indicating that lacustrine and mire conditions were established over large parts of the basin at certain times, for example during the mid–late Westphalian B (Fig. 5). Individual units can usually be sub-divided into a lower and an upper part separated by a maximum gamma ray peak which corresponds with an organic-rich shale (e.g. marked 'X' on Fig. 4). Below this gamma ray peak, the shale line in the mud-dominated unit often shows a subtle fining-upwards trend and above it an overall coarsening-upwards trend (e.g. see Fig. 5). For this reason the lower and upper parts of mud-dominated units are defined as 'fining-upwards sub-units' and 'coarsening-upwards sub-units', respectively. The fining upwards sub-units would seem to have good sealing potential but intra-formational traps have not been discovered in any of the study wells. Fine-grained mouth bar sandstones occur towards the top of coarsening-upwards sub-units and randomly interrupt the shale line on the gamma ray log, but these generally have low reservoir potential and can rarely be correlated between wells.

No core in the study wells was available from the middle of mud-dominated intervals to determine whether marine fossils occur at the level of a maximum gamma ray peak. However, in general, marine fossils are extremely rare in the northern part of the Southern North Sea (Quirk 1993) although exceptions have been reported south of the study area by Besly *et al.* (1993) and Ritchie & Pratsides (1993). Furthermore, Knowles (1964) has shown that in the Upper Carboniferous onshore UK only some marine bands are radioactive and that radioactivity within individual marine bands can be highly variable laterally. Persistent radioactivity in shales is more likely to correspond with bituminous matter and fine clays. Maximum gamma ray peaks are therefore interpreted to represent condensed horizons rather than indicating marine influence. However, palaeontological evidence for occasional marine incursions is present towards the centre of the basin where strata are predominantly shaley.

Locally, coarsening-upwards sub-units are either partially or wholly missing beneath certain sand-rich units. Where this occurs, particularly at the base of the reflective lower Westphalian B package, erosional truncation can sometimes be observed on high quality seismic data (e.g. Fig. 3b). However, the amount of truncation amounts to no more than 50m of missing section and is significantly less than the thickness of the overlying sand-rich unit. The surface of erosion is also rather flat and does not have the form of an incised valley on 3D seismic data. The recognition of erosional truncation is, however, particularly useful for identifying and mapping the bases of potential reservoir-prone sand-rich units on seismic.

*Cycle stratigraphy*
By dividing the succession into units as described above, it can be shown that the Westphalian consists of six large-scale sand-rich/mud-dominated cycles. Three of these cycles are shown in Fig. 5 and, although their ages cannot be constrained precisely using palynology, they are approximately upper Westphalian A (Langsettian), Westphalian B (Duckmantian) and lower Westphalian C (Bolsovian) in age. The other three cycles are assigned lower Westphalian A, upper Westphalian C/lower Westphalian D and upper Westphalian D/early Stephanian ages, although the upper two cycles are almost completely barren of palynomorphs in the study area. Locally the Westphalian B cycle, which consists of the lower Westphalian B reflective package and the mid-upper Westphalian B shaly interval, can be sub-divided into two higher order sand-rich/mud-dominated cycles (Fig. 5).

Regional well correlations show that sand-rich/mud-dominated cycles in the Westphalian show predictable lithological trends. In the southern part of the study area, they contain thicker mud-dominated units but thinner sand-rich units and in the north most cycles contain thinner mud-dominated units but thicker sand-rich units (Figs 5 & 6a). Consequently, most cycles in the northern part of the southern North Sea have rather uniform total thicknesses. Minimal thickness variations are also observed on seismic data within the study area (e.g. Fig. 2) which is explained by Quirk & Aitken (this volume) as due to the relatively large size of the foreland basin. For example, an interval comprising two sequences with a total thickness of approximately 300 m shows less than 20% variation in two-way time over an area of some 500 km$^2$ on a 3D seismic survey within the study area. At least 10% of this variation is due to differences in seismic velocity and the

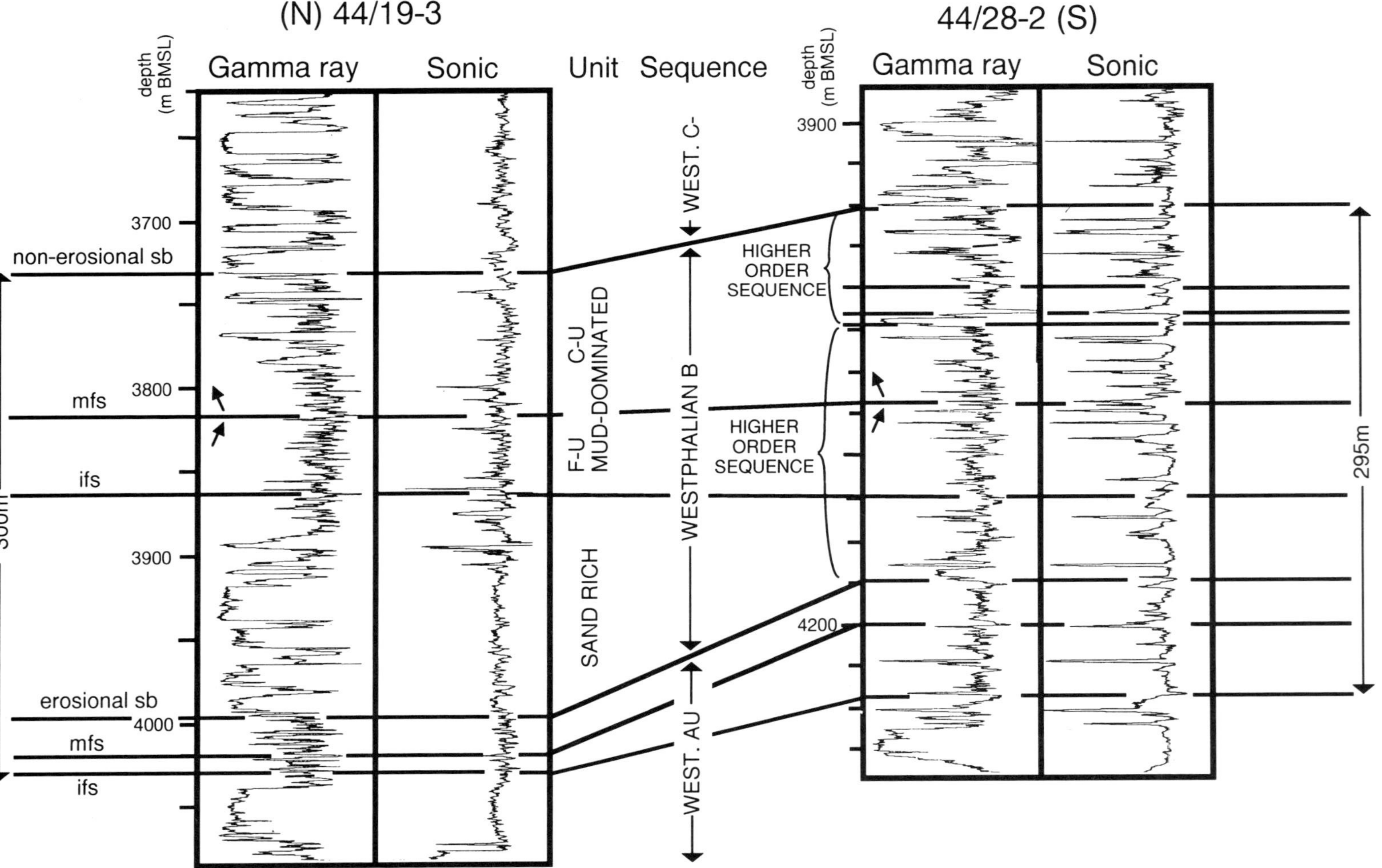


**Fig. 5.** Well log correlation of two wells in the north and south of the study area which form part of a regional correlation of 40 wells based on seismic, lithostratigraphy and palynological data. The approximate distance between the wells is 50 km. Note how an increase in the thickness of sand-rich units from south to north is balanced by a decrease in the thickness of mud-dominated units. sb, sequence boundary; mfs, maximum flooding surface; ifs, initial flooding surface; F–U, fining-upwards; C–U, coarsening-upwards; arrows indicate fining-upwards and coarsening-upwards trends on gamma ray logs.

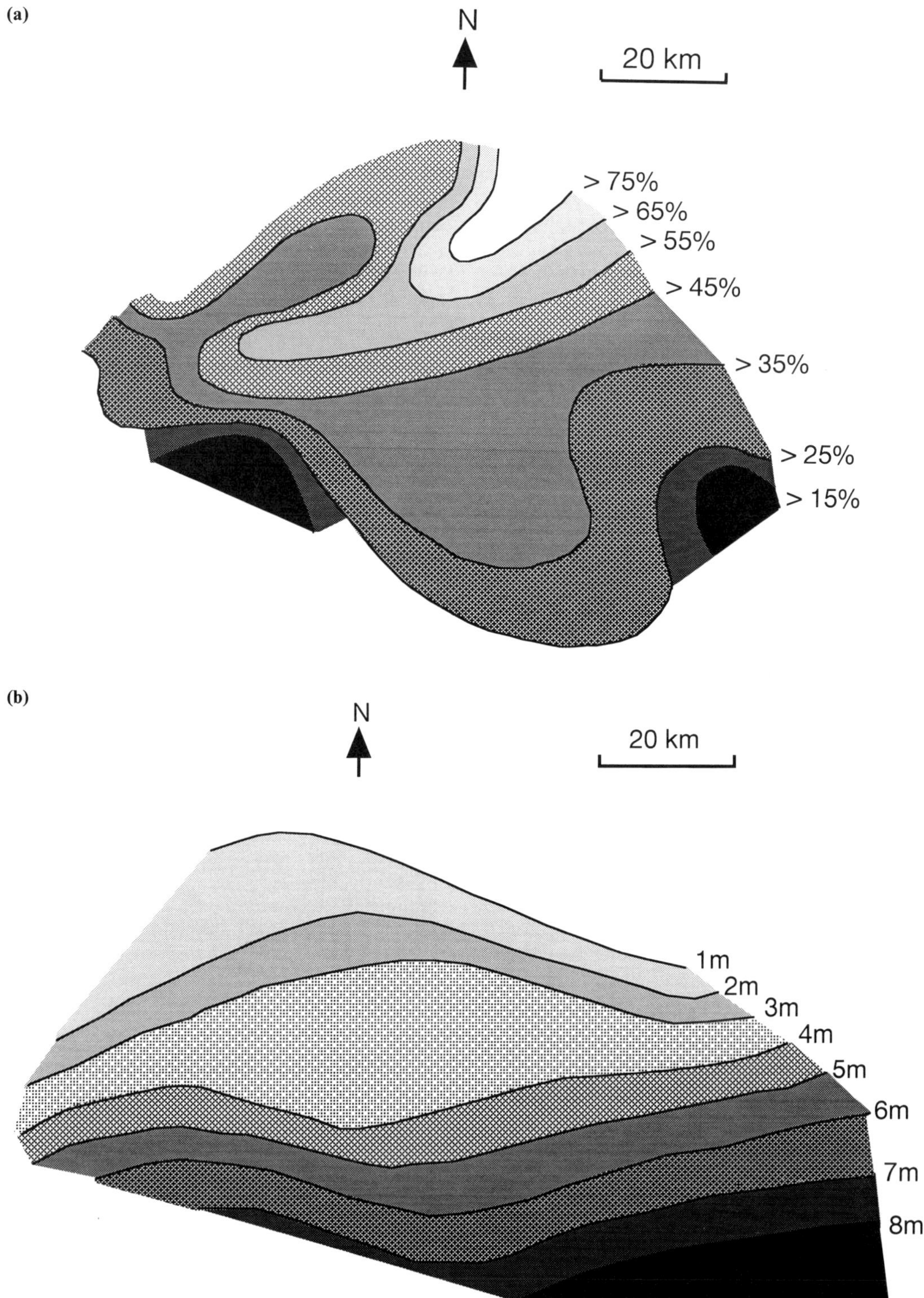


**Fig. 6.** Map of (**a**) net sandstone (in percent) and (**b**) net coal (in metres) across part of the study area for a sand-rich unit (early Westphalian B) based on 29 wells (location removed for reasons of confidentiality).

precision of the seismic picks at the top and bottom of the interval is ±25 m; hence, the amount of thickness variation that can be attributed to geological factors is less than 10%.

Coal is present within all cycles of Westphalian A/B age, particularly within fining-upwards mud-dominated sub-units, although there are significantly fewer coals in sand-rich units in the north than in the south. For example, there is little coal present in the lower Westphalian B reflective package in the north (Fig. 6b). However, coals throughout the study area are thin and, although well spacing is not ideal, it appears that many coals are of limited lateral extent (cf. Quirk 1993). Coal is virtually absent from cycles of Westphalian C/D age, which are generally red, except in the extreme south of the study area. In most of the study area, the change from coal-bearing to barren red strata occurs fairly abruptly at the top of the Westphalian B. This may reflect a change to a more arid climate as has been described during the Westphalian D by Besly (1987), although Besly *et al.* (1993) have pointed out that this change is diachronous. Carbonaceous rooted horizons have been observed in cores of mud-dominated Westphalian C/D strata in the study area indicating that vegetation was still present but was later mostly oxidized. Caliche or calcretes are confined to the Stephanian. Apart from the presence or absence of coal, cycles throughout the Westphalian succession have a similar well log expression (e.g. Fig. 4).

## Sequence stratigraphic interpretation

The observation that erosional truncation is occasionally present at the base of sand-rich units is used to redefine the sand-rich/mud-dominated cycles as sequences. Hence, truncation surfaces are classified as erosional sequence boundaries (e.g. Fig. 5), analogous to the type 1 sequence boundaries of Posamentier & Vail (1988) except that peneplanation rather than incision appears to have been the main form of erosion. Where truncation has not been observed, the bases of sand-rich units are instead interpreted as non-erosional sequence boundaries, similar to type 2 sequence boundaries. The assumption has therefore been made here that any erosional truncation below the resolution of seismic and well log correlations is due to normal fluvial processes such as channel scour.

Deposition within any sequence began with fluvial (sand-rich) strata on top of the sequence boundary. The fluvial strata consist predominantly of braided or low sinuousity channel deposits (Quirk 1993). In the north, fluvial sandstones are thick and laterally extensive indicating that aggradation and channel avulsion were important processes during their deposition. It has also been observed within most sequences that the thickness of each sand-rich unit is greater in the north of the study area than in the south and is inversely proportional to the thickness of the overlying mud-dominated unit (e.g. Fig. 5). The number of sandstone bodies within any sand-rich unit is proportional to its total thickness and hence, within each sequence, lacustrine strata are interpreted to overstep fluvial strata from south to north due to transgression. At the top of any sand-rich unit, fluvial deposition was terminated when mud-dominated lake conditions became established. The dark mudstone which occurs at the top of each sand-rich unit is interpreted to represent the surface of transgression or initial (lake) flooding surface (Figs 4 & 5) which is diachronous. It appears therefore, that fluvial sandstones were deposited north of the contemporary lake margin during periods of transgression, i.e. coarse-grained sediment was trapped on the coeval alluvial plain due to rising lake levels. Quirk (1996) has shown that such a situation occurs because the graded profile of a river must rise as the relative height of the coastline increases thus creating accommodation space for coarse-grained sediment on the alluvial plain.

The presence of coal within the fining-upwards sub-units indicates that shallow water conditions prevailed during transgression. This was probably because the rate of fine-grained sediment supply and organic productivity were sufficiently high to almost balance the rate of creation of accommodation caused by rising relative lake level. Consequently, a true shelf break did not develop and high energy shoreface sandstones are absent. The top of each fining-upwards sub-unit is marked by a maximum gamma ray peak corresponding with an organic-rich shale. This is interpreted as a maximum (lake) flooding surface (Figs 4 & 5).

Above the maximum flooding surface, the lacustrine strata gradually coarsen upwards as a result of deltaic progradation until alluvial conditions became once more established. The upper part of the coarsening-upwards interval has in some cases been removed by erosion because of a relative fall in lake level prior to deposition of the succeeding sequence. In other cases, fluvial deposition followed lacustrine deposition without interruption as aggradation succeeded progradation without any apparent intervening period of erosion.

A significant proportion of the coal in any one sequence occurs towards the top of the sand-rich unit and in the lower, fining-upwards part of the mud-dominated unit (Fig. 4). Hence, most coal accumulated during periods of lake transgression with mires forming along the margins of the lake, basinward (south) of the alluvial plain where coeval fluvial strata were deposited. The reasons for this are probably that:

(1) the amount of coarse-grained sediment reaching the coast during transgression was low because of the effects of fluvial aggradation on the coeval alluvial plain;
(2) shallow, low energy, fresh water conditions prevailed;
(3) climatic conditions were appropriate for the development of mires;
(4) the rate of creation of accommodation was optimum for preservation of peat (Cross 1988).

## Sequence stratigraphic model

The Westphalian deposits in the study area do not conform with existing sequence stratigraphic models which tend to assign fluvial strata to incised valley fills (e.g. Dalrymple *et al.* 1994) or highstand (progradational) systems tracts (e.g. Posamentier & Vail 1988). The fluvial sandstones are neither laterally nor vertically constrained within the confines of an incised valley and they are markedly retrogradational rather than progradational. Instead, a model is presented in Fig. 7 for a typical Westphalian A–B sequence in the Southern North Sea. Westphalian C–D sequences have a similar origin but lack coal due to drier conditions. The model has been developed on the basis of (i) detailed observations made in the study area; (ii) general observations made to the south; and (iii) by extrapolation to the north of the study area where the Westphalian is mostly eroded. It illustrates that sand-rich units, which form the main reservoir target in the northern part of the Southern North Sea, were deposited due to fluvial aggradation (and avulsion) during periods of lake transgression. Coarse-grained fluvial sediment was trapped on the alluvial plain as river profiles attempted to keep up with rising lake levels. Coals and fining-upwards lacustrine strata were deposited above sand-rich units following initial flooding (Fig. 7). Both the sand-rich unit and the fining-upwards mud-dominated sub-unit together are equivalent to a transgressive systems tract (cf. Posamentier *et al.* 1988). The end of transgression is marked by a maximum gamma ray peak which corresponds to the maximum (lake) flooding surface. The

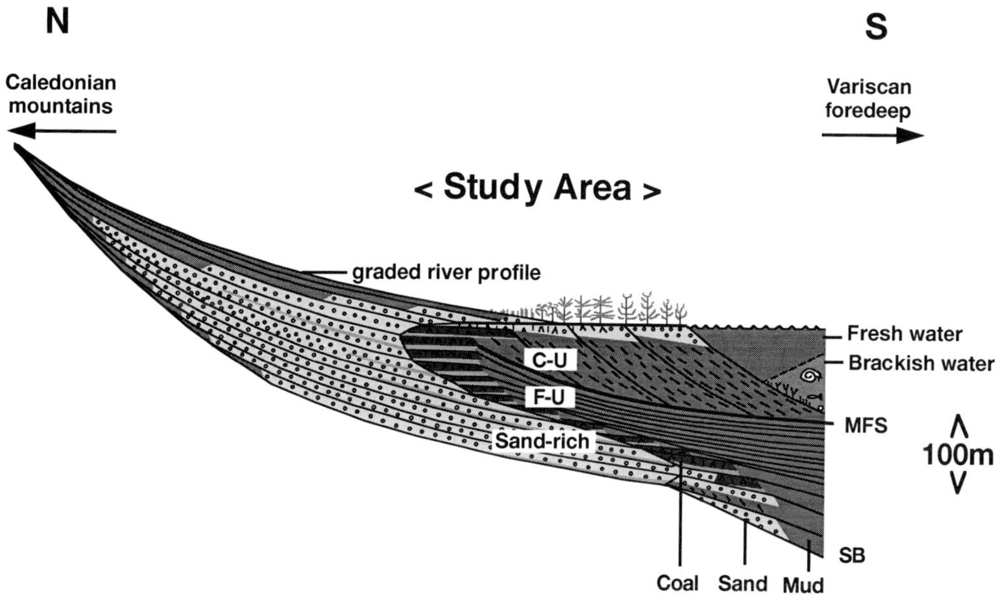


**Fig. 7.** Schematic model of a typical Westphalian sequence in the Southern North Sea. F–U, fining-upwards; C–U, coarsening-upwards; SB, sequence boundary; MFS, maximum flooding surface.

coarsening-upwards sub-unit, which overlies the maximum flooding surface, is progradational and this can be assigned to the highstand systems tract of standard sequence stratigraphic models (e.g. Posamentier & Vail 1988). Where erosional truncation has occurred at the base of a sequence, a lowstand systems tract may have been deposited in the deeper part of the basin farther to the south. Where a non-erosional sequence boundary is present (e.g. Fig. 7), the lowermost part of the sand-rich unit probably does not retrograde and is therefore equivalent to the shelf margin systems tract of Posamentier et al. (1988).

The six sequences identified in the Westphalian of the study area span a period of time equivalent to approximately 10 Ma (Hess & Lippolt 1988), although new SHRIMP zircon age dates may reduce this interval (Riley et al. 1994). This suggests that each sequence was deposited within an interval of 1.5 Ma or less. This sort of periodicity is typical of third order eustatic cycles as defined by Vail et al. (1977). The presence of marine bands in the Westphalian in NW Europe indicates that the foreland basin was tenuously connected to open ocean and may therefore have been affected by eustatic cycles. However, at present it is difficult to rule out the possibility that additional factors led to changes in relative lake level, such as Variscan tectonism or variations in sediment input (cf. Quirk 1996).

Little internal variation in thickness can be detected within any one sequence either on seismic (e.g. Fig. 3b) or in well log correlations (e.g. Fig. 5) despite a marked change in sandstone, mudstone and coal content from north to south across the study area. This can be explained by the large lateral and vertical scale of each sequence (Fig. 7) and the time span involved. Thus, because of high sediment-supply rates, local effects such as differential compaction are insignificant compared with regional processes such as subsidence and eustasy that led to the creation (and removal) of accommodation space at the third order, basin-wide scale.

It is perhaps surprising to note that the effects of the change to a more arid climate in Westphalian C/D times (Besly 1987) has little obvious effect on the overall sequence stratigraphy of the Westphalian apart, that is, from the presence or absence of coal. The explanation is thought to lie in the scale of the river system supplying sediment to the basin. This was probably sourced from a major highland area a few hundred kilometres north of the study area (Quirk 1993) which could have remained relatively unaffected by climate changes in the study area.

## Future prospects

The 'Exxon' sequence stratigraphic models of Vail et al. (1977) have been used widely in the prediction and correlation of shallow marine strata (e.g. Cross et al. 1993; O'Byrne & Flint 1993; Wehr & Brasher 1996). However, sequence stratigraphy has rarely been applied to alluvial–lacustrine successions with the notable exception of Legarreta et al. (1993) and McKie & Garden (1996). The model developed here for Westphalian alluvial–lacustrine sequences helps explain certain unusual aspects of the lithostratigraphy such as the widespread occurrence of retrogradational fluvial reservoirs and the absence of shoreface sandstones. However, more importantly, it has improved the resolution and confidence of seismic and well log interpretation in the northern part of the Southern North Sea.

Quirk & Aitken (this volume) have shown that potential exists for future gas discoveries in structures already drilled in the study area due to the angular nature of the Saalian unconformity; i.e. most wells that have failed to discover gas in the Westphalian have encountered a mud-dominated unit within base Rotliegend structure. The prediction of where reservoir-prone (sand-rich) intervals lie within structure depends on mapping reflective Westphalian intervals and tying these in to nearby wells. The tramline-like nature of the Westphalian strata allows thicknesses to be added or subtracted from mapped horizons to precisely locate the position of sand-rich units with proven or possible reservoir potential. The use of colour enhancement and horizon flattening on seismic workstations improves the accuracy of any interpretation. The results of seismic facies analysis have, in contrast, proved disappointing.

Reservoir thickness and quality within sand-rich units may be estimated from regional trends based on detailed correlation of pre-existing wells (e.g. Fig. 6a). In a general sense, the number, thickness and quality of reservoirs are predicted to increase towards the north. Preferably wells should be targetted on up-faulted blocks and where sand-rich units with high net to gross ratios are interpreted to lie directly below the base Rotliegend (Quirk & Aitken this volume); the objective need not lie at the mapped structural culmination.

Even where there is a high degree of confidence in the seismic interpretation, a contingency side-track well should be planned in the event that additional or better reservoir zones encountered in the initial well are projected to lie within structure elsewhere below the Saalian unconformity. It is also worth noting that the effects of weathering and diagenesis below the Saalian unconformity may lead to localised areas of porosity enhancement or porosity reduction. For correlation and mapping purposes, it is recommended that any well should be drilled to a minimum depth of 300 m below the base Rotliegend so that at least two sequences are penetrated which can then be integrated with existing interpretations and help improve the regional picture.

Although mud-dominated units would appear from well logs to have good sealing capacity, to date no intra-Westphalian traps have been proved in the study area, probably because of seal breaches caused by the large number of early Permian faults that affect the Westphalian (Quirk & Aitken this volume). Therefore, deeper targets are best tested as secondary rather than primary objectives.

## Conclusions

The Westphalian in the northern part of the southern North Sea contains sequences of uniform thickness which can be mapped with confidence on good quality seismic data. Regional well correlations allow sandstone and coal percentages to be predicted. Each individual sequence consists of a sand-rich and a mud-dominated unit. The sand-rich unit occurs in the lower part of any sequence and comprises stacked fluvial sandstones. These sandstones are predominantly coarse-grained, laterally extensive, braided or low sinuousity channel deposits that developed by fluvial aggradation and avulsion during periods of relative lake level rise and transgression. The mud-dominated unit overlies the sand-rich unit and comprises mostly lacustrine strata with coal in its lower part (fining-upwards sub-unit) and pro-delta mudstones and mouth bar sandstones in its upper part (coarsening-upwards sub-unit). The lower and upper parts of the mud-dominated unit are separated by a maximum flooding surface. Although Westphalian sequences are different to those described in marine basins, the sand-rich unit and the fining-upwards sub-unit together are equivalent to the transgressive systems tract of Posamentier & Vail (1988) and the coarsening-upwards sub-unit is equivalent to the highstand systems tract. Each sequence was deposited over a period of approximately 1.5 Ma or less suggesting that they were controlled to some degree by third order eustatic processes.

Confidential well data and proprietary seismic data used in this study were provided by Fina and Block 44/29 partners, Goal Petroleum (now Talisman Energy), Statoil and Intera (now Schlumberger Geo-Quest) following on from earlier work carried out by D.G.Q. whilst working at the Nederlandse Aardolie Maatschappij (NAM) and reported in Quirk (1993). The author is grateful for the assistance and advice provided by J. Aitken, J. Kunka, D. Lawton, P. Mason, B. Evans, G. Plint and N. Jones and to L. Hill, who drafted most of the diagrams.

## APPENDIX: *Wells used in the study*

| UK Sector | DUTCH Sector |
| --- | --- |
| 44/14-2 | D12-3 |
| 44/18-1 | D12-4 |
| 44/18-2 | D15-2 |
| 44/19-3 | D15-3 |
| 44/22-1 | D15-4 |
| 44/22-3 | E10-1 |
| 44/22-4 | E13-1 |
| 44/23-4 | E13-2 |
| 44/23-5 | E16-3 |
| 44/23-6 | E18-2 |
| 44/23-7 | F10-2 |
| 44/23-8 | J3-1 |
| 44/24-2 | J3-2 |
| 44/24-3 | K1-2 |
| 44/24-4 | K2-1 |
| 44/26-4 | K3-1 |
| 44/28-1 | K5-2 |
| 44/28-2 | |
| 44/28-3 | |
| 44/28-4 | |
| 44/29-1 | |
| 44/29-3 | |
| 44/29-4 | |

## References

BESLY, B. M. 1987. Sedimentological evidence for Carboniferous and early Permian palaeoclimates of Europe. *Annales de la Société Géologique du Nord*, **106**, 131–143.

——1988. Palaeogeographic implications of late Westphalian to early Permian red-beds, Central England. *In*: BESLY, B. M. & G. Kelling (eds.) *Sedimentation in a synorogenic basin complex: the Carboniferous of northwest Europe*. Blackie, Glasgow, 200–221.

——, BURLEY, S. D. & TURNER, P. 1993. The late Carboniferous 'Barren Red Bed' play of the Silver Pit area, Southern North Sea. *In*: PARKER, J. R. (ed.) *Petroleum Geology of Northwest Europe: Proceedings of the 4th Conference*. Geological Society, London, 727–740.

CALVER, M. A. 1969. Westphalian of Britain. *Compte Rendu du Sixiéme Congrès International de Stratigraphie et de Géologie du Carbonifère*, Sheffield 1967, **1**, 233–254.

CAMERON, T. D. J. 1993. Carboniferous and Devonian of the Southern North Sea. *In*: KNOX, R. W. O'B. & CORDEY, W. G. (eds) *Lithostratigraphic nomenclature of the UK North Sea*, **5**. British Geological Survey, Nottingham.

COLLINSON, J. D., JONES, C. M., BLACKBOURN, G. A., BESLY, B. M. ARCHARD, G. M. & MCMAHON, A. H. 1993. Carboniferous depositional systems in the Southern North Sea. *In*: PARKER, J. R. (ed.) *Petroleum Geology of Northwest Europe: Proceedings of the 4th Conference*. Geological Society, London, 677–687.

COPE, J. C. W., GUION, P. D., SEVASTOPULO, G. D. & SWAN, A. R. H. 1992. Carboniferous. *In*: COPE, J. C. W., INGHAM, J. K. & RAWSON, P. F. (eds) *Atlas of palaeogeography and lithofacies*. Geological Society, London, Memoirs, **13**, 67–86

CROSS, T. A. 1988. Controls on coal distribution in transgressive-regressive cycles, Upper Cretaceous, Western Interior, U.S.A. *In*: WILGUS, C. K., HASTINGS, B. S., KENDALL, C. G. St. C., POSAMENTIER, H. W., ROSS, C. A. & VAN WAGONER, J. C. (eds) *Sea-level changes: an integrated approach. Society of Economic Paleontologists and Mineralogists*, Special Publications, **42**, 371–380.

——, BAKER, M. R., CHAPIN, M. A., CLARK, M. S., GARDNER, M. H., HANSON, M. S., LESSENGER, M. A., LITTLE, L. D., MCDONOUGH, K.-J., SONNENFELD, M. D., VALASEK, D. W., WILLIAMS, M. R. & WITTER, D. N. 1993. Applications of high-resolution sequence stratigraphy to reservoir analysis. *In*: ESCHARD, R. & DOLIGEZ, B. (eds) *Subsurface Reservoir Characterization from Outcrop Observations*. Éditions Technip, Paris, 11–34.

COWARD, M. P. 1990. The Precambrian, Caledonian and Variscan framework to NW Europe. *In*: HARDMAN, R. F. P. & BROOKS, J. (eds.) *Tectonic Events Responsible for Britain's Oil and Gas Reserves*. Geological Society, London, Special Publications, **55**, 1–34.

DALRYMPLE, R. W., BOYD, R. & ZAITLIN, B. A. 1994. History of research, valley types and internal organization of incised-valley systems: introduction to the volume. *In*: DALRYMPLE, R. W., BOYD, R. & ZAITLIN, B. A. (eds) *Incised-valley systems: origin and sedimentary sequences*. SEPM (Society for Sedimentary Geology), Special Publications, **51**, 3–10.

DROZDZEWSKI, G. 1993. The Ruhr Coal Basin (Germany): structural evolution of an autochthonous foreland basin. *International Journal of Coal Geology*, **22**, 231–250.

FIELDING, C. R. 1984. A coal depositional model for the Durham Coal Measures of NE England. *Journal of the Geological Society*, London, **141**, 919–931.

——1986. Fluvial channel and overbank facies from the Westphalian of the Durham coalfield, NE England. *Sedimentology*, **33**, 119–140.

FRASER, A. J. & GAWTHORPE, R. L. 1990. Tectono-stratigraphic development and hydrocarbon habitat of the Carboniferous in northern England. *In*: HARDMAN, R. F. P. & BROOKS, J. (eds) *Tectonic events responsible for Britain's oil and gas reserves*. Geological Society, London, Special Publications, **55**, 49–86.

——, NASH, D. F., STEELE, R. P. & EBDON, C. C. 1990. A regional assessment of the intra-Carboniferous play of northern England. *In*: BROOKS, J. (ed.) *Classic Petroleum Provinces*. Geological Society, London, Special Publications, **50**, 417–440.

GAYER, R. A., COLE, J. E., GREILING, R. O., HECHT, C. & JONES, J. A. 1993. Comparative evolution of coal bearing foreland basins along the Variscan northern margin in Europe. *In*: GAYER, R. A., GREILING, R. O. & VOGEL, A. K. (eds) *Rhenohercynian and subvariscan fold belts*. Vieweg Publishing, Braunschweig/Weisbaden, 47–82.

GUION, P. D., BANKS, N. L. & RIPPON, J. H. 1995. The Silkstone Rock (Westphalian A) from the East Pennines, England: implications for sand body genesis. *Journal of the Geological Society, London*, **152**, 819–832.

HASZELDINE, R. S. 1984. Muddy deltas in freshwater lakes, and tectonism in the Upper Carboniferous Coalfield of NE England. *Sedimentology*, **31**, 811–822.

—— & ANDERTON, R. 1980. A braidplain facies model for the Westphalian B Coal Measures of northeast England. *Nature*, **284**, 51–53.

HESS, J. C. & LIPPOLT, H. J. 1986. $^{40}Ar/^{39}Ar$ ages of tonstein and tuff sanidines: new calibration points for the improvement of the Upper Carboniferous time scale. *Chemical Geology (Isotope Geoscience Section)*, **59**, 143–154.

KNOWLES, B. 1964. The radioactive content of the coal measures sediments in the Yorkshire-Derbyshire coalfield. *Proceedings of the Yorkshire Geological Society*, **34**, 413–450.

LEEDER, M. R. & HARDMAN, M. 1990. Carboniferous geology of the Southern North Sea Basin and controls on hydrocarbon prospectivity. *In*: HARDMAN, R. F. P. & BROOKS, J. (eds) *Tectonic events responsible for Britain's oil and gas reserves*. Geological Society, London, Special Publications, **55**, 87–105.

——, RAISWELL, R., AL-BIATTY, H., MCMAHON, A. & HARDMAN, M. 1990. Carboniferous stratigraphy, sedimentation and correlation of well 48/3–3 in the southern North Sea Basin: integrated use of palynology, natural gamma/sonic logs and carbon/sulphur geochemistry. *Journal of the Geological Society*, London, **147**, 287–300.

LEGARRETA, L., ULIANA, M. A., LAROTONDA, C. A. & MECONI, G. R. 1993. Approaches to nonmarine sequence stratigraphy – theoretical models and examples from Argentine Basins. *In*: ESCHARD, R. & DOLIGEZ, B. (eds) *Subsurface Reservoir Characterization from Outcrop Observations*. Éditions Technip, Paris, 125–143.

McKie, T. & Garden, I. R. 1996. Hierarchical stratigraphic cycles in the non-marine Clair Group (Devonian) UKCS. *In*: Howell, J. A. & Aitken, J. F. (eds) *High Resolution Sequence Stratigraphy: Innovations and Applications.* Geological Society, London, Special Publications, **104**, 139–157.

O'Byrne, C. J. & Flint, S. S. 1993. High-resolution sequence stratigraphy of Cretaceous shallow marine sandstones, Book Cliffs Outcrops, Utah, USA – application to reservoir modelling. *First Break*, **11**, 445–459.

Posamentier, H. W. & Vail, P. R. 1988. Eustatic controls on clastic deposition II – sequence and systems tract models. *In*: Wilgus, C. K., *et al.* (eds) *Sea-level changes: an integrated approach.* Society of Economic Paleontologists and Mineralogists Special Publications, **42**, 125–154.

Posamentier, H. W., Jervey, M. T. & Vail, P. R. 1988. Eustatic controls on clastic deposition I – conceptual framework. *In*: Wilgus, C. K. *et al.* (eds) *Sea-level changes: an integrated approach.* Society of Economic Paleontologists and Mineralogists Special Publications, **42**, 109–124.

Quirk, D. G. 1993. Interpreting the Carboniferous of the Dutch Cleaver Bank High. *In*: Parker, J. R. (ed.) *Petroleum Geology of Northwest Europe: Proceedings of the 4th Conference.* Geological Society, London, 697–706.

——1996. Base profile: a unifying concept in alluvial sequence stratigraphy. *In*: Howell, J. A. & Aitken, J. F. (eds) *High Resolution Sequence Stratigraphy: Innovations and Applications.* Geological Society, London, Special Publications, **104**, 37–50.

—— & Aitken, J. F. 1997. The structure of the Westphalian in the northern part of the Southern North Sea. *This volume.*

Ramsbottom, W. H. C., Calver, M. A., Eagar, R. M. C., Hodson, F., Holliday, D. W., Stubblefield, C. J. & Wilson, R. B. 1978. *A correlation of Silesian rocks in the British Isles.* Geological Society, London, Special Reports, **10**.

Rijkers, R. H. B. & Duin, Ed J. Th. 1994. Crustal observations beneath the southern North Sea and their tectonic and geological implications. *Tectonophysics*, **240**, 215–224.

Riley, N. J., Claoué-Long, J., Higgins, A. C., Owens, B., Spears, A., Taylor, L. & Varker, W. J. 1994. Geochronometry and geochemistry of the European mid-Carboniferous boundary global stratotype proposal, Stonehead Beck, North Yorkshire, UK. *Annales de la Société géologique de Belgique*, **116**, 275–289.

Ritchie, J. S. & Pratsides, P. 1993. The Caister Fields, Block 44/23a, UK North Sea. *In*: Parker, J. R. (ed.) *Petroleum Geology of Northwest Europe: Proceedings of the 4th Conference.* Geological Society, London, 759–769.

Rothwell, N. R. & Quinn, P. 1987. The Welton Oilfield. *In*: Brooks, J. & Glennie, K. (eds) *Petroleum Geology of North West Europe.* Graham & Trotman, London, 181–189.

Storey, M. W. & Nash, D. F. 1993. The Eakring-Dukeswood oil field: an unconventional technique to describe a field's geology. *In*: Parker, J. R. (ed.) *Petroleum Geology of Northwest Europe: Proceedings of the 4th Conference.* Geological Society, London, 1527–1537.

Tubb, S. R., Soulsby, A. & Lawrence, S. R. 1986. Palaeozoic prospects on the northern flanks of the London Brabant massif. *In*: Brooks, J., Goff, J. C. & van Hoorn, B. (eds) *Habitat of Palaeozoic gas in NW Europe.* Geological Society, London, Special Publications, **23**, 55–72.

Vail, P. R., Mitchum, R. M., Todd, R. G., Widmier, J. M., Thompson, S., Sangree, J. B., Bubb, J. N. & Hatlelid, W. G. 1977. Seismic stratigraphy and global changes of sea level. *In*: Payton, C. E. (ed.) *Seismic stratigraphy – applications to hydrocarbon exploration.* American Association of Petroleum Geologists Memoirs, **26**, 49–212.

Wehr, F. L. & Brasher, L. D. 1996. Impact of sequence-based correlation style on reservoir model behaviour, North Cormorant Field, UK North Sea. *In*: Howell, J. A. & Aitken, J. F. (eds) *High Resolution Sequence Stratigraphy: Innovations and Applications.* Geological Society, London, Special Publications, **104**, 115–128.

Whittaker, A., Holliday, D. W. & Penn, I. E. 1985. *Well logs in British stratigraphy.* Geological Society, London, Special Reports, **18**.

Ziegler, P. A. 1990. *Geological atlas of Western and Central Europe* (2nd edition). Shell Internationale Petroleum Maatschappij B.V., The Hague.

# Modelling of sandbody connectivity in the Schooner Field

F. C. J. MIJNSSEN

*Shell UK Exploration and Production, 1 Altens Farm Road, Nigg, Aberdeen AB9 2HY, UK*
*Present address: Shell International Exploration and Production, Postbus 60,*
*2280 AB Rijswijk, The Netherlands*

**Abstract:** The Schooner field will be one of the first stand-alone developments of a gas accumulation in the Carboniferous Barren Red Measures Group of the UK Sector of the Southern North Sea. It illustrates the potential of Barren Red Measures Group as a future play in this area.

The Schooner Field, located some 150 km (94 miles) east of the South Yorkshire coastline, is a NW–SE-oriented faulted dip closure. The reservoir consists predominantly of fluvial channel sandstones of the Barren Red Measures Group, with small amounts of gas located in the underlying Coal Measures Group. The sandstone channels are embedded in flood plain mudstones, which comprise 65–70% of the stratigraphic section. Because of the relatively low proportion of sandstone to mudstone, sandbody connectivity has been identified as the critical factor affecting recovery of gas.

In order to asses the impact of sandbody connectivity on recovery efficiency, reservoir modelling work was performed based on eight different geological models and three well design options. The factors that were varied in order to build the geological models included (1) presence of minor faults, (2) the width of the channels, and (3) channel orientation. Well design options considered included (1) wells with deviations between 10 and 63°, (2) sub-horizontal wells drilled along the strike of the structure, and 3) sub horizontal wells drilled along the dip of the structure.

The results of the modelling work showed that sandbody orientation had the largest influence on connectivity, with connectivity being the highest when the channels were oriented along the strike of the structure. The width of the channels and the presence of minor faults had some impact on connectivity, while varying well design made little or no difference. Close inspection of the models did, however, show that the major faults compartmentalize the reservoir, resulting in large quantities of gas being by-passed if wells are not properly positioned. Consequently, more work is needed to evaluate the optimal location of the development wells.

The Schooner Field lies in Blocks 44/26a and 43/30 of the UK Sector of the southern North Sea (Fig. 1). This is on the southern margin of the Carboniferous producing trend. The field was discovered with Well 44/26–2, which was drilled in 1987 to test a large faulted dip closure at the Top Carboniferous Level (Fig. 2). The well found a total of 99 m (325 ft) of gas pay in the Upper Carboniferous Barren Red Measurers Group. The areal extent of the accumulation was appraised by drilling two additional wells, 44/26–3 and 44/26–4, and acquiring a 3D seismic survey. The field will be one of the first stand-alone developments in the Barren Red Measures Group of the southern North Sea, hence illustrating the potential of this group as a future play in this area.

The general stratigraphic sequence in the Schooner Field is shown in Fig. 3. The reservoir consists of fluvial sandstones in the Barren Red Measures Group and Coal Measures Group. The Barren Red Measures Group contains 99 % of the gas, with the remaining 1% being located in the Coal Measures Group. Desert-lake mudstones of the Rotliegend Group form the top seal for the reservoir.

Reservoir quality in the sandstones of the Barren Red Measures Group is generally good to excellent, with and average porosity of 12% and permeabilities that range from 10 to 1000 mD. The sandstones are, however, embedded in impermeable flood plain mudstones that comprise 65–70% of the stratigraphic section. Consequently, significant concern exists about sandbody connectivity and the impact it will have on the recovery of gas.

This paper presents the results of a study undertaken to asses the range of uncertainty in reserves resulting from different sand body connectivity models. In the study, eight three-dimensional reservoir models were constructed, and three different well design options were considered. The results of the study have been utilized to optimize the design of development wells in the Schooner Field, and may also be applicable to other Carboniferous Fields in the southern North Sea.

*From* Ziegler, K., Turner, P. & Daines, S. R. (eds), 1997, *Petroleum Geology of the Southern North Sea: Future Potential*, Geological Society Special Publication No. 123, pp. 169–180.

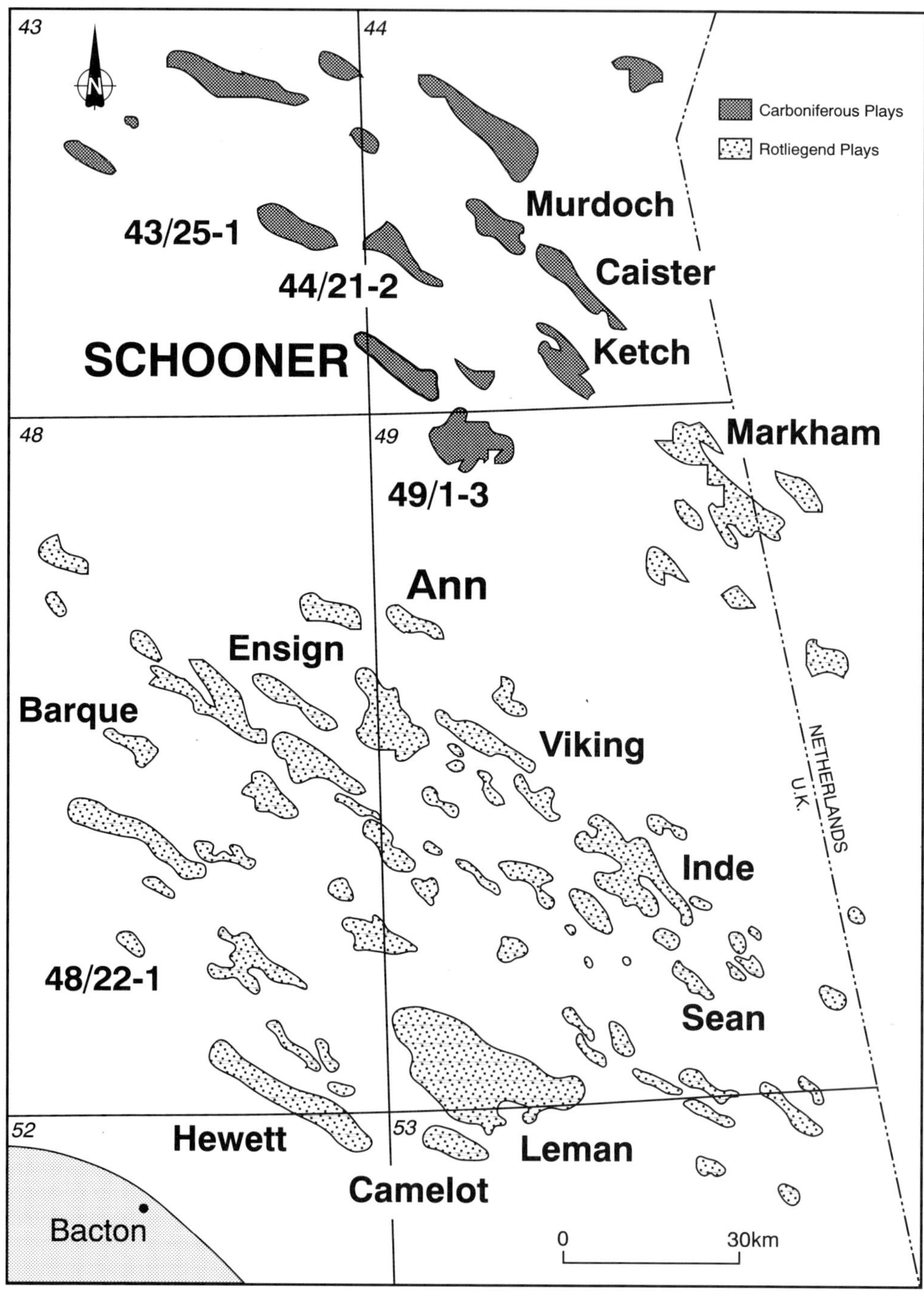



**Fig. 1**. Location of the Schooner Field.

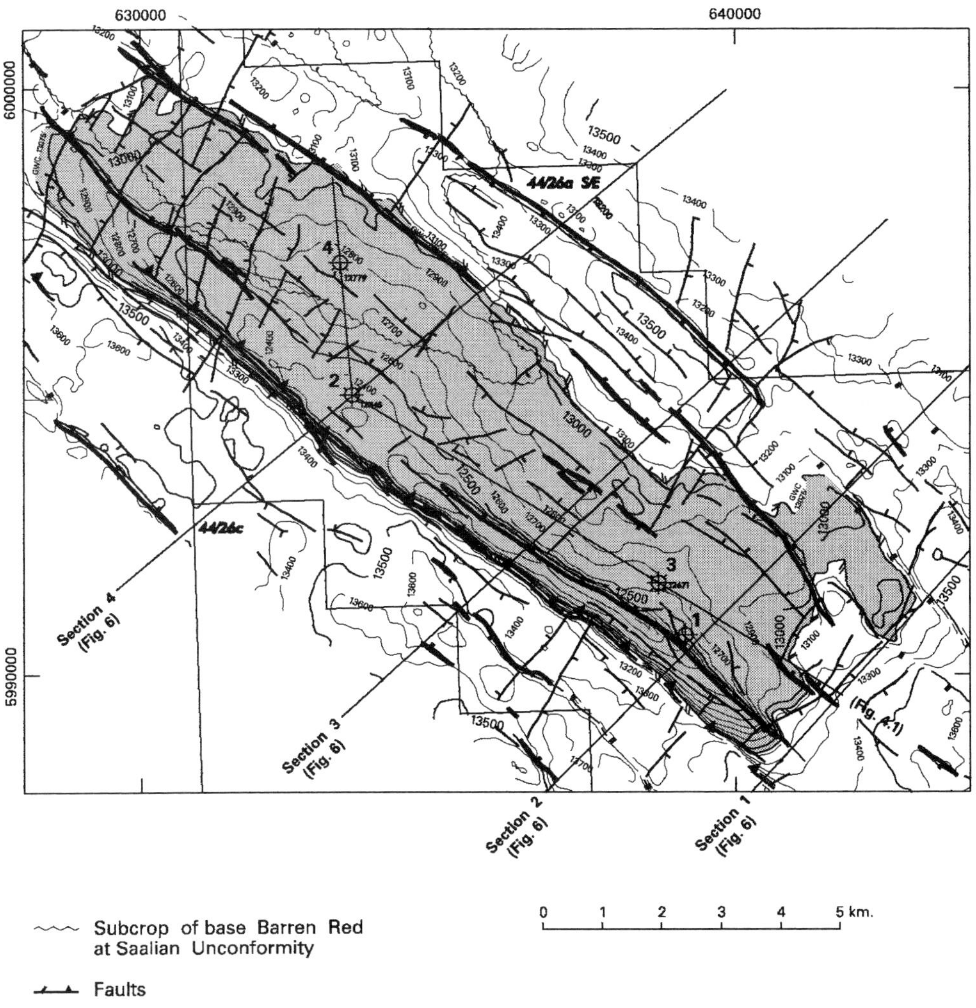



~~~ Subcrop of base Barren Red
at Saalian Unconformity

⊥⊥⊥ Faults

⊕ Well Location

**Fig. 2.** Structure map of the Top Carboniferous in the Schooner Area showing location of the discovery and appraisal wells.

## Reservoir sedimentology

### Genetic units

*Barren Red Measures Group.* Based on evaluation of core and log data, the following genetic units are differentiated in the Barren Red Measures Group.

*Composite low-sinuosity channel.* Poorly stratified conglomerates, festoon cross-bedded sandstones and ripple laminated sandstones characterise this genetic unit. Often numerous reactivation surfaces are present. Usually there is no obvious grain size trend. A blocky gamma ray (GR) response and a clear density/neutron (FDC/CNL) log separation characterize this genetic

unit. Eberth & Miall (1991) interpreted comparable deposits from the Cutler Formation in New Mexico, USA, as braided river channels.

*Simple low-sinuosity channel.* Festoon cross-bedded sandstones and ripple laminated sandstones represent this genetic unit. Usually this genetic unit shows a fining-upwards sequence resulting in a bell shaped GR response. This genetic unit is interpreted as the remains of small channels that cross cut the flood plain sediments between the main braided channels (Fig. 4). This interpretation is based on sandbody thickness (up to 4 m), the presence of erosional lower surfaces, and the lack of reactivation surfaces.
~~~

                    F. C. J. MIJNSSEN

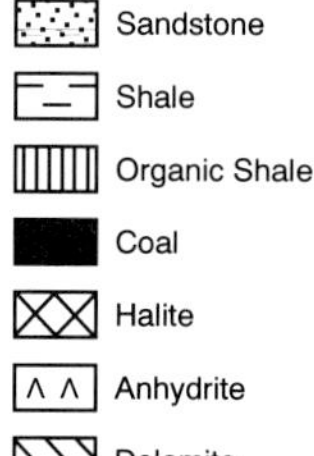

| SYSTEM | GROUP | FORMATION | LITHOLOGY |
|---|---|---|---|
| Permian | Zechstein | Zechstein IV | |
| | | Zechstein III | |
| | | Zechstein II | |
| | | Zechstein I | |
| | | Kupferschiefer | |
| | Rotliegend | Silverpit | |
| Carboniferous | Barren Red Measures | | |
| | Coal Measures | | |
| | Millstone Grit | | |
| | Yoredale | | |



**Fig. 3.** General chrono- and lithostratigraphy of the Schooner Area.

*Proximal overbank deposit.* Ripple laminated sandstones and bioturbated and root mottled sandstone characterise this genetic unit. The individual sand layers are usually thin (15–30 cm) and often homogeneous. These sands are thought to be formed during flooding events as a result of channels over spilling their banks. A spiky GR response and a vague FDC/CNL separation characterize these units.

*Flood plain.* Bioturbated sandstones, bioturbated mudstones and horizontally laminated mudstones, as well as a high GR response, characterise this genetic unit. Kerr & Jirik (1990) interpret comparable deposits in the Frio Formation of Texas, USA, as flood-plain sediments.

*Coal Measures Group.* Based on core and log evaluation the following genetic units are differentiated in the Coal Measures Group.

*Composite low-sinuosity channel.* Festoon cross-bedded sandstones and ripple-laminated sandstones characterize this genetic unit, which has been encountered only in the 44/26-4 Well. Due to the numerous reactivation surfaces and the resemblance of this unit to the composite channels in the Barren Red Measures Group, this channel is interpreted as a braided river channel. Since it occurs at the top of the Coal Measures Group, it probably represents the change from the Coal Measures depositional system to the Barren Red Measures

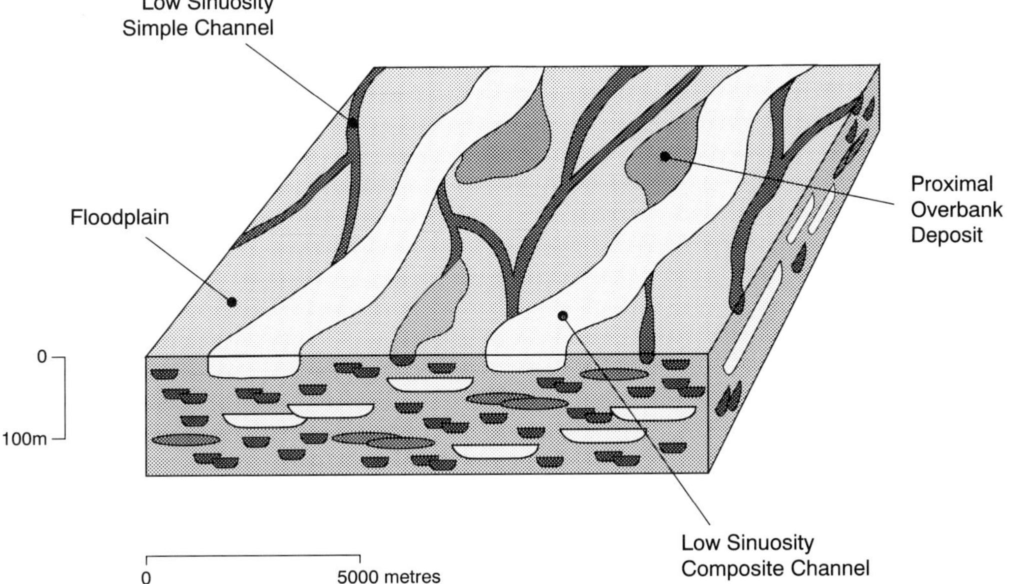


**Fig. 4.** Depositional model postulated for the Barren Red Measures Group in the Schooner Field. Because channel dimension is uncertain, the horizontal scale should read 0 to 500 m for the narrow channel scenarios.

depositional system. A low GR response and clear FDC/CNL separation characterize this genetic unit.

*Simple meandering channel.* Festoon cross-bedded sandstones and ripple-laminated sandstones characterize this genetic unit. A clear fining upward sequence is often visible. This genetic unit has a smooth bell shaped GR response and a clear FDC/CNL separation. Puigdefabregas & van Vliet (1978) interpret comparable deposits in the Montlobatt Formation of the Southern Pyrenees, Spain, as meandering river channels.

*Crevasse splay.* Ripple laminated sandstones and bioturbated sandstones characterize this genetic unit. No obvious grain-size trend occurs. These sediments often show alternating sand and shale drapes towards the top indicative of waning flow. Considering the coexistence of this genetic unit with the simple meandering channels described above, it is interpreted to be a crevasse splay deposit. In well logs this genetic unit has an intermediate GR response and FDC/CNL separation.

*Flood plain.* Bioturbated sand, bioturbated mudstone and horizontally laminated mudstone characterise this genetic unit. Due to the bioturbated and muddy nature of this genetic unit, these sediments are interpreted as representing flood plain deposits. This genetic unit usually exhibits a high GR response.

## Depositional environment

Based on Eberth & Miall (1991), who describe a facies association for the Cutler Formation of New Mexico, USA, which is comparable with the Barren Red Measures Group, the overall depositional setting is interpreted to be a braided to anastomosing river system in a proximal alluvial plain (Fig. 4). Within this system major composite low-sinuosity channels developed. The simple channels formed small subsidiaries flowing between the large composite channels. Proximal overbank deposits formed adjacent to the main channel areas during flooding events.

Besly *et al.* (1993) propose an alternative depositional model for the Barren Red Measures Group in which fluctuations in base level result in an alternation of braided river systems formed during low stands and meandering-river systems formed during high stands. In this model, the low-stand braided-river systems incise into the sediments formed during the high base-level periods. From a reservoir geological point of view, the depositional model of Besly *et al.* (1993) would result in a similar reservoir architecture to that advocated by the model proposed in this paper.

To explain the juxtaposition of the genetic units in the Coal Measures Group, the overall depositional environment is interpreted to

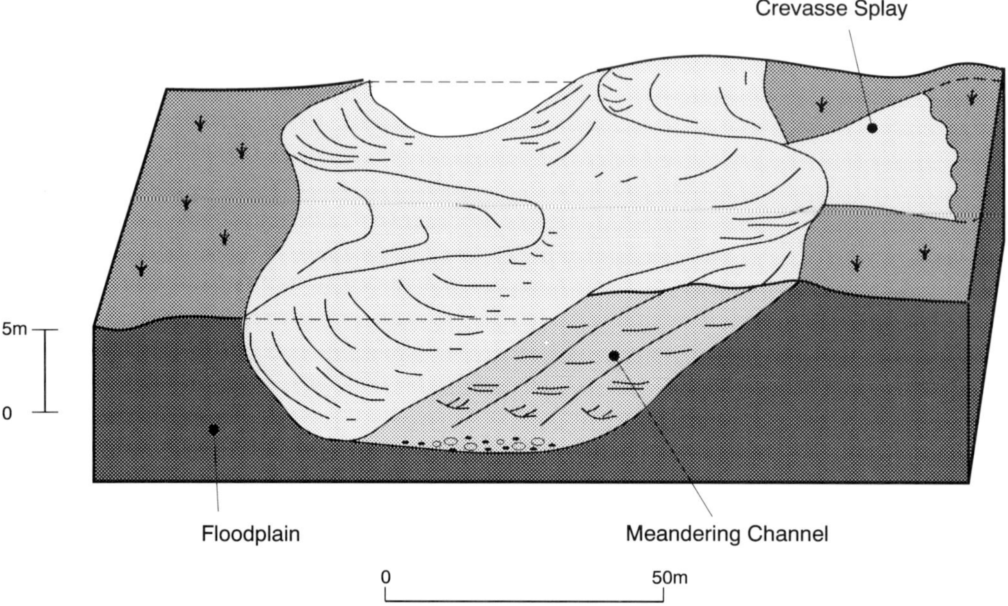



**Fig. 5.** Depositional model postulated for the Coal Measures Group in the Schooner Field.

represent a swampy alluvial plain cross cut by meandering rivers (Fig. 5). The crevasse splays were probably formed adjacent to the meandering channels during flooding periods. At the top of the sequence, the channels had a more braided character probably indicating a shift to a more proximal fluvial style. This could have been caused by a tectonic pulse in the hinterland or a change in climate.

## Reservoir structure

The Schooner Field is an elongate NW–SE-trending inversion-related horst block bounded by high-angle transpressional oblique-slip faults. The structure is 16 km (10 miles) long by 4 km (2.5 miles) wide, with the crest of the structure slightly offset to the southwest. Within the structural closure, the Carboniferous strata have been deformed into a broad southeast-plunging anticlinal swell. The main reservoir unit, the Barren Red Measures Group, forms a southeasterly thickening wedge within the closure due to the progressive truncation of this unit beneath the Saalian Unconformity (Fig. 6). The gas column is some 300 m (1000 ft), with the free water level at 3985 m ss (13075 ft ss).

The Barren Red Measures Group is seismically transparent. Consequently minor faults are not visible in this zone. However, numerous minor faults are visible in the underlying Coal

Measures Group. These faults abut against the base of the Barren Red Measures Group. If these faults extend upwards into the Barren Red Measures Group, they could cause compartmentalisation of the reservoir in this zone.

## Three-dimensional reservoir modelling

### Uncertainties

*General.* Sandbody connectivity has been identified as the most important attribute to model in order to optimize development of the Schooner Field. The main uncertainties associated with sand body connectivity are the presence or absence of minor faults, channel width, and channel orientation (Fig. 7). In order to quantify the impact of these uncertainties on the recovery of gas, several 3D reservoir models were constructed using Shell's propriety 3D modelling package MONARCH. This package allows 3D modelling of reservoir architecture based on detailed well log correlation. It can be used for volumetrics calculation, as well as well planning. It can also generate input for reservoir simulation (Budding *et al.* 1992; 1995).

*Minor faults.* As already mentioned, the Barren Red Measures Group is seismically transparent. Consequently it is uncertain if the numerous minor faults that occur in the underlying Coal

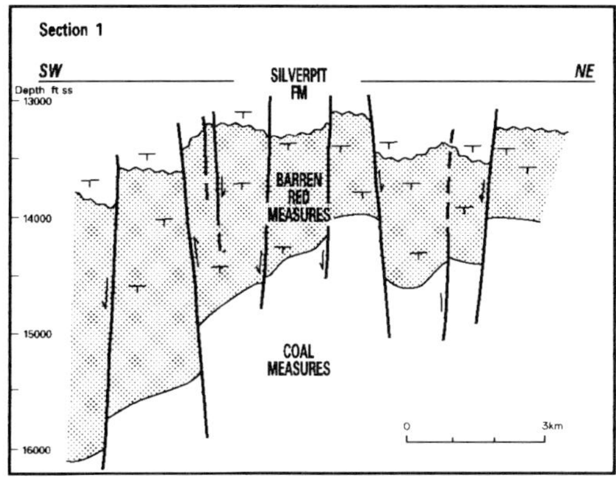


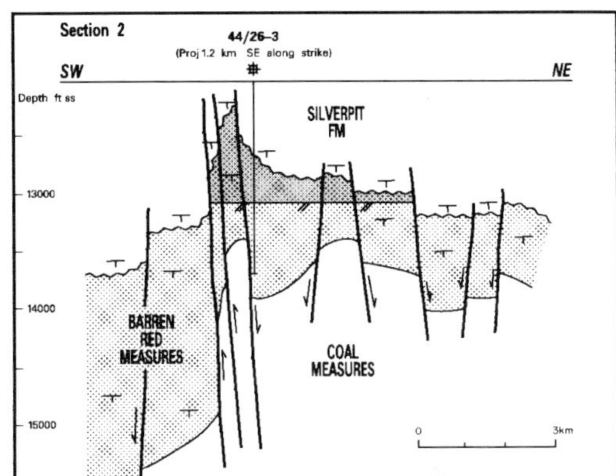


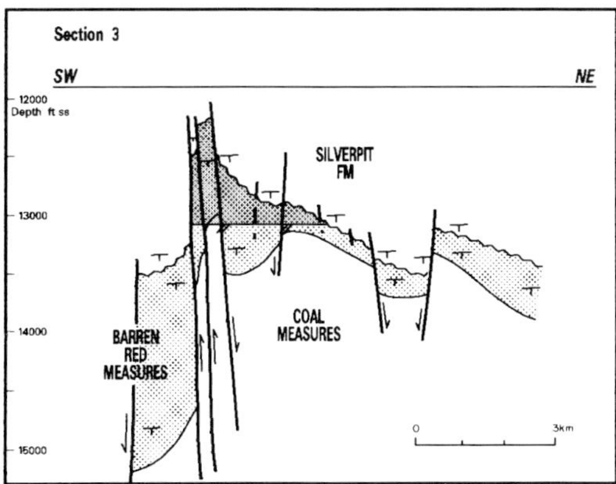


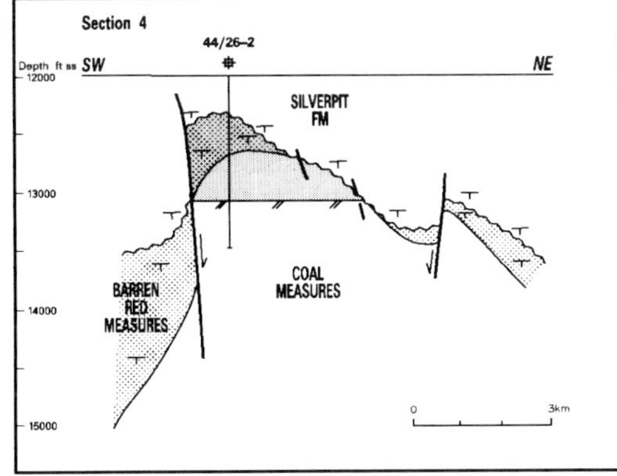


**Fig. 6.** Dip sections through the Schooner Field. Fig. 2 shows the location of the sections.

Measures Group extend upwards into the Barren Red Measures Group. Because of this uncertainty, two fault scenarios were considered during the 3D modelling: (1) all faults seen in the Coal Measures Group extend upwards into the Barren Red Measures Group, and (2) only the major faults extend upwards (Fig. 7). In both scenarios faults were considered to be sealing. A third scenario, in which all faults extend upwards into the Barren Red Measures Group but with non-sealing minor faults, was not modelled since it would give essentially the same results as scenario 2.

*Channel orientation.* Regional well log correlations and isopach mapping suggest that channels in the Barren Red Measures Group and Coal Measures Group are predominantly oriented NE–SW. However, there are no indications that local trends could not be different. For example channel orientations could have been controlled by pre-existing fault trends, in which

case they might be oriented SE–NW. There is no dipmeter or other dip information that uniquely defines a particular channel orientation in the Schooner Field Area. Therefore two trends, one oriented NE–SW and the other SE–NW, were considered in the reservoir modelling to test the effect of channel orientation on reservoir connectivity (Fig. 7). For the modelling work, it was assumed that there is no difference in channel orientation between the Barren Red and Coal Measures Group.

*Channel width.* Barren Red Measures Group. Data obtained from outcrop studies reported in the open literature (see Bibliography 1) and closely spaced bore holes drilled by the British Coal Board were used to model channel geometries in the Barren Red Measures Group. These data show a bimodal distribution for the composite and simple channels. Hence, it was decided to model two scenarios with respect to channel widths:

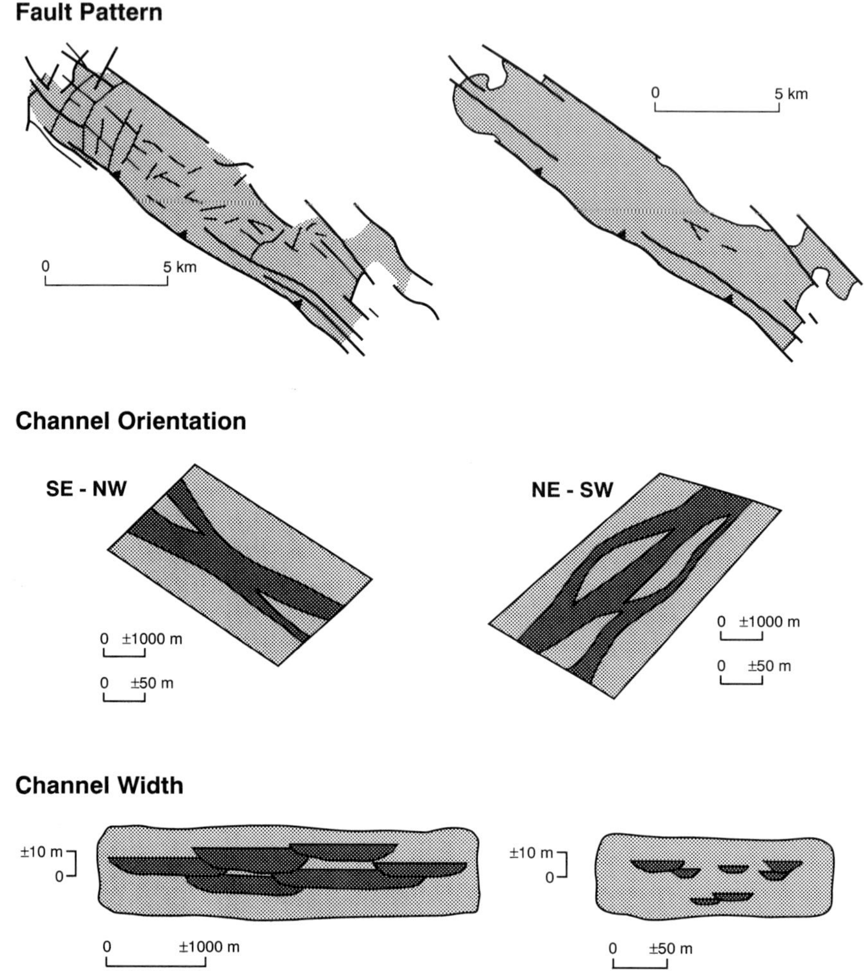


**Fig. 7.** Main uncertainties with respect to the reservoir geology of the Schooner Field.

(1) narrow composite low sinuosity channels ($c.\,70$ m) and narrow simple low sinuosity channels ($c.\,20$ m),

(2) wide composite low sinuosity channels ($c.\,1000$ m) and wide simple low sinuosity channels ($c.\,300$ m).

The narrow channel option seems quite extreme, especially for the composite channels. However, well test information can be interpreted to indicate that the these channels have such dimensions.

If the depositional model proposed by Besly *et al.* (1993) is the more appropriate model for the Barren Red Measures Group, the width of the simple channels will have been underestimated in the reservoir model given that width/thickness data for simple meandering channels obtained from the open literature (see Bibliography 2) generally show somewhat larger widths. This could result in underestimation of connectivity. However for the thickness range found in the Schooner Field, analogue data for both the simple low sinuosity and simple meandering channels show similar width ranges.

*Coal Measures Group.* Width/thickness data for simple meandering channels obtained from outcrop studies reported in the open literature (see Bibliography 2) were used during the 3D modelling of the Coal Measures Group. The composite low sinuosity channel at the top of the Coal Measures Group in well 44/26-4 was modelled using the parameters used for the Barren Red Measures Group.

*Reservoir connectivity*

Based on the aforementioned uncertainties, eight geological scenarios were modelled (Table 1). Because the intention was to assess connectivity in terms of gas initially in place (GIIP) connected to well bores, the geological models were transformed into 'volume' models. In these models each model element was characterized by a gas volume. This gas volume was determined by the porosity assigned to the model element, a saturation height function and a gas expansion factor.

For each geological scenario three well options were considered:

(1)  deviated wells (10–63°);
(2)  horizontal wells parallel to the strike of the structure;
(3)  horizontal wells perpendicular to the strike of the structure.

The horizontal well options were considered in order to test if horizontal wells could improve connectivity in comparison to the much easier to drill deviated wells. The horizontal wells were 'drilled' either perpendicular or parallel to the strike of the structure to test if the drilling direction would influence connectivity in relation to channel orientation.

Figure 8 shows the results of the connectivity analysis. This figure shows the following.

- Channel orientation has the largest impact on connected GIIP, especially in the wide channel scenarios.
- For all model scenarios there are no major differences in connected GIIP between the well design options considered. This probably reflects the limited vertical connectivity in the Schooner Field when compared to horizontal connectivity.
- Channel dimensions do not influence connected GIIP significantly, except when the channels are oriented SE–NW and minor faults are present. In this scenario, connected GIIP is reduced considerably.

Based on these observations it can be concluded that in order to estimate connected GIIP more accurately, the most important uncertainty to reduce is channel orientation. This can be done by obtaining channel orientation data using dipmeter or borehole imaging tools or oriented core. If these data show that the channels are oriented SE–NW, then the next major uncertainties to reduce are channel width and the occurrence of minor faults. Well tests and interwell seismic techniques might achieve this.

Since the modelled well design options did not have an impact on connected GIIP, there would appear to be little value in acquiring additional data to constrain sandbody geometry in order to optimize well trajectories. However, close inspection of the connectivity models shows that for all tested development options, the major faults isolate large blocks of the reservoir. If no wells are drilled in these blocks, they will not be drained effectively, resulting in a suboptimal field development. Consequently more work is needed to evaluate the optimal locations of the development wells.

## Conclusions and recommendations

(1) The Schooner Field will be one of the first stand-alone developments of a gas accumulation in the Barren Red Measures Group of the Southern North Sea and hence illustrates the potential of this group as a future play in this area.

(2) Based on core and wireline log information, the Barren Red Measures Group within the Schooner Field is interpreted to represent a braided to anastomosing river system formed on a proximal alluvial plain (Fig. 4). The sediments of the Coal Measures Group are interpreted to be a meandering river system flowing through a swampy alluvial plain (Fig. 5).

(3) Because net-to-gross is low in the Schooner Reservoir, sand body connectivity will be the key geological factor influencing reservoir behaviour.

**Table 1.** *Scenarios modelled with* MONARCH

Wide channels, oriented NE–SW with all faults present
Wide channels, oriented NE–SW with only major faults present
Wide channels, oriented SE–NW with all faults present
Wide channels, oriented SE–NW with only major faults present
Small channels, oriented NE–SW with all faults present
Small channels, oriented NE–SW with only major faults present
Small channels, oriented SE–NW with all faults present
Small channels, oriented SE–NW with only major faults present

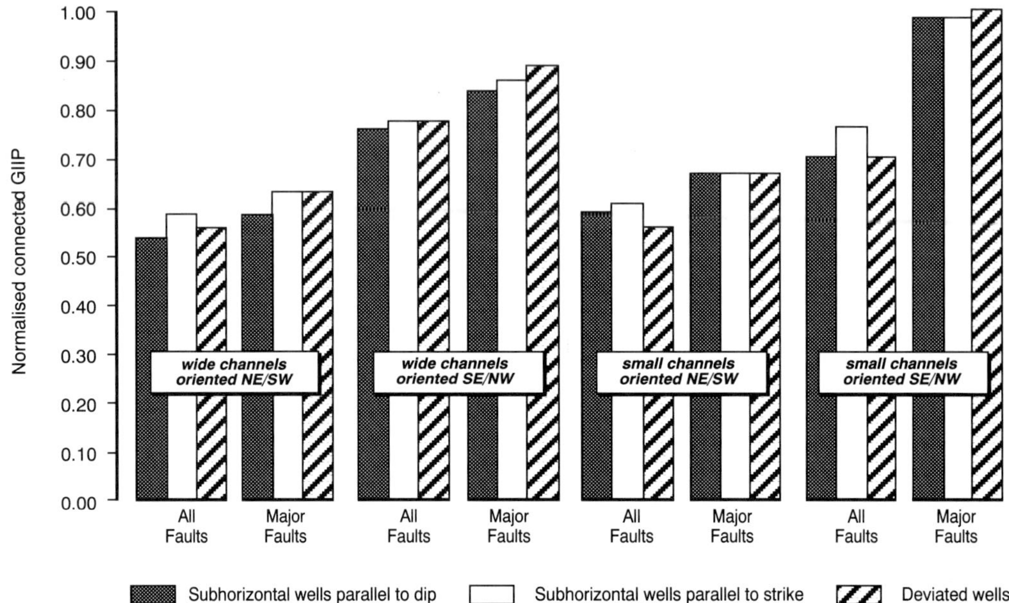



**Fig. 8.** Connected GIIP Volumes for the geological scenarios studied and well design options considered. Volumes are normalised to the scenario with the largest connected GIIP Volume, i.e. small channels, oriented SE/NW and all faults present using deviated wells.

The main uncertainties that will effect connectivity are (i) the presence of minor faults in the Barren Red Measures Group, (ii) channel orientation, and (iii) channel width (Fig. 7). Eight geological scenarios (Table 1) were modelled to evaluate the effect of these uncertainties on sand body connectivity.

(4) Channel orientation has the largest impact on connected GIIP, especially in the wide channel scenarios.

(5) For all scenarios, there are no major differences in connected GIIP between the three well design options.

(6) Channel dimensions do not influence connected GIIP significantly, except when the channels are oriented SE–NW and minor faults are present. In this scenario, connected GIIP is reduced considerably.

(7) Close inspection of the connectivity models shows that for all tested development options, the major faults isolate large blocks of the reservoir. If no wells are drilled in these blocks, they will not be drained effectively.

(8) In order to more accurately determine relative connected GIIP, the most important uncertainty to eliminate is channel orientation.

(9) Since the well design options do not have an impact on connected GIIP, there is no need to acquire additional data to optimize well trajectories.

(10) To be able to drain isolated fault blocks effectively, more work is needed to evaluate the optimal locations of the development wells.

The work described in this paper is based on a number of studies performed by employees of Shell U.K. Exploration and Production, ESSO UK Exploration and Production and various contractors. I want to thank my colleagues in the Southern Field Unit and the Reservoir Geology Section of Shell U.K. Exploration and Production for their support. I also want to thank my colleagues at Shell Research, especially R. Bennett, for the help and support during the 3D modelling. I would like to thank K. Ziegler, M. Sweet and especially G. Leveille for reviewing this paper. I want to thank the management of Shell U.K. Exploration and Production, ESSO UK Exploration and Production, Sovereign Exploration and Amoco UK Exploration for permission to publish this paper.

## References


BESLY, B. M., BURLEY, S. D. & TURNER, P. 1993. The Late Carboniferous 'Barren Red Bed' play of the Silver Pit area, Southern North Sea. *In*: PARKER, J. R. (ed.) *Petroleum geology of Northwest Europe: Proceedings of the 4th Conference*. Geological Society, London, 727–740.

BUDDING, M. C., DAVIES, A. H., DAVIES, P., FOWLES, J. D. & VAN DER VORST, J. M. C., 1995. 3-D reservoir modelling with GEOLOGIX and MONARCH (abs.). *Technical programme and abstracts of papers and posters, Annual Convention of the American Association of Petroleum Geologists*, Houston, 5–8 March, 13.

——, PAARDEKAM, A. H. M. & VAN ROSSEM, S. J. 1992. 3D connectivity and architecture in sandstone reservoirs. *Society of Petroleum Engineers Paper 22342, Proceedings of the Society of Petroleum Engineers International Meeting*, Beijing, China, 24–27 March, 131–139.

EBERTH, D. A. & MIALL, A. D. 1991. Stratigraphy, sedimentology and evolution of a vertebrate-bearing braided to anastomosed fluvial system, Cutler Formation (Permian-Pennsylvanian), north central New Mexico. *Sedimentary Geology*, 72, 225–252.

KERR, D. R. & JIRIK, L. A. 1990. Fluvial architecture and reservoir compartmentalisation in the Oligocene Middle Frio Formation, South Texas. *Transactions of the Gulf Coast Geological Societies*, XL, 373–380.

PUIGDEFABREGAS, C. & VAN VLIET, A. 1978. Meandering stream deposits of the Southern Pyrenees. *In*: MIALL, A. D. (ed.) *Fluvial sedimentology*, Canadian Society of Petroleum Geologists Memoirs, 10, 469–485.


## Bibliography 1: Publications from which width/thickness data were obtained for composite channels


ALLEN, P. M. & MATTER, A. 1982. Oligocene meandering stream sedimentation in the eastern Ebro Basin, Spain. *Ecologae Geologie Helvetica*, 75, 33–49.

ATKINSON, C. D. 1983. *Comparative sequence of ancient fluvial deposition in the Tertiary south Pyrenees, Northern Spain*. PhD thesis, University of Wales.

BHRENSMEYER, A. K. & TAUXE, L. 1982. Isochronous fluvial systems in Miocene Deposits of Northern Pakistan. *Sedimentology*, 29, 331–352.

BONYTHON, C. W. 1963. Further light on river floods reaching Lake Eyre. *Proceedings of the Royal Geographical Society of Australia, South Australia Branch*, 64, 9–22.

DELUCA, J. L. & ERIKSSON, K. A. 1989. Controls on synchronous ephemeral and perennial river sedimentation in the middle sandstone member of the Triassic Chinle Formation, north-eastern New Mexico, U.S.A. *Sedimentary Geology*, 61, 155–175.

EBERTH, D. A. & MIALL, A. D. 1991. Stratigraphy, sedimentology and evolution of a vertebrate-bearing braided to anastomosed fluvial system, Cutler Formation (Permian–Pennsylvanian), north central New Mexico. *Sedimentary Geology*, 72, 225–252.

EVANS, J. E. 1991. Facies relationships, alluvial architecture and paleohydrology of a Palaeocene, humid-tropical alluvial-fan system: Chumstick Formation, Washington State, U.S.A. *Journal of Sedimentary Petrology*, 61, 732–755.

FRIEND, P. F., HIRST, J. P. P. & NICHOLS, G. J. 1986. Sandstone body structure and river process in the Ebro Basin of Aragon, Spain. *Cuadernos Geologia Iberica*, 10, 9–30.

——, SLATER, M. J. & WILLIAMS, R. C. 1979. Vertical and lateral building of river sandstone bodies, Ebro Basin, Spain. *Journal of the Geological Society, London*, 136, 39–36.

GIBLING, M. R. & RUST, B. R. 1990. Ribbon sandstones in the Pennsylvanian Waddens Cove Formation, Sydney Basin, Atlantic Canada: the influence of siliceous duricrusts on channel body geometry. *Sedimentology*, 37, 45–65.

HILL, G., 1989. Distal alluvial fan sediments from the Upper Jurassic of Portugal: controls on their cyclicity and channel formation. *Journal of the Geological Society, London*, 146, 539–555.

HUBERT, J. F. & HYDE, M. G. 1982. Sheet-flow deposits of graded beds and mudstones on an alluvial sandflat-playa system: Upper Triassic Blomidon redbeds, St Mary's Bay Nova Scotia. *Sedimentology*, 29, 457–474.

JOHNSON, E. A. & PIERCE, F. W. 1990. Variations in fluvial deposition on an alluvial plain: an example from the Tongue River Member of the Fort Union Formation (Palaeocene), southern Powder River Basin, Wyoming U.S.A. *Sedimentary Geology*, 69, 21–36.

LAWRENCE, D. A. & WILLIAMS, B. P. J. 1987. Evolution of drainage systems in response to Arcadian deformation: the Devonian Battery Point Formation, eastern Canada. *In*: ETHRIDGE, F. G., FLORES, R. M. & HARVEY, M. D. (eds) *Recent developments in fluvial sedimentology*, SEPM Special Publications, 39, 287–300.

MAIZELS, J. K. 1990. Raised channel systems as indicators of palaeohydrologic change: a case study from Oman. *Palaeogeography, Palaeoclimatology, Palaeoecology*, 76, 241–277.

MOODY-STUART, M. 1966. High and low sinuosity stream deposits with examples from the Devonian of Spitsbergen. *Journal of Sedimentary Petrology*, 36, 1102–1117.

ORI, G. G. 1982. Braided to meandering channel patterns in humid region alluvial fan deposits, River Reno, Po Plain (northern Italy). *Sedimentary Geology*, 32, 231–248.

QUI, Y., XUE, P. & XIAO, J. 1987. Fluvial sandstone bodies as hydrocarbon reservoirs in lake basins. *In*: ETHRIDGE, F. G., FLORES, R. M. & HARVEY, M. D. (eds) *Recent developments in fluvial sedimentology*, SEPM Special Publications, 39, 329–342.

SOPENA, A., RAMOS, A. & PEREZ-ARLUCEA, M. 1989. *Permian and Triassic fluvial systems in central Spain*. Excursion guidebook, 4th international conference on fluvial sedimentology.

STEAR, W. M. 1983. Morphological characteristics of ephemeral stream channel and overbank splay sandstone bodies in the Permian Lower Beaufort Group, Karoo Basin, South Africa. *In*: COLLINSON, J. D. & LEWIN, J. (eds) *Modern and ancient fluvial systems*. I.A.S. Special Publications, 6, 405–420.

STEER, B. L. & ABBOTT, P. L. 1984. Palaeohydrology of the Eocene Ballena Gravels, San Diego County, California. *Sedimentary Geology*, **38**, 181–216.

TUNBRIDGE, I. P. 1981. Old Red Sandstone sedimentation – an example from the Brownstones (highest Lower Old Red Sandstone) of south central Wales. *Geological Journal*, **16**, 111–124.

——1984. Facies models for a sandy ephemeral stream and playa complex; the Middle Devonian Trentishoe Formation of North Devon. *Sedimentology*, **31**, 697–716.

WEISS, M. A. 1988. *Sedimentology and depositional environments of the Permian Lower Cutler Group: Turks Head, Canyonlands National Park, Utah.* MSc thesis, University of Utah.

## Bibliography 2: Publications from which width/thickness data were obtained for meandering channels

ALLEN, P. M. & MATTER, A. 1982. Oligocene meander-ing stream sedimentation in the eastern Ebro Basin, Spain. *Ecologae Geologie Helvetica*, **75**, 33–49.

ATKINSON, C. D. 1983. *Comparative sequence of ancient fluvial deposition in the Tertiary south Pyrenees, Northern Spain.* PhD thesis, University of Wales.

BESLY, B. M. & COLLINSON, J. D. 1991. Volcanic and tectonic controls of lacustrine and alluvial sedimentation in the Malpas-Sort Basin, Catalonian Pyrenees. *Sedimentology*, **38**, 3–26.

EBANKS, W. J. & WEBER, J. F. 1982. Development of a shallow heavy oil deposit in Missouri. *Oil and Gas Journal*, September, 221–234.

EVANS, J. E. 1991. Facies relationships, alluvial architecture and paleohydrology of a Palaeocene, humid-tropical alluvial-fan system: Chumstick Formation, Washington State, U.S.A. *Journal of Sedimentary Petrology*, **61**, 732–755.

KERR, D. R. & JIRIK, L. A. 1990. Fluvial architecture and reservoir compartmentalisation in the Oligocene Middle Frio Formation, South Texas. *Transactions of the Gulf Coast Geological Societies*, **XL**, 373–380.

MARZO, M., NIJMAN, W. & PUIGDEFABREGAS, C. 1988. Architecture of the Castissent fluvial sheet sandstone, Eocene, South Pyrenees, Spain. *Sedimentology*, **35**, 719–738.

NAMI, M. & LEEDER, M. 1978. Changing channel morphology and magnitude in the Scalby Formation (Middle Jurassic) of Yorkshire, England. *In*: MIALL, A. D. (ed.) *Fluvial sedimentology*. Canadian Society of Petroleum Geologists Memoirs, **10**, 431–440.

OLSEN, H. 1989. Sandstone-body structures and ephemeral stream processes in the Dinosaur Canyon Member, Utah, U.S.A. *Sedimentary Geology*, **32**, 231–248.

ORI, G. G. 1982. Braided to meandering channel patterns in humid region alluvial fan deposits, River Reno, Po Plain (northern Italy). *Sedimentary Geology*, **32**, 231–248.

PADGET, G. V. & EHRLICH, R. 1976. Paleohydrologic analysis of a Late Carboniferous Fluvial System, southern Morocco. *Geological Society of America Bulletin*, **87**, 1101–1104.

PUIGDEFABREGAS, C. 1973. Miocene Pointbar deposits in the Ebro Basin, Northern Spain. *Sedimentology*, **20**, 133–144.

—— & VAN VLIET, A. 1978. Meandering stream deposits of the Southern Pyrenees. *In*: MIALL, A. D. (ed.) *Fluvial sedimentology*, Canadian Society of Petroleum Geologists Memoirs, **10**, 469–485.

QUI, Y., XUE, P. & XIAO, J. 1987. Fluvial sandstone bodies as hydrocarbon reservoirs in lake basins. *In*: ETHRIDGE, F. G., FLORES, R. M. & HARVEY, M. D. (eds) *Recent developments in fluvial sedimentology*. SEPM Special Publications, **39**, 329–342.

RICCI-LUCHI, F., COLELLA, A., ORI, G. G. & OGLIANI, F. 1982. *Pliocene Fan Deltas of the Intra-Appenninic Basin, Bologna.* Excursion Guidebook, 2nd I.A.S. European Regional Meeting, 162pp.

STEAR, W. M. 1980. Channel sandstone and bar morphology of the Beaufort Group Uranium District near Beaufort West. *Transactions of the Geological Society of South Africa*, **83**, 391–398.

——1983. Morphological characteristics of ephemeral stream channel and overbank splay sandstone bodies in the Permian Lower Beaufort Group, Karoo Basin, South Africa. *In*: COLLINSON, J. D. & LEWIN, J. (eds) *Modern and ancient fluvial systems*. I.A.S. Special Publications, **6**, 405–420.

# Application of high-frequency cycle analysis in high-resolution sequence stratigraphy


C. S. YANG[1] & Y. A. BAUMFALK[2]

IGC (International Geo Consultants BV), Kapelstraat 5B, 2161 HD Lisse, The Netherlands
[1] Present address: Torenwacht 5, 2353DB Leiderdorp, The Netherlands
[2] Present address: Javastraat 178HS, 10965CN Amsterdam, The Netherlands



**Abstract:** High-resolution sequence stratigraphy analysis is based on the integration of third-order sequence analysis and high-frequency (Milankovitch) cyclicity analysis. Sliding window spectral analysis of wireline logs reveals wavelength changes of the high-frequency cycles along the bore hole. The abrupt wavelength change, shown as the break of the sliding window spectrum bands, indicates a stratigraphic discontinuity. The gradual wavelength change, detected by the gradual shift of the spectrum bands, reflects a gradual change in the rate of net sediment accumulation. These cyclicity patterns reflect the thickness variations of preserved depositional cycles, controlled by basin subsidence, eustatic sea-level/base-level fluctuations, and sediment supply. Observations from the subsurface of the southern North Sea indicate that the breaks of spectrum bands as well as the trend of spectrum band shift can be correlated between wells to help detailed sequence correlation. Cyclicity analysis of the synthetic logs generated from a two-dimensional sequence stratigraphic modelling confirms the lateral correlatability of the cyclicity pattern between wells. Integration of cyclicity analysis with traditional methods will help accurate and reliable sequence stratigraphic correlations in various settings, including biostratigraphically barren strata, thick intervals of fine-grained or coarse-grained sediments, and chalk sections. Such correlations are essential for the establishment of a proper geological model in the hydrocarbon exploration, development and production.


## High-frequency depositional cycles

Sequence-stratigraphic concepts deal with the stratigraphic response to the interaction between sediment flux on the one hand and the space that is made available for those sediments to fill on the other. The accommodation space is a function of eustatic change and total sea-floor subsidence (due to tectonics, loading, sediment compaction etc.) (Posamentier & Vail 1988; Posamentier *et al.* 1988). Due to the cyclic changes of the eustatic sea level and climate, the cyclic nature of stratigraphic successions is a basic attribute of the sequence stratigraphic approach (Posamentier & James 1993). These sequence stratigraphic principles apply at all scales, both temporal and spatial (Posamentier & James 1993).

The hierarchy of cycles is summarized in Fig. 1. The first-, second- and third-order cycles have the periods of 10–70 Ma, 3–10 Ma, and 0.5–3 Ma respectively (Vail *et al.* 1977*a*, *b*; Haq *et al.* 1987). They are tectonic cycles with an episodic nature, but may also reflect long-term climate changes related to the long-period Milankovitch cycles (Fig. 1; Schwarzacher 1993). They have a high preservation potential. The fourth-, fifth- and sixth-order cycles are Milankovitch cycles with the periods of 20–400 ka (Berger *et al.* 1989). They have a true periodic nature and a high-medium preservation potential.

In this paper the fourth-, fifth- and sixth-order cycles are referred to the high-frequency cycles. They are the building blocks of the third-order sequences, and can be used as quantitative parameters to evaluate the third-order sequences. The recognition and determination of these cycles within the third-order sequences are the basis of the high-resolution sequence stratigraphy analysis.

Figure 2 illustrates an example of the hierarchy of cycles in the Upper Rotliegend Group in well G17–1 of the Netherlands offshore. The Upper Rotliegend Group in the Southern Permian Basin is a Lower Permian lithostratigraphic unit. It consists of a barren siliciclastic succession, with evaporites in the basin centre (Van Wijhe *et al.* 1980; Glennie 1990; Yang & Nio 1994). The Upper Rotliegend Group in well G17-1 is characterized by cyclic alternations of halite and lacustrine claystone. Three orders of cycles can be differentiated (Yang & Nio 1994; Yang & Kouwe 1995). There are five second-order cycles (super-sequence RO1 to RO5), which have a thickness of 100–200 m. They are interpreted to reflect long-term (1.3–2.3 Ma) climatic changes from arid to relatively humid conditions. Cycles RO2, RO3 and RO4 can be subdivided into third-order cycles (sequence RO2.1 to 2.2, RO3.1 to 3.4, RO4.1 to 4.4), which have a thickness of 25–75 m and represent medium-term


*From* Ziegler, K., Turner, P. & Daines, S. R. (eds), 1997, *Petroleum Geology of the Southern North Sea: Future Potential*, Geological Society Special Publication No. 123, pp. 181–203.

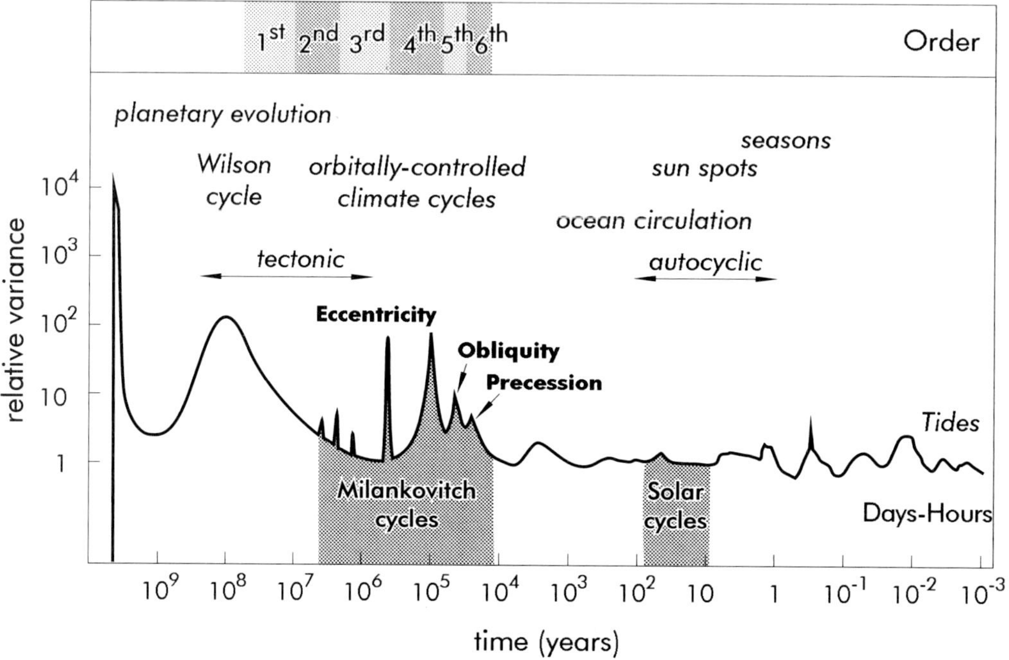


**Fig. 1.** Geological cycles and their order. The schematic spectrogram illustrates the cyclic phenomena occurring at geologically relevant time scale (modified after Berger 1980 and Schwarzacher 1991). The cycle orders are after Vail *et al.* (1977*a, b*) and Haq *et al.* (1987).

(0.5–0.8 Ma) climatic changes. The high-order cycles have a thickness of less than 10 m. These cycles reflect climatic changes interpreted as being related to the 100 ka Milankovitch cycles (Yang & Nio 1994; Yang & Kouwe 1995).

Milankovitch cycles reflect variations in the orbital parameters of the Earth. As a consequence, the amount of solar radiation reaching a certain point on Earth surface varies significantly. The main astronomical parameters include the eccentricity of the Earth's orbit (100 ka and 400 ka periods), the obliquity of the Earth's rotation axis (54 ka and 41 ka periods), and the precession of the equinoxes (23 ka and 19 ka periods) (Berger *et al.* 1989). In addition, there are several long eccentricity cycles with the periods of 1290, 2030 and 3400 ka (Schwarzacher 1993).

The variations of the Earth's orbital parameters result in changes in insolation (solar radiation) at the surface of the Earth, both in absolute terms and relatively over the seasons. These changes have an important effect on widespread climatic changes and sea-level/base-level fluctuations, which control sequence development and sedimentary facies variations recorded in the wire-line log data (Hays *et al.* 1976; Imbrie & Imbrie 1979; Fischer & Bottjer 1991; de Boer & Smith 1994). How

these climatic changes are reflected in the sedimentary record will depend on the latitude and the importance of seasonality in the sediment supply, the sea level/base level and the general depositional setting, and the depositional processes in the receiving basin.

In the marine pelagic realm the changes in insolation may influence the (biological) primary production (and therefore sedimentation) directly or indirectly (e.g., mediated by changes in fertility of oceanic water, Thunell *et al.* 1991; Fischer *et al.* 1991). In continental sedimentary settings, prevailing wind direction, weathering and erosion in the hinterland, and the type of fluvial system may change with climate. In either case, the deposits may show rhythms with the same periods, although the intensity of the contribution of each orbital parameter may vary from place to place.

Studies in the Quaternary show that these astronomical parameters have a major effect on global climate and are the main control on the fluctuation of continental ice-caps and the changes in depositional systems (Sarnthein 1978; Talbot 1985; Kocurek *et al.* 1991; Fischer & Bottjer 1991). There is convincing evidence that orbital forcing cycles of climate, causing truly cyclic patterns in sedimentation, existed

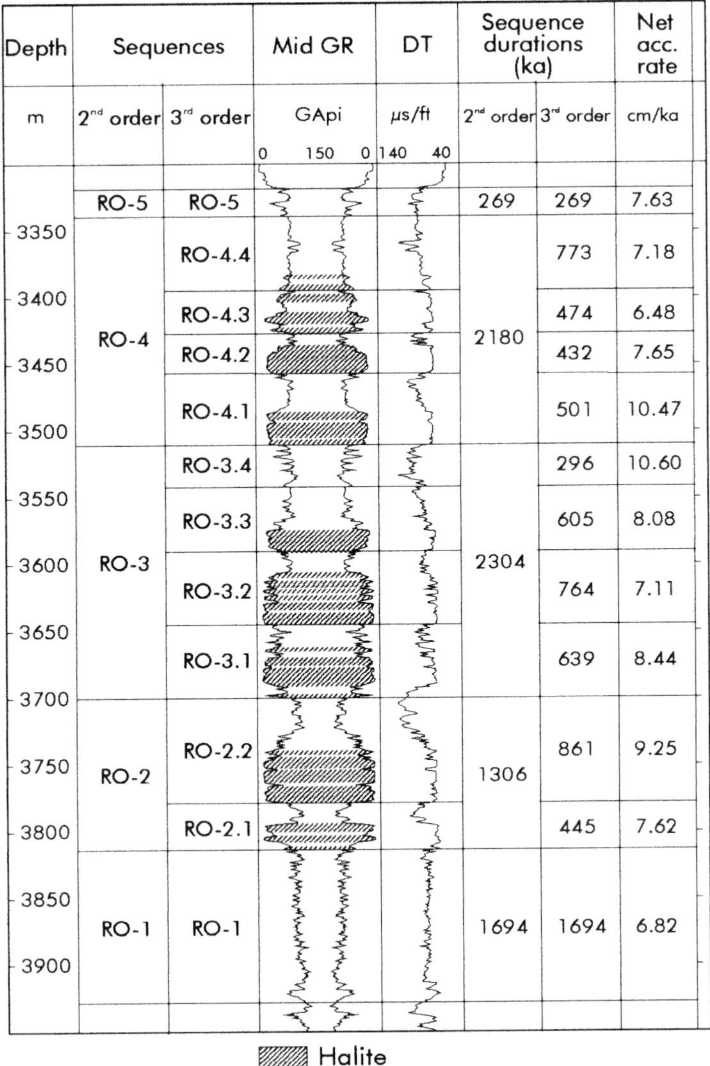



**Fig. 2.** An example of the hierarchy of cycles in the Upper Rotliegend in well G17-1 of the Netherlands offshore (modified after Yang & Kouwe 1995; see Fig. 3b for location of the well). The alternations of halite and lacustrine claystone show three orders of cycles: the second-order cycles (super-sequence RO1 to RO5) with a thickness of 100–200 m, the third-order cycles (sequence RO2.1 to 2.2, RO3.1 to 3.4, RO4.1 to 4.4) with a thickness of 25–75 m, and the high-order cycles with a thickness of less than 10 m. These cycles reflect long-term, medium-term and short-term climatic changes respectively. The third-order sequence duration and net sediment accumulation rate are estimated from Interval Spectral Analysis of wireline log. Mid GR refers to Mirror image display of gamma ray log.

also in the pre-Quaternary. Truly cyclic patterns which are related to precisely measurable orbital events have been recognized in pre-Quaternary deposits, especially from wire-line data in subsurface studies (Smith 1989; Melnyk & Smith 1989; Worthington 1990; Nio *et al.* 1993; Yang & Baumfalk 1994).

The recent periods of orbital parameters mentioned above are valid for the last few hundred thousand years. There are strong indications that in the more remote past these values were different, although of the same order of magnitude. For example, the gravitational effects of the other bodies in the solar system may have caused the variation of the Milankovitch periods over the Earth's history. Calculations for several periods in geologic time have been given by Berger *et al.* (1989) and Berger & Loutre (1994). For example, the periods of the Milankovitch cycles in the Early Permian were estimated as 100 ka (eccentricity), 44.3 ka and 35.1 ka (obliquity) and 21 ka and 17.6 ka (precession) (Berger *et al.* 1989).

Various methods have been developed to analyse the Milankovitch cycles. Recent research follows mainly two lines: (1) wireline-log cyclicity analysis to detect and evaluate Milankovitch cycles recorded in sedimentary sections; (2) high-resolution sequence-stratigraphic modelling to produce synthetic sedimentary successions using sequence stratigraphic principles and Milankovitch parameters. These two lines of research are closely related to each other, forming an integrated approach for high-resolution sequence-stratigraphy study. Wireline-log cyclicity analysis reveals Milankovitch parameters, which can be used to calibrate the high-resolution sequence stratigraphic modelling, whereas the modelling provides important knowledge which help us to understand and interpret the cyclicity patterns in wireline logs. In this paper, results from the cyclicity analysis of the synthetic wells generated in the one-dimensional and two-dimensional sequence stratigraphic modelling will be used to help the interpretation of the cyclicity patterns in the subsurface of the southern North Sea.

## Wireline-log cyclicity analysis

Spectral analysis of wireline logs is an efficient way to detect cycles because wireline logs meet the essential requirements for time-series analysis, such as effective sampling, data continuity and resolution. Among various wireline logs, gamma ray logs are often used in the cyclicity analysis because they record the lithological features of depositional conditions (Yang & Kouwe 1995).

The wireline-log cyclicity analysis uses different methods, such as the fast Hartley (Fourier) transform and the maximum entropy techniques. The advantages and disadvantages of these two techniques have been discussed by ten Kate & Sprenger (1992). Weedon (1991) and Schwarzacher (1991) have also given good summaries of spectral analysis principles.

The Fourier model consists of the sum of harmonically related sinusoids of fixed frequency, called Fourier harmonics, i.e. a series of data can be considered as being composed of a set of sine waves, each with its own wavelength and amplitude. The amplitude and phase of each harmonic are estimated from the data. If the original signal contains cyclic patterns, the power spectrum derived from the Fourier transform will show pronounced peaks corresponding with the frequencies of these cyclicities. In the fast Fourier transform (FFT) algorithm, the data are directly transformed into a power spectrum. This method has the advantage of accurate estimation of spectral amplitudes (ten Kate & Sprenger 1992).

The maximum entropy spectral analysis is based on the autocorrelations calculated from the data. Maximum entropy has the advantages of high-resolution in the low-frequency range. It produces continuous spectrum. In addition this method can be used to analyse short sequences. Presence of a few cycles in the 'original' data is sufficient to get a reliable estimation of the cycle length and power (ten Kate & Sprenger 1992).

In the spectral analysis of a stratigraphic interval we are interested mainly in the ratios between the dominant frequencies in the data. These ratios are then compared to the ratios of the Milankovitch cyclicities. Our basic assumption in a spectral analysis of wireline-log data is that the cyclicities present in the log data represent Milankovitch cycles, if their ratios are close to the ratios of the predicted and calculated Milankovitch cyclicities. The periods of the Milankovitch cycles and their changes through geological time have been calculated by Berger *et al.* (1989). Therefore the duration of Milankovitch cycles are known. This allows us to estimate the time duration of the analysed segment with a resolution of some tens of thousands of years. Once the duration of the segment is estimated, the average accumulation rate can be calculated as well.

Figure 3 shows a spectrogram of part of the Westphalian D in well 44/21-3 from the Silver Pit area of the Southern North Sea Basin. The Silesian in the basin was deposited in the foreland basin of the Variscan Orogen (Ziegler 1989). Coarse-grained fluvial sedimentation was widespread in the Late Westphalian C. During the Early Westphalian D however, the sand-dominated deposition was replaced by fine-grained flood-plain sedimentation with some sandstones interpreted as fluvial-channel or sheet-flood sandstones. A semi-arid climate prevailed, causing a widespread primary reddening of the sediments (Besly *et al.* 1993). The analysed interval consists mainly of fine-grained sediments. The spectrogram shows several peaks, called A, A′, B, C and E. They have a wave length of 21.4 m 14.8 m 9.6 m 7.8 m and

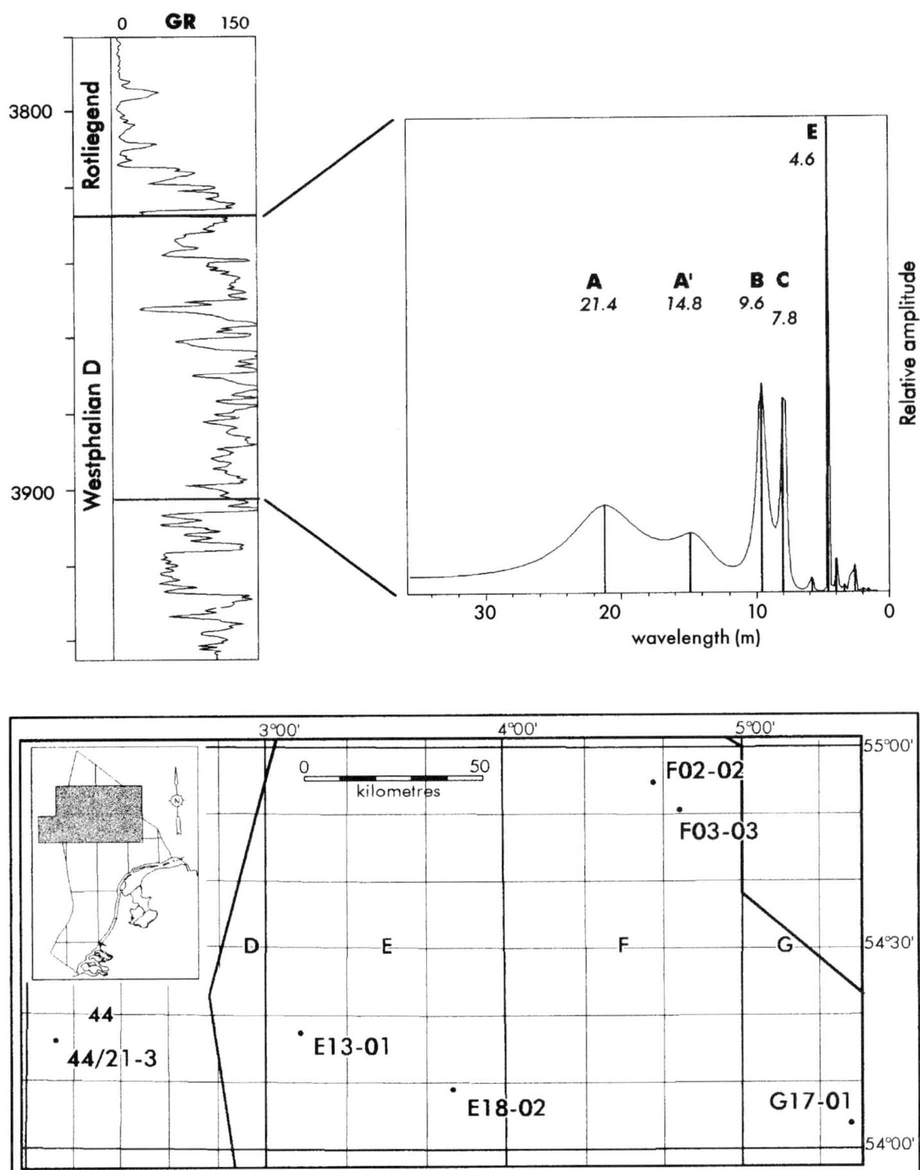



**Fig. 3.** (a) Spectrogram from the interval spectral analysis of the Westphalian D in well 44/21-3, UK sector of the Southern North sea. Peaks A, B, C and E are recognized as the 100 ka, 43 ka, 34 ka and 20 ka Milankovitch cycles. The numbers under the horizontal axis of the power spectrum represent the wavelength. See text for detailed explanation. (b) Location of well 44/21-3 and other studied wells mentioned in this paper.

4.6 m. The ratios between the A, A', B, C and E peaks are $21.4 : 14.8 : 9.6 : 7.8 : 4.6 = 1 : 0.69 : 0.45 : 0.36 : 0.21$. The ratios of A, B, C and E are very close to the ratios between the eccentricity (100 ka), two obliquities (43 and 34 ka) and the precession (20 ka) during the Carboniferous time (Berger *et al.* 1989). There-fore these peaks are interpreted as Milankovitch cycles. In addition to these principal Milanko-vitch cycles, secondary cycles can also be recognized (A' at 69 ka). The eccentricity cycle (100 ka) in the analysed interval has a wave-length of 21.4 m. Therefore the average net accumulation rate for this interval is 21 cm/ka.

Because the thickness is known, it is possible to calculate the duration time of this interval. For the analysed interval this results in a duration time of 336 ka.

Usually the ratios of the peaks in the wire line log spectrum show several possible matches with the ratios of the Milankovitch cycles. In order to reduce the uncertainty, the different solutions in the recognition of the Milankovitch cycles are checked using known biostratigraphic data. Such independent checks are important also due to the fact that erosion/non-deposition may occur during part of the lower order cycle. As a result the ratio between the different order cycles in the log data may differ from the theoretical ratio. This will be discussed later in the cyclicity analysis of the synthetic logs generated from a two-dimensional sequence stratigraphic modelling.

## Sliding-window spectrum patterns of the synthetic GR logs from one-dimensional modelling

Sliding-window spectral analysis of wireline logs is used to detect abrupt or gradual changes in the frequency and wavelength of the high-frequency cycles along the borehole. The former represent the discontinuities related to the stratigraphic boundaries of lower-order cycles. The latter reflect a gradual change in the rate of net sediment accumulation or in the rate of sediment-accommodation development.

In order to have a better understanding of the sliding-window spectral analysis patterns, a series of synthetic logs are generated under the condition of known frequency parameters. Sliding-window spectral analyses are then applied to these synthetic logs to investigate the sliding-window spectrum band patterns corresponding to the input parameters. The modelling is one-dimensional because it generates the sedimentary section at one location, without considering the lateral variations.

Figure 4a shows the simplest situation of a synthetic log produced using one constant frequency at harmonic number 23. The log shows a very regular sinusoid pattern (Fig. 4a). The along-borehole variation in the frequencies or wavelengths of this synthetic log is investigated using the sliding window spectral analysis. In this analysis, a window of fixed size (200 m in this case) is moved down the synthetic log by steps of a fixed length (e.g. 6 m). In each step a power spectrum is calculated for the window using spectral analysis. The high-frequency part (50%) of the spectrogram is plotted alongside the synthetic log and positioned at the middle depth point of the window. Only peaks with their amplitudes above a certain threshold are shown. Spectrograms of successive windows may partly overlap. Since the synthetic log in Fig. 4a is generated by a constant frequency, the sliding-window spectral analysis reveals a continuous spectrum band at the same frequency as the input signal (Fig. 4a).

An additional frequency is added in the synthetic logs of Fig. 4b. The input signals include two constant frequencies at harmonic numbers 23 and 42 respectively. The synthetic log is composed of these two sinusoids and displays an interference pattern of the two (Fig. 4b). The sliding-window spectral analysis of this synthetic log shows two continuous spectrum bands at the same frequencies as the original signals (Fig. 4b).

An abrupt frequency (and wavelength) change is introduced into the synthetic log in Fig. 4c. The input frequencies are displayed in the power spectrum of the entire synthetic log. The lower half of the synthetic log is generated by two constant frequencies at peaks 2 and 4. In the overlying synthetic log the two input frequencies are lower at peaks 1 and 3 (Fig. 4c). The frequency change occurred at 1012 m. Consequently the lower half of the synthetic log shows the interference of the two sinusoids with relatively small wavelengths, whereas the upper half shows an interference of the two sinusoids with relatively large wavelengths. A stratigraphic discontinuity is present at 1012 m (Fig. 4c). The sliding-window spectral analysis reveals the details of the frequency variation along the log: two continuous frequency bands at relatively high frequencies in the lower half, and two continuous frequency bands at relatively low frequencies in the upper half. A break and interference of the spectrum bands occurs around 1012 m (Fig. 4c).

The synthetic log in Fig. 5 also displays an abrupt frequency change, similar to that in Fig. 4c. This synthetic log, however, is generated using three Milankovitch parameters: the eccentricity (100 ka), the obliquity (41 ka) and the precession (23 ka) cycles. Although the Milankovitch cycle periods (in terms of time) remain constant over the entire synthetic log, they will be recorded as different wavelengths (in terms of thickness) if the net sediment accumulation rate changes. In Fig. 5 the net sediment accumulation rate is $20\,cm\,ka^{-1}$ below 95 m decreases to $5\,cm\,ka^{-1}$ above 95 m. The abrupt change of the net sediment accumulation rate at 95 m causes a sharp shift in the wavelengths of the input Milankovitch cycles. The input

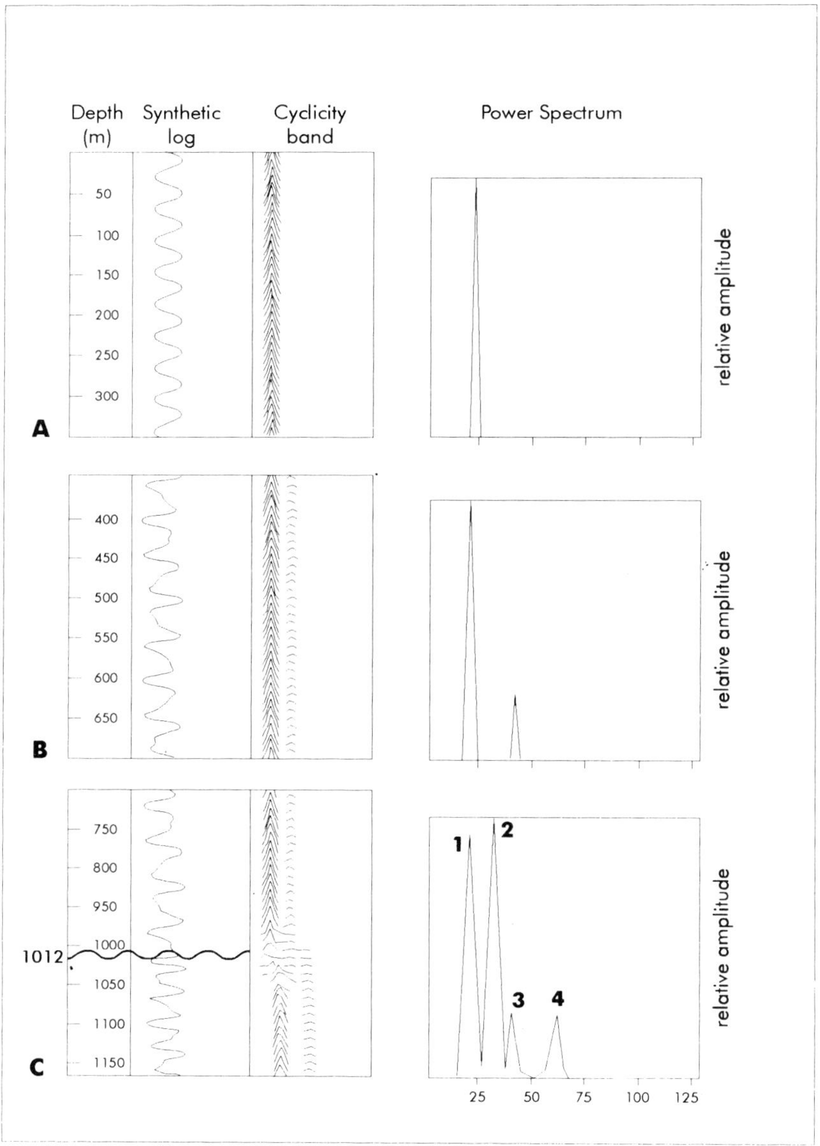



**Fig. 4.** Sliding window spectral analysis of the synthetic logs (modified after Nio *et al.* 1993). (**a**) The synthetic log generated using a constant frequency. (**b**) The synthetic log generated using two constant frequencies. (**c**) The synthetic log generated using two frequencies with an abrupt frequency (and wavelength) change at 1012 m. The numbers under the horizontal axis of the power spectrum represent the harmonic number, or the number of cycles per basic interval. The vertical axis of the power spectrum indicates the relative amplitude.

Milankovitch cycles in the lower part of the synthetic log have wavelengths 18.3 m (eccentricity), 7.5 m (obliquity) and 4.2 m (precession). In the overlying synthetic log the input eccentricity and obliquity cycles have wavelengths 4.65 m and 1.8 m respectively (Fig. 5). Consequently the lower part of the synthetic log shows the interference of the three Milankovitch cycles with relatively large wavelengths, whereas the upper part shows an interference of the Milankovitch cycles with relatively small wavelengths. A stratigraphic discontinuity is present at 95 m (Fig. 5). The sliding-window spectral analysis reveals the details of the frequency variation along the log: two continuous frequency bands at wavelengths of 7.5 m

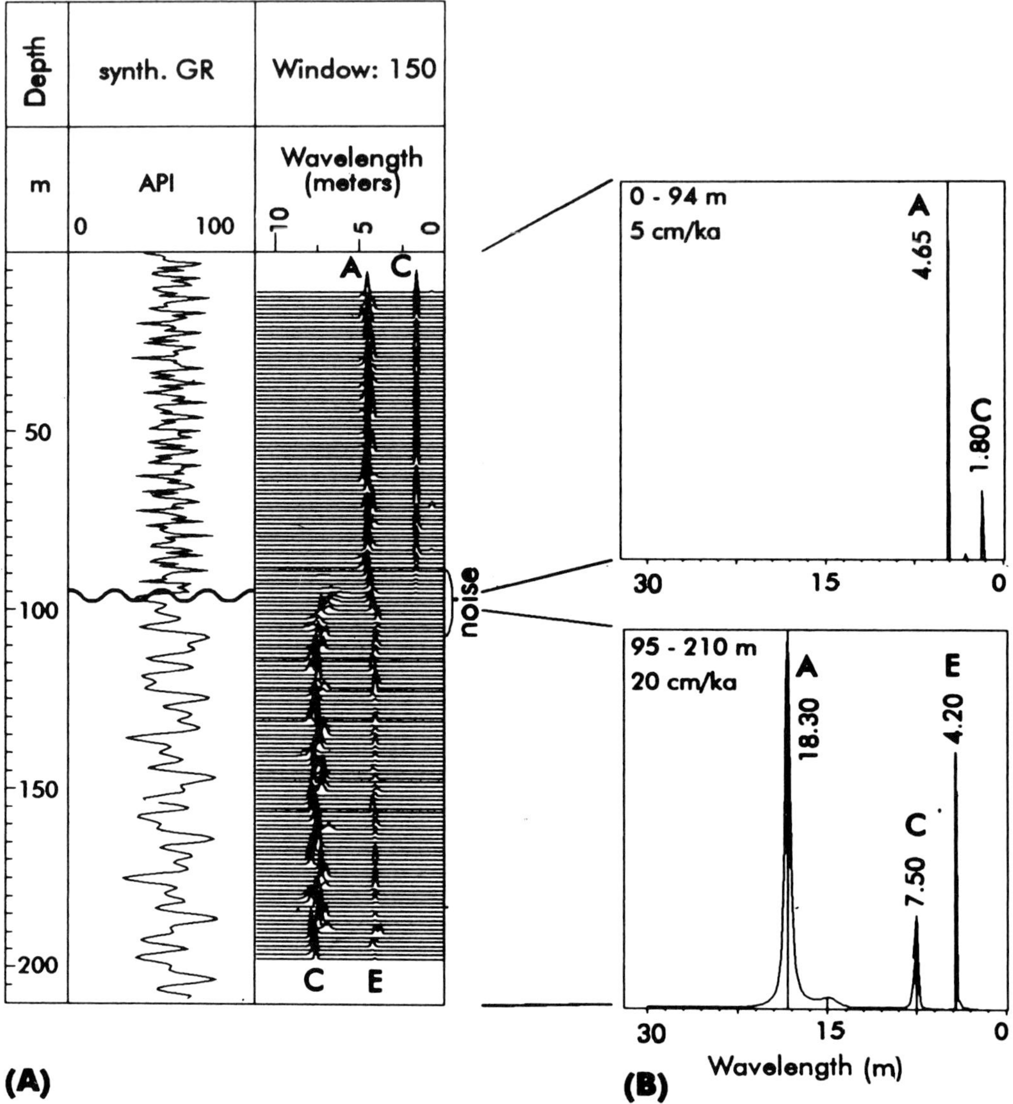



**Fig. 5.** Sliding-window spectral analysis of a synthetic log generated using three Milankovitch cycles (modified after Nio *et al.* 1993). The abrupt change of the net sediment accumulation rate at 95 m causes a sharp shift in the wavelengths of the Milankovitch cycles. A, C and E in the power spectrum indicate the eccentricity, obliquity and precession cycles. The window contains 150 data points.

(obliquity) and 4.2 m (precession) in the low part, and two continuous-frequency bands at wavelengths of 4.65 m (eccentricity) and 1.8 m (obliquity) in the upper part. A break and interference of the spectrum bands occurs around 95 m (Fig. 5). The eccentricity cycle (A) is not shown in the lower part because its wavelength (18.3 m) is too large to be covered by the window. The precession cycle in the upper part is not displayed in the power spectrum because it is thin (1 m) and does not contain enough data points to be defined. The sliding-window spectral analysis in this example shows a stacking of the spectrum peaks (Fig. 5).

The synthetic log in Fig. 6 displays abrupt as well as gradual frequency (and wavelength) changes. This synthetic log is generated by three constant Milankovitch periods (100 ka, 41 ka and 23 ka) and a changing net sediment accumulation rate. The net sediment accumulation

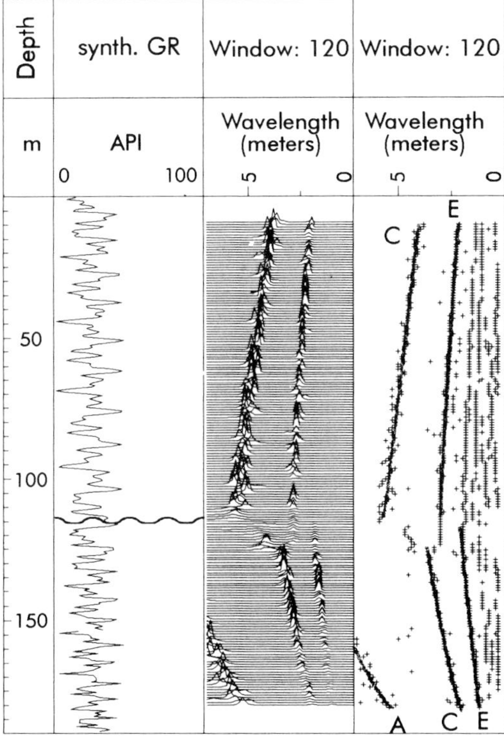

Fig. 6. Sliding-window spectral analysis of a synthetic log generated using three Milankovitch cycles with abrupt as well as gradual wavelength changes. This synthetic log is generated by three constant Milankovitch periods (100 ka, 41 ka and 23 ka) and a changing net sediment accumulation rate. The window contains 120 data points.

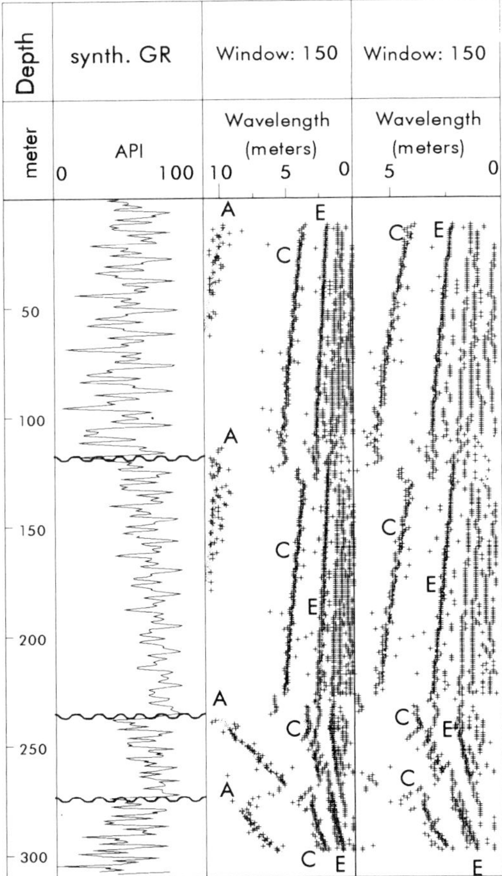

Fig. 7. Sliding-window spectral analysis of a synthetic log showing wavelength changes and lithological variations (modified after Yang & Kouwe 1995). The lithological changes have no direct influence on the spectrum band patterns. The window contains 150 data points. The middle column gives an overview of the spectrum peaks, including eccentricity (A), obliquity (C) and precession (E). In the right column the high-frequency part (40%) of the spectrogram is enlarged to show details of the obliquity and precession peaks.

rate first increases gradually in the up-borehole direction, changes abruptly across 115 m and then decreases gradually in the up-borehole direction. This produces a thickening-upward succession in the lower part of the synthetic log, followed by a thinning-upward succession in the upper part. A stratigraphic discontinuity is present at 115 m. The sliding window spectral analysis reveals two continuous spectrum bands representing the obliquity (C) and precession (E) cycles. In the lower part of the synthetic log the two spectrum bands shift gradually towards longer wavelength in up-borehole direction, whereas in the upper part they shift gradually towards shorter wavelength in up-borehole direction. A break and interference of the spectrum bands occurs around 115 m. The eccentricity cycle (A) is only partly shown in the spectrum because its wavelengths are too large to be covered by the window. The sliding window spectral analysis in this example shows

a stacking of the spectra peaks as well as the crosses indicating the wavelengths of the spectrum peaks.

The synthetic log in Fig. 7 shows not only frequency changes, but also lithological variations. The high gamma ray values are considered to reflect stronger gamma ray signals in fine-grained sediments. This synthetic log consists of four sections: a thickening-upward and fining-upward succession in the lowermost part of the synthetic log, followed by a thickening-upward and coarsening-upward succession in the lower part, a thinning-upward and coarsening-upward succession in the middle part, and a thinning-upward and fining-upward succession in the upper part. Three stratigraphic discontinuities are present. The sliding window spectral analysis of this synthetic log reveals spectrum bands representing the eccentricity (A), obliquity (C) and precession (E) cycles. The distinct breaks and interference of these spectrum bands are related to the stratigraphic discontinuities, whereas the gradual shift in the wavelengths of these spectrum bands reflect the gradual thickening or thinning of the cycles which is caused by a gradual increase or decrease of the net sediment accumulation rate. It is interesting to note that the lithological changes have no direct influence on the spectrum band patterns. The spectrum bands highlight the frequency or wavelength variations along the borehole (i.e., thickening-upward or thinning-upward trends). They do not show any difference between the coarsening-upward and fining-upward successions. In reality, however, the lithological changes may influence differential compactions, and therefore causing shifts in frequency and wavelength.

*Summary of sliding window spectral analysis*

(1) Discontinuities in the stratigraphic section are often marked by an abrupt wavelength change of high-frequency cycles, therefore causing a break or interference of the spectrum bands.

(2) A gradual increase or decrease of net sediment accumulation rates will cause a gradual increase or decrease in the wavelength of high-frequency cycles, which will be reflected in the gradual shift of the spectrum band.

(3) Lithological changes have no direct influence on the spectrum band patterns. Although, indirectly, the lithological changes may influence differential compactions, and therefore cause shifts in wavelength.

(4) The selection of the window size is based on experience in which considerations of required interval accuracy and the good definition of a stable spectrum play a role. A larger window contains more data points, and therefore may give a more reliable spectrum, but has a greater likelihood to contain discontinuities. A smaller window does not contain enough data points to cover the lower-frequency cycles. The good ranges of the window size are often determined by trial and error to obtain the best trade-off between interval accuracy and spectrum reliability.

## Example from the Tertiary of the southern North Sea

Figure 8 shows the results of sliding-window spectral analysis of the gamma-ray log in the Palaeogene interval of well F02–02 from the Dutch sector of the southern North Sea Basin. This large Tertiary basin is a thermal sag basin, conformable with the present-day North Sea Basin. The Palaeogene succession in well F02-02 is dominated by claystones with some marl deposited in a deep water setting near the basin centre. The sequence stratigraphic zonation is based on wireline log interpretation, wireline log cyclicity analysis and biostratigraphic evaluation. The figure shows several interesting features.

*Interpretation of the sliding-window spectrum bands*

The gamma ray-log of sequences TA3.5, TA4.2, TA4.3 and TA4.4 shows only subtle variations. The spectrum peaks, however, show a very distinct pattern of variations in net accumulation rates with obvious breaks. These breaks could be sequence boundaries.

Sequence TA3.5 initially shows an upward shift of the spectrum peaks to the right (to a higher-frequency range, i.e. an upward-thinning of units), followed by an upward shift to the left (to a lower frequency range, i.e. an upward-thickening of the units). This sequence was deposited in a deep water setting with sufficient accommodation space. The upward thinning/thickening of the beds can be interpreted to indicate the decrease/increase of sediment supply, related to the retrogradation/progradation of depositional systems during the rise/fall of the eustatic sea-level. These patterns of variations in the net accumulation rate can be used to help well-to-well correlations.

*Calibration of sequence boundaries*

An example is the sequence boundary at the base of sequence TA4.2. The gamma ray-log sliding-window spectral peaks reveals an obvious break in the variation of net accumulation rate at 1720 m. Below this boundary, the net accumulation rate increases upwards until this boundary is reached. Above the boundary the net accumulation rate is high in the lower part of sequence TA4.2, and then decreases upwards. In the case of deep water sedimentation, this can be interpreted to have occurred due to the progradation

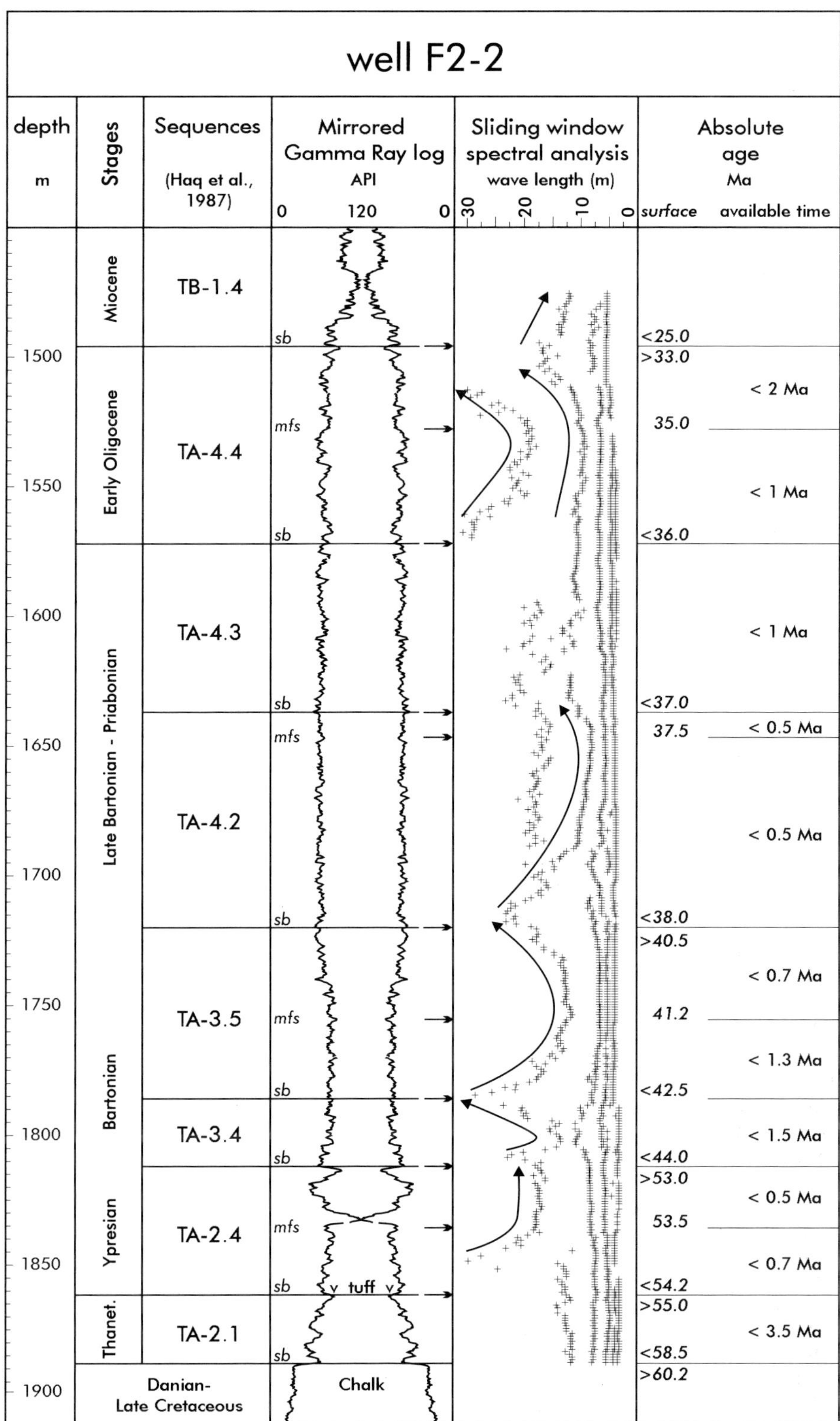



**Fig. 8.** Sliding-window spectral analysis of the Palaeogene GR log from well F02-02 of The Netherlands offshore (see Fig. 3b for location of the well).

of depositional systems during the eustatic sea-level fall, the aggradation during the lowstand and the retrogradation during the eustatic sea-level rise. Similar examples can be found at about 1812 m (TA3.4 s.b.), 1785 m (TA3.5 s.b.), 1572 m (TA4.4 s.b.), and 1496 m (TB1.4 s.b.).

### Calibration of maximum-flooding surfaces

An example of a maximum-flooding surface can be observed at 1749 m (TA3.5 mfs). The GR-log sliding-window spectral peaks reveal the variation of the net accumulation rate. Below this surface the net accumulation rate decreases upwards until this surface, then the net accumulation rate increases upwards. Therefore the maximum-flooding surface is marked by a minimal rate of net accumulation. Two similar examples of the maximum-flooding surface can be recognized at approximately 1527.5 m (TA4.4 mfs) and 1647 m (TA4.2 mfs).

Based on available biostratigraphic data, absolute ages are assigned to the Palaeogene sequences of F2-2. As the biostratigraphic data were obtained from cutting samples, these ages should be viewed only as an approximate indication. Nevertheless these data give some ranges of the maximum durations of the sequences. Sequences TA-4.2 and TA-4.3, e.g., have a maximum duration of 1 Ma each. In these two sequences the wavelengths of the most prominent cycles are approximately 10 m. If they represent the eccentricity cycle, this implies that the recorded time durations of sequences TA-4.2 and TA-4.3 are 830 and 640 ka respectively, close to the maximum durations. For other sequences, the durations estimated from Milankovitch cycles are shorter than the maximum durations based on sequence ages. This may reflect the incomplete stratigraphic record of these sequences.

In summary, the sliding-window spectral analysis of the gamma-ray log from well F02-02 reveals a pattern of vertically continuous bands of spectrum peaks in intervals where cyclicities are pronounced and a blank pattern in intervals where cyclicities are missing or blurred. The spectrum bands shift gradually towards lower frequencies in an up-borehole direction in a thickening-upward succession (upward net accumulation rate increase), and shift gradually towards higher frequencies in an up-borehole direction in a thinning-upward succession (upward net accumulation rate decrease). When the window includes a discontinuity, the pattern is interrupted for some length and a break or interference pattern is visible.

## Sliding-window spectrum patterns of the synthetic logs from two-dimensional sequence-stratigraphic modelling

Our observations from the subsurface of the southern North Sea indicate that the cyclicity patterns are often correlatable between wells. In order to understand the similarity of the cyclicity patterns between wells, synthetic GR logs are generated from a two-dimensional sequence stratigraphic model. Sliding-window spectral analysis is then applied to these logs to investigate the lateral correlatability of the cyclicity pattern between the wells.

There is a rapid development in the forward computer modelling to simulate the interaction between tectonics, eustacy and sedimentation (see e.g. Bitzer & Harbaugh 1987; Rivenæs 1995; among many others). The purpose of this two-dimensional sequence-stratigraphic modelling is to generate synthetic GR logs for further cyclicity analysis. The modelling uses a computer program developed by IGC. It is a geological process-based computer model. The model can be used to simulate and predict sequence development, sedimentary facies variation and reservoir sand distribution in space and time based on sequence stratigraphic principles. The basic input parameters include basin subsidence, sea-level/base-level changes, amplitudes and periods of Milankovitch cycles, sediment influx and dispersal, and hinterland area. Most parameters are derived from basic geological data or from well studies (e.g. the duration and net accumulation rate estimated in cyclicity analysis).

Based on these parameters the computer model starts from a topographic section crossing the study area from proximal position (basin margin) to distal position (basin). It then simulates the changes (deposition or erosion) along this cross-section at a fixed time step. At each step the deposition or erosion is controlled by topography, accommodation space development (basin subsidence and sea-level/base-level changes), sediment supply, and climatic changes related to Milankovitch cycles. In this way the model can simulate the sequence development, sedimentary facies variations and sand distributions along the cross section as a response to basin subsidence, sea-level/base-level fluctuations, climatic changes, and variations in sediment supply.

The results can be presented as a compacted geological cross-section, showing sequence boundaries and internal strata stacking patterns, sedimentary facies variation, and sand distribution (Fig. 9). Such geological cross-section can

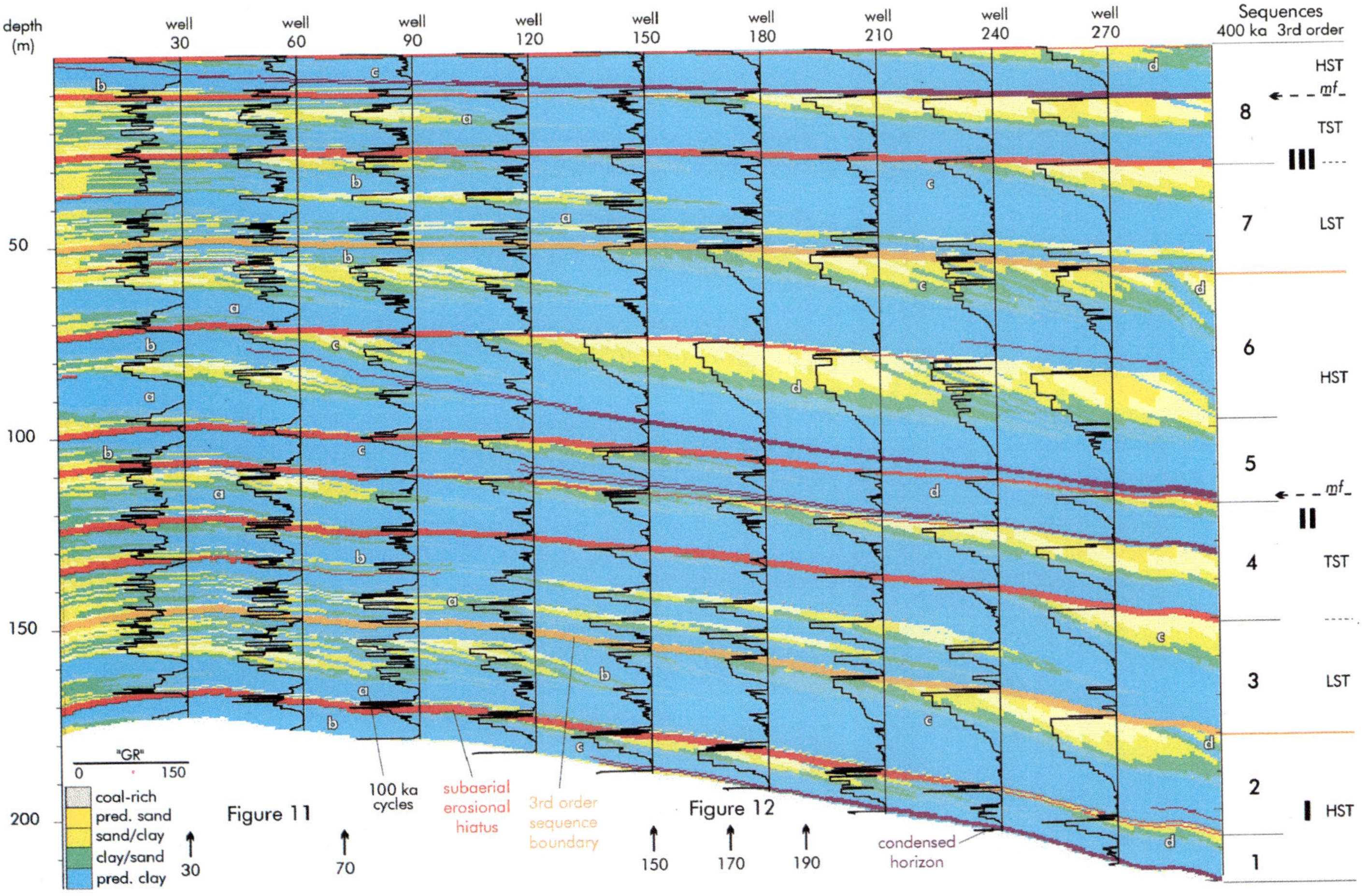


**Fig. 9.** Compacted geological cross-section from a two-dimensional high-resolution sequence-stratigraphic modelling. This dip-section simulated a delta progradation during a period of 3.2 Ma in the Carboniferous time (Late Namurian to Early Westphalian). The GR logs of selected synthetic wells are also shown in this section. The third-order and the 400 ka sequences are shown in the column to the right, whereas the 100 ka cycles are marked by letters a, b, c and d. The locations of the synthetic logs in Figs 11 and 12 are indicated at the bottom of the section.

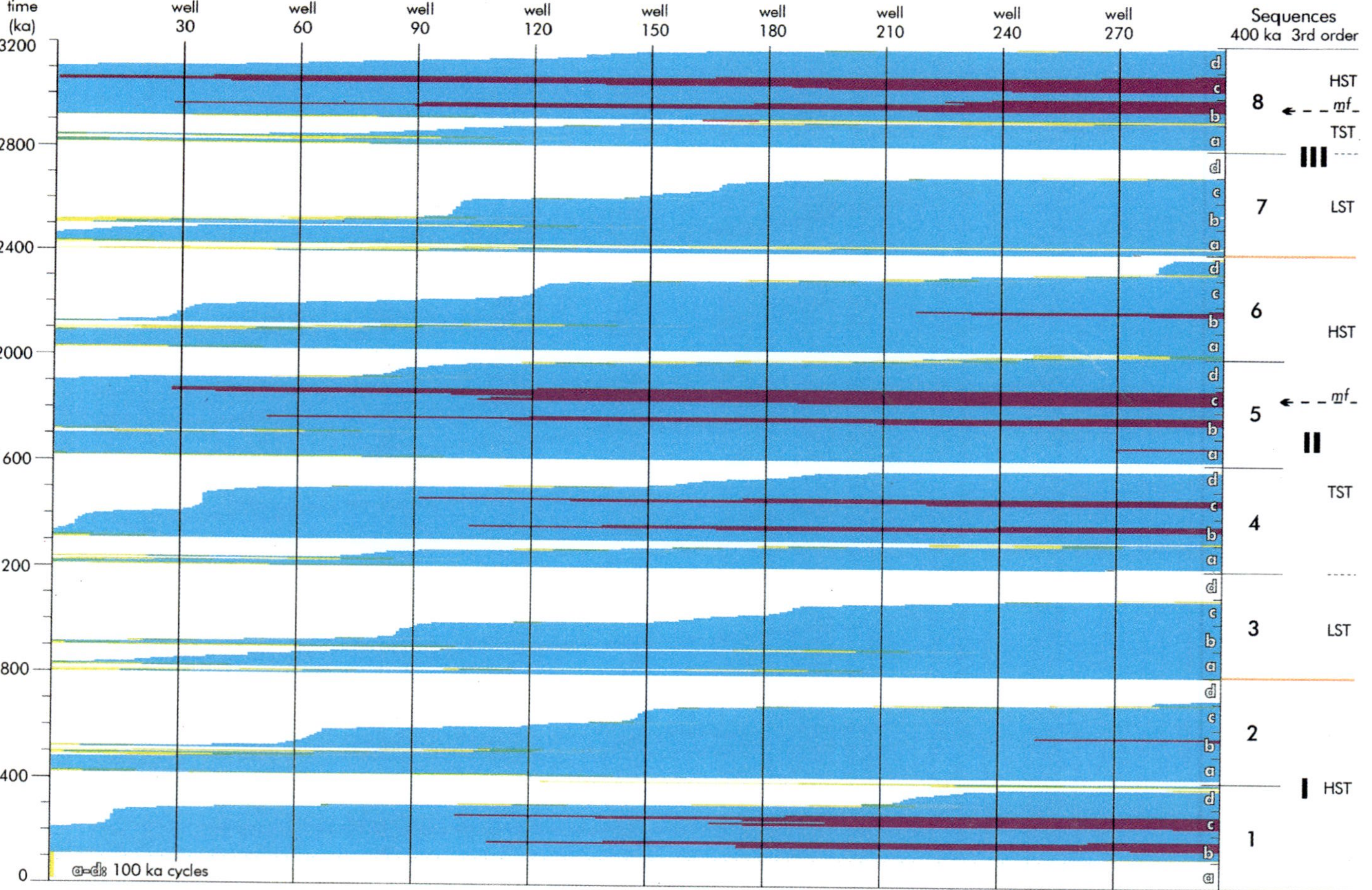



**Fig. 10.** Time section from a two-dimensional high-resolution sequence-stratigraphic modelling, illustrating the chronostratigraphic development of sequences during the delta progradation of Fig. 9. The third-order and the 400 ka sequences are shown in the column to the right, whereas the 100 ka cycles are marked by letters a, b, c and d.

be compared to, and further calibrated with, real well data. The modelling results can also be presented as a time section, showing a chronostratigraphic development of sequences (Fig. 10).

Furthermore, the modelling can generate a number of synthetic wells along the cross-section. Such data can be used to produce a synthetic seismic section, which can then be compared to, and calibrated with, real seismic sections. In addition, the modelling results can be displayed as a 'movie', showing the details of sequence deposition or erosion in response to sea-level/base-level fluctuations.

After the comparison and calibration with well and seismic data, the model can be adjusted in an iterative and interactive process. In this way, the model can simulate and predict reservoir sand distribution, which will shed light on detailed reservoir simulation studies.

Figure 9 shows a compacted geological cross-section from a two-dimensional high-resolution sequence-stratigraphic modelling for the delta progradation during the Late Namurian to Early Westphalian time in the southern North Sea Basin. The Silesian in this basin represents an overall regressive process, filling the foreland basin of the Variscan Orogen (Ziegler 1989). During the Namurian, the basin was gradually filled by cyclic delta progradations. In the Late Namurian C and Early Westphalian A delta plain and fluvial plain became an important stage in the delta progradation.

The modelled dip section is 300 km long, deposited during a time duration of 3.2 Ma. The basic input parameters are listed in Table 1. The basin subsidence rate is 15 cm ka$^{-1}$ in the proximal position, and 19 cm ka$^{-1}$ in the distal position. A maximum amplitude of 100 m is assumed for the eustatic sea-level changes. Three Milankovitch cycles are used in the model, the 400 ka and 100 ka eccentricity cycles and the 34 ka obliquity cycle. The 'climate influence' in Table 1 is a parameter describing the ranges of variations in the erosional and depositional rates during Milankovitch climate cycles. In Table 1 a climate influence 0.5 means that the erosion rate changes from 200 cm ka$^{-1}$ in the climate maximum, to 100 cm ka$^{-1}$ in the climate minimum. Similarly the depositional rates vary from 50 cm ka$^{-1}$ (clay) and 10 cm ka$^{-1}$ (sand) during the climate maximum to 25 cm ka$^{-1}$ (clay) and 5 cm ka$^{-1}$ (sand) during the climate minimum.

In the compacted geological cross-section, sediment transport is from left to right. The fluvial sandstones at the left side of the cross-section change to a delta-plain setting in the central part, where clayey sandstone, sandy claystone and claystone increase significantly.

**Table 1.** *Basic input parameters for the high-resolution sequence-stratigraphic modelling of the Carboniferous section*

| | |
|---|---|
| *Subsidence* | |
| Proximal | 15 cm ka$^{-1}$ |
| Distal | 19 cm ka$^{-1}$ |
| *Maximum sea-level amplitude* | 10 000 cm |
| *Milankovitch cycles* | |
| Eccentricity | 400 ka |
| Eccentricity | 100 ka |
| Obliquity | 34 ka |
| *Erosion* | |
| Maximum erosion rate | 200 cm ka$^{-1}$ |
| Climatic influence | 0.5 |
| *Deposition* | |
| Fine sediments | 50 cm ka$^{-1}$ |
| Coarse sediments | 10 cm ka$^{-1}$ |
| Climatic influence | 0.5 |
| *Dimension* | |
| Number of cells | 300 |
| Cell width | 1 km |
| Number of slices | 3200 |
| Slice time duration | 1 ka |
| Shore slope | 0.1 |
| Hinterland width | 400 km |

Further basinward are delta-lobes, which show periods of distinct progradation, alternated with periods of transgression. The depositional sequences are divided by subaerial erosion/hiatus (red or orange line in Fig. 9) and subaqueous condensed section (purple line in Fig. 9).

The time section (Fig. 10) shows the chronostratigraphic development of sequence. On this time section colours represent periods of sedimentation, including sand (yellow), clayey sand (greenish yellow), sandy clay (green) and clay (blue). Blank areas reflect the periods of erosion or non-deposition, whereas purple areas indicate the periods of subaqueous hiatus.

Three order of sequences can be recognized in the modelled section (Figs 9 and 10). The most distinct sequence corresponds to the 400 ka cycle. There are eight 400 ka sequences. They are bounded by subaerial erosional surfaces (1 to 8 in Figs 9 and 10), with a rather uniform thickness over the section. Each 400 ka sequence consists of four 100 ka sequences, which show a clear strata stacking pattern characterised by a rapid retrogradation of the lowermost 100 ka sequence over the basal erosional boundary, followed by sedimentation in the proximal position and subaqueous hiatus (condensed section) in the distal position in the middle

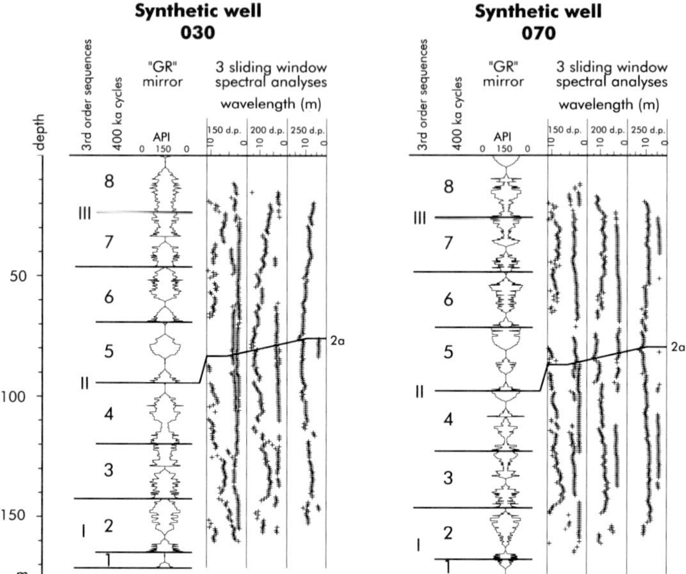

**Fig. 11.** Correlation of the sliding-window spectrum patterns of two synthetic GR logs from wells 030 and 070 in the proximal part of the cross-section shown in Fig. 9. The locations of the wells are indicated in Fig. 9.

part of the 400 ka sequence (e.g., cycle 5 in Figs 9 and 10). The upper part of the 400 ka sequence displays a basinward progradation.

The stacking pattern of the 400 ka sequences, in turn, shows a large-scale progradation–retrogradation, corresponding to the third-order sequence. In a complete third-order sequence (II in Figs 9 and 10), the LST shows extensive erosion in the proximal area and the most basin-ward progradation of the delta (cycle 3 in sequence II). The TST is marked by a rapid transgression at the base, and the most extensive condensed section at the top. The HST displays a progradation of the delta towards the basin.

Figure 9 displays also the synthetic GR logs generated along the modelled dip section. Synthetic wells in the proximal position show a succession dominated by fining-upward cycles, whereas those in the distal position display a succession of coarsening-upward cycles. This is interpreted to reflect the transition from fluvial dominated sedimentation in the proximal position to delta sedimentation in the distal position. These strata stacking and log patterns are very similar to those observed in the Carboniferous (Late Namurian to Early Westphalian) in the Silver Pit–Cleaver Bank area of the southern North Sea Basin.

Figure 11 shows the GR logs of two synthetic wells from the proximal position of the modelled

dip section. Well 030 and 070 are 40 km apart. Both wells display a succession dominated by fining-upward cycles, interpreted to have been deposited in a fluvial-delta plain setting. The sliding window spectral analysis reveals a distinct spectrum band at the wavelength of about 10 m. This spectrum band reflects the 100 ka eccentricity cycle. The spectrum band displays an increasing wavelength in the up-borehole direction until the base of sequence 5 and then a decreasing wavelength in the up-borehole direction in the sequences above. This pattern is comparable between the two wells, although the log patterns do show some difference (e.g. cycles 4 and 5 in the two wells). The cycle thickness trend is interpreted to reflect the strata stacking pattern in the proximal position, where the thickest deposition cycles (the high sedimentation rate) are expected to occur in the late TST and the early HST.

Figure 12 shows the GR logs of three synthetic wells from the middle part of the modelled dip section. Well 170 is 20 km from well 150, whereas well 190 is 20 km further away. These wells are dominated by successions of coarsening-upward cycles, interpreted to represent delta progradations. The sliding window spectrum bands with the wavelength of 7–8 m correspond to the 100 ka eccentricity cycles. The spectrum band between break 1b and 1a displays a thickening-upward trend. A thickening-upward trend can

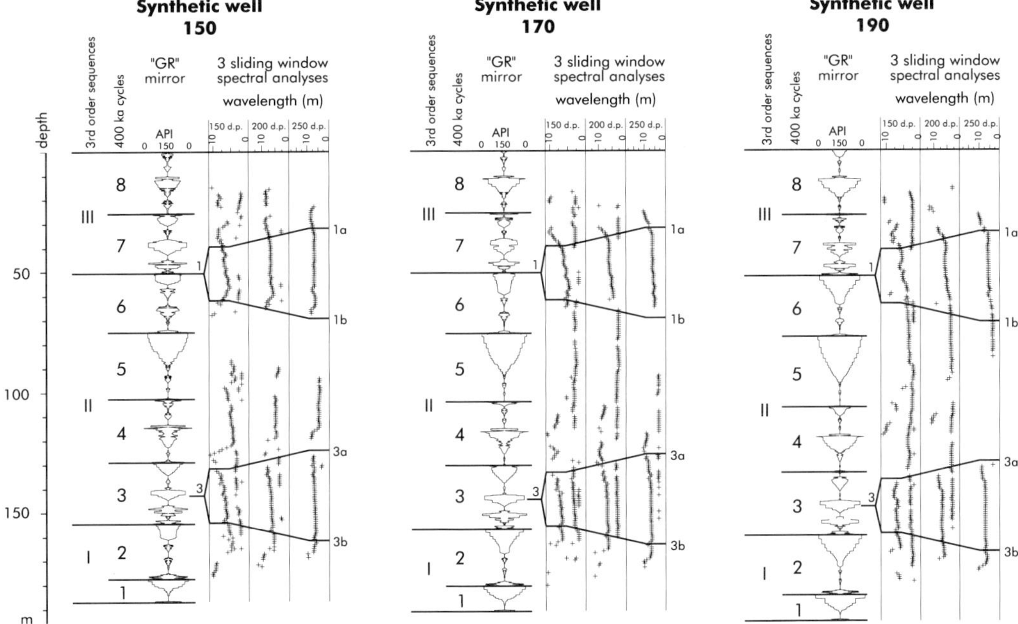

**Fig. 12.** Correlation of the sliding-window spectrum patterns of three synthetic GR logs from wells 150, 170 and 190 in the middle part of the cross-section shown in Fig. 9. The locations of the wells are indicated in Fig. 9.

also be recognised in the spectrum band between breaks 3b and 3a. Break 1 and 3 as well as the thickening-upward trend are comparable among the three wells.

The results from this two-dimensional sequence stratigraphic modelling indicate that the sliding-window spectrum band patterns can be correlated over a distance of at least 40 km along the modelled dip section. The difference in the spectrum band patterns of the synthetic wells between the proximal and the middle parts of the dip section reflects the different sedimentation in the two positions during the delta progradation. It can be expected that the sliding window spectrum band patterns may be correlatable over even a longer distance along a strike section. Such cases have been observed from the subsurface data of the southern North Sea, and can be investigated with the cyclicity analysis of synthetic wells generated in a three-dimensional high-resolution sequence stratigraphic modelling.

As shown in Figs 9 and 10, within a cycle, deposition often takes place during part of the cycle only, whereas the rest of the cycle is marked by erosion and non-deposition. Nevertheless, the preserved sequence can be viewed as a record of the amount of increased accommodation

space due to basin subsidence and eustasy during the whole cycle, which was filled by sediments and preserved. In this sense the preserved sequence can be viewed as the stratigraphic record of the whole cycle. This applies for different orders of cycles. The lower order depositional cycles (e.g. the 400 ka cycles in Figs 9 and 10) have thicker sequences with a higher preservation potential. They can be correlated over a long distance with rather stable thickness. The higher order cycles (e.g. the obliquity and precession cycles) have thinner sequences with more lateral variations. In reality they may be overprinted by autocyclic processes, such as channel avulsion in a fluvial system, or lobe switching on a delta.

Due to the erosion/non-deposition during part of a 400 ka cycle, its stratigraphic record may consists of four, three, two or even only one 100 ka cycles, depending on the position from the basin towards the margin (Figs. 9 and 10). This implies that the ratio between the wavelengths of different orders of cycles in the stratigraphic record may differ from the theoretical ratio. Such a difference can be used to evaluate the amount of erosion in the lower order cycles and to infer the position of the well during the deposition.

## Examples of sliding window spectrum pattern correlation from the southern North Sea

### Neogene in wells F2-2 and F3-3 of the Dutch offshore

Wells F2-2 and F3-3 are located in the Dutch sector of the North Sea Basin, a Tertiary thermal sag basin conformable with the present-day North Sea Basin. Wells F2-2 and F3-3 are approximately 8.5 km from each other. The Neogene succession in the two wells is dominated by fine-grained sediments deposited in a shelf setting. Sands can be recognised in a few intervals (base TB-3.1 and TB-3.4 in Fig. 13), which are interpreted as the LST sediments deposited during eustatic sea-level falls. The sliding-window spectral analysis of the Neogene gamma ray logs from the two wells shows a number of cyclicity breaks (Fig. 13). At least 13 spectrum band breaks can be correlated between the 2 wells. These cyclicity breaks have been numbered 1 to 13 (Fig. 13). The spectrum bands between the breaks are also comparable in the two wells, such as:

- the spectral bands above break 2a show a slight thinning-up trend in both wells;
- the spectral bands between breaks 3b and 2a show an overall thinning-up trend in both wells;
- the spectral bands between breaks 5b and 3b display a slight thinning-upward trend in the lower part, and a thickening-upward trend in the upper part;
- the spectral bands between breaks 8a and 6a show a distinct thickening-up trend in both wells;
- the spectral bands between breaks 10a and 9a are similar in the two wells.
- The spectral bands between breaks 12b and 13a are similar in the two wells.

The comparison of these spectrum-band breaks as well as the patterns of the spectrum-band shift can be used to help detailed sequence correlation. The fact that cyclicity patterns can be correlated between different wells indicate that the spectrum band patterns are not random variations. They reflect net accumulation rate variations controlled by accommodation (basin subsidence and eustatic sea-level fluctuations) and sediment supply. This provides a sound base for the application of cyclicity analysis to support sequence zonation and well-to-well correlation.

The correlation shown in Fig. 13 is based on wireline-log interpretation and cyclicity analysis. The sequence ages are interpreted from the biostratigraphic data of F2-2. This correlation is different from the biostratigraphic correlation. The biostratigraphic data of F3-3 are considered not very reliable due to the limited cutting samples with large spacing. The correlation based on cyclicity results has been confirmed by seismic data. This correlation suggests that the two major sand intervals at the base of TB-3.1 and TB-3.4 are correlatable between the two wells. They are interpreted to have been deposited during the eustatic sea-level falls (Haq et al. 1987).

Sequence-stratigraphic concepts have been applied to siliciclastic rock successions that display an alternation of coarse-grained intervals with fine-grained intervals. However, thick successions comprising fine-grained sediments (e.g. the claystone and siltstone successions in this example), or a more-or-less monotonous lithofacies (e.g. chalk), are problematical in sequence stratigraphic analysis. The establishment of a sequence-stratigraphic framework for fine-grained rock succession is essential in exploration. Observations from wells show a consistent occurrence of high-frequency cycles in thick fine-grained rock successions. The climate-forcing changes of high-frequency cycles caused changes in the physical and chemical properties in marine and continental basins, which were recorded as minor lithofacies changes in the fine-grained successions. The cyclicity analysis of well F02-02 and F03-03 shows a good example that sliding window spectral analysis is a useful tool to support well-to-well sequence correlations in fine-grained successions.

### Carboniferous in wells E13-1 and E18-2 of the Dutch offshore

Wells E13-1 and E18-2 are located in the Cleaver Bank area of the southern North Sea Basin, a foreland basin of the Variscan Orogen during the Carboniferous (Ziegler 1989). The basin was gradually filled by cyclic delta progradations in the Namurian and Westphalian A time. Subsequently the area was dominated by two stages of widespread coarse-grained fluvial sedimentation in the Early Westphalian B and the Late Westphalian C alternated with stages of extensive fine-grained sedimentation in the Late Westphalian B to Early Westphalian C (mainly shallow lacustrine facies) and the Early Westphalian D (flood plain facies).

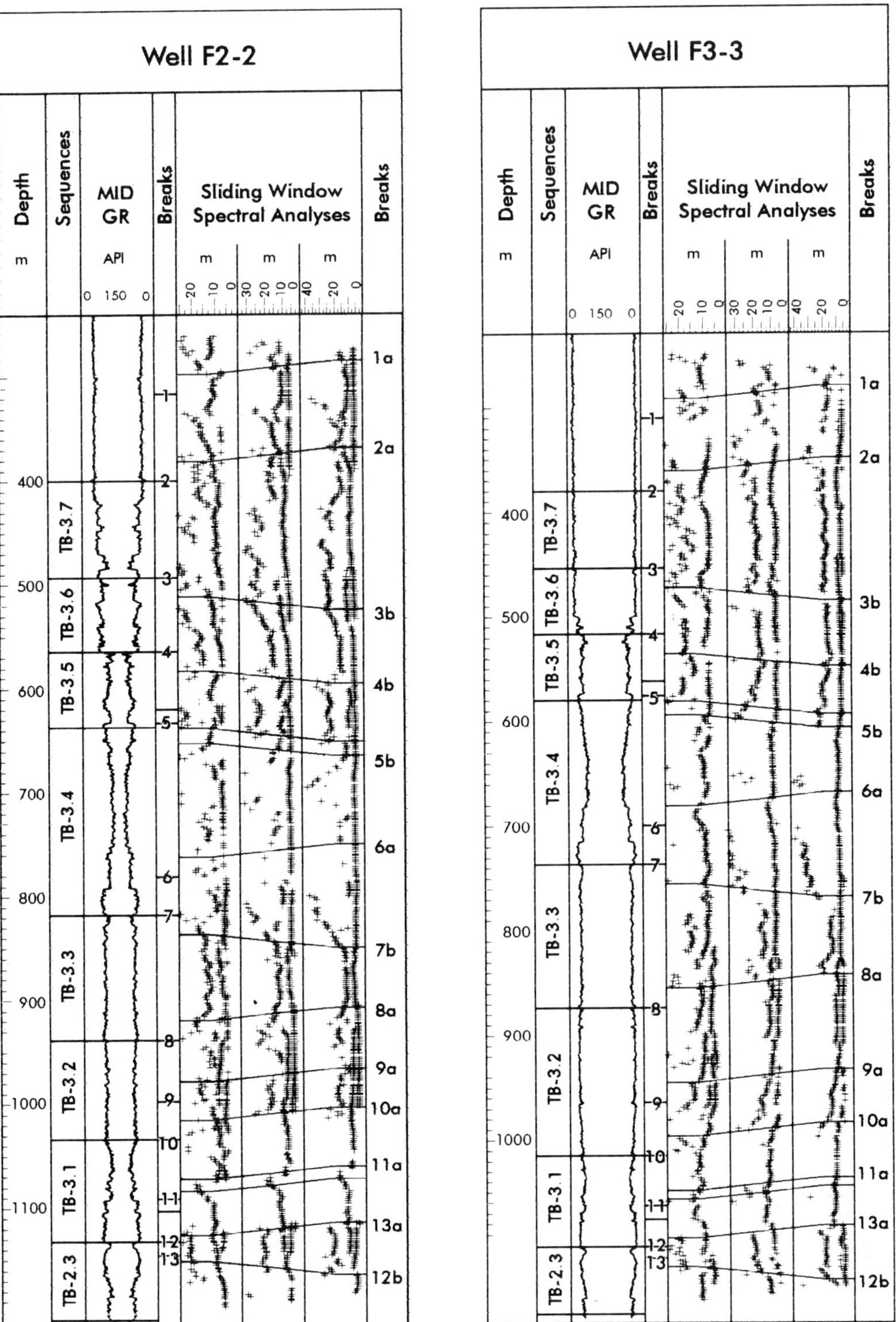

**Fig. 13.** Sliding-window spectral analysis of the Tertiary in wells F2-2 and F3-3 in the Netherlands offshore (see Fig. 3b for location of the wells). MID GR refers to Mirror image display of gamma ray log. The spectral peaks were produced with the maximum entropy spectral analysis. Sequence names follow Haq *et al.* (1987). The patterns of spectrum band shift and spectrum band breaks can be correlated between the two wells to help detailed well-to-well sequence correlation.

Wells E13-1 and E18-2 are approximately 44.5 km apart. The sliding-window spectral analysis of the Westphalian section in the two wells shows distinct spectrum bands interrupted by breaks (Fig. 14). At least six major breaks can be correlated between the two wells. The spectral bands between breaks 2a and 1a show a thickening-up trend in both wells. The spectral bands between breaks 3b and 4a are similar in the two wells. The spectral bands between breaks 6b and 3b show an overall thickening-up trend in both wells. These cyclicity breaks are correlated between the two wells to help sequence correlations. In biostratigraphically barren sections such as the Westphalian C and D sequence stratigraphic correlations are

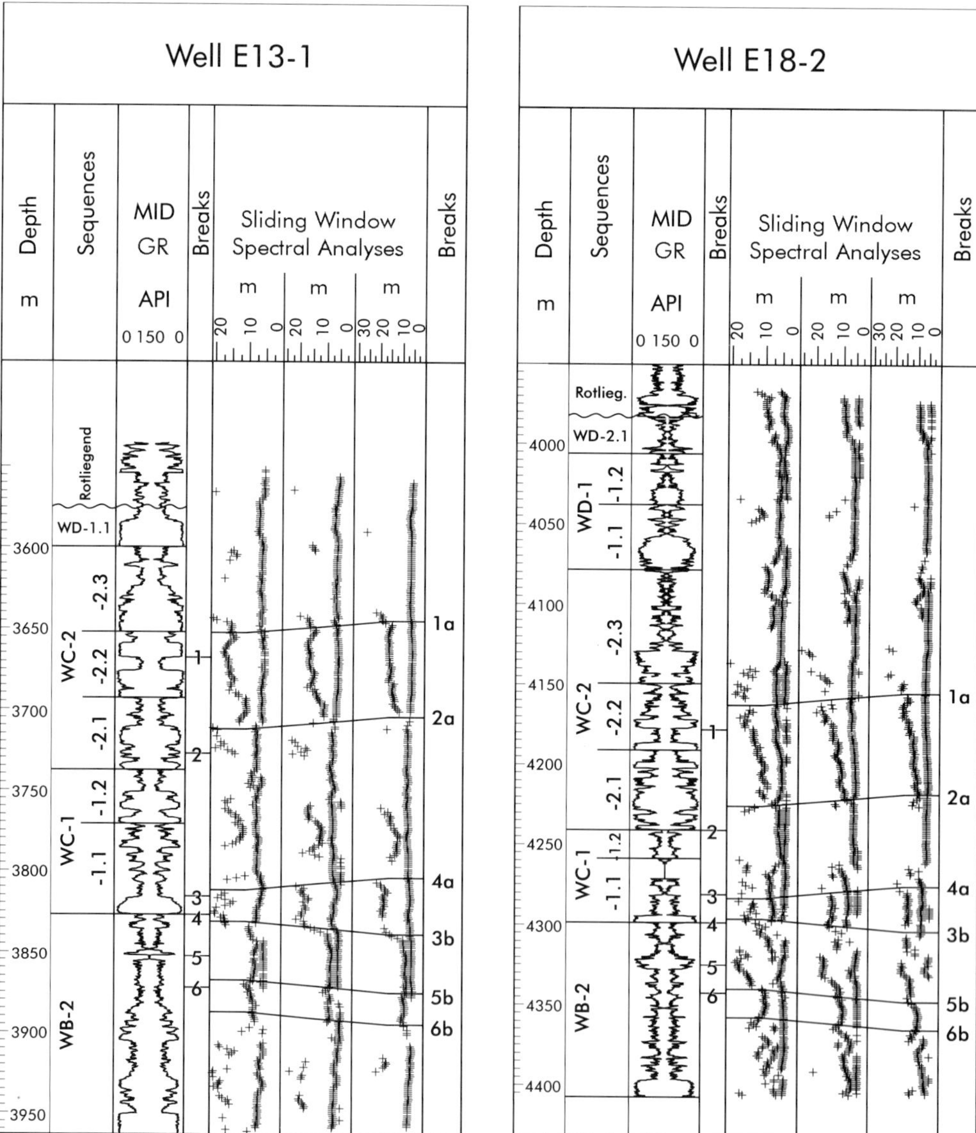

**Fig. 14.** Sliding-window spectral analysis of the Westphalian section in wells E13-1 and E18-2 of the Netherlands offshore (see Fig. 3b for location of the wells). The analysed interval includes Late Westphalian B (sequence WB-2), Westphalian C (sequences WC-1.1 to WC-2.3) and Early Westphalian D (sequences WD-1.1 to WD-2.1). MID GR refers to Mirror image display of gamma ray log. The spectral peaks were produced with the maximum entropy spectral analysis. The breaks and shift of the spectrum bands are comparable between the two wells.

often difficult. The pattern recognition and comparison of high-frequency spectrum bands from wireline log cyclicity analysis provides an additional tool for well-to-well sequence stratigraphic correlations.

## Conclusions

(1) Sliding window spectral analysis can be used to detect wavelength changes of the high-frequency cycles along the bore hole. The break or interference of the spectrum bands indicates an abrupt wavelength change of the high-frequency cycles, which represents a discontinuity related to the stratigraphic boundaries. Such breaks can be used to recognize and calibrate sequence boundaries and maximum-flooding surfaces. The gradual shift of the spectrum bands reflects a gradual wavelength change of the high-frequency cycles, which represents a gradual change in the rate of net sediment accumulation or in the rate of the development of sediment accommodation space.

(2) Observations from the subsurface of the southern North Sea indicate that the cyclicity patterns can be correlated between wells. This suggests that the cyclicity patterns are not random features in individual wells. They reflect the thickness variations of preserved depositional cycles, controlled by basic geological processes, i.e. basin subsidence, eustatic sea-level/base-level fluctuations, and sediment supply. Therefore the breaks of spectrum bands as well as the pattern of spectrum band shift can be correlated between wells to help detailed sequence correlation.

(3) Cyclicity analysis of the synthetic logs generated from a two-dimensional sequence stratigraphic modelling confirmed the lateral correlatability of the cyclicity pattern between wells. The 400 ka sequence can be correlated over the entire length (300 km) of the modelled dip section with a rather uniform thickness. The 100 ka sequences do show more variations in thickness and lithology along the dip section, but are still correlatable over a distance of at least 40 km (Fig. 9).

(4) It is crucial to realise that the mathematical treatment and computer analysis of wireline logs is not a 'black box', but an additional tool to use ignored, yet valuable, information in wire-line logs and to look at the geology from a new dimension. Integrated with traditional methods (e.g., log correlation, biostratigraphic and seismic interpretation), this approach will reduce the uncertainty and improve the resolution in correlations. The method has been used to establish accurate and reliable sequence stratigraphic correlations in different settings, including red beds of the Westphalian C and D in Late Carboniferous, the biostratigraphically barren strata of the Lower Permian Upper Rotliegend Group, the chalk section of the Upper Cretaceous, and the thick intervals of fine-grained or coarse-grained sediments. Such correlations are essential for the establishment of a proper geological model to understand and predict the reservoir sandstone distribution and source rock sedimentation.

The Authors thank the Geological Survey of The Netherlands for providing the wireline logs of released wells for this study, and thank IGC B.V. for permission to publish this paper. Appreciation and thanks to W. Kouwe, D. Koelewijn, H. v.d. Veen and E. Scheele for help in preparing the manuscript. H. Jansen kindly provided the biostratigraphic data for the Tertiary in wells F2-2 and F3-3 of the Dutch offshore. Appreciation and thanks also to K. Ziegler, P. Sansom and J. Howell for their thorough and constructive reviews which led to significant improvements in the paper.

## References

BERGER, A. L. 1980. The Milankovitch astronomical theory of palaeoclimate: a modern review. *Vistas in Astronomy*, **24**, 103–122.
—— & LOUTRE, M. F. 1994. Astronomical forcing through geological time. *In*: DE BOER, P. L. & SMITH, D. G. (eds) *Orbital forcing and cyclic sequences*. International Association of Sedimentologists, Special Publications, **19**, 15–24.
——, —— & DEHANT, V. 1989. Astronomical frequencies for pre-Quaternary palaeoclimate studies. *Terra Nova*, **1**, 474–479.
BESLY, B. M., BURLEY, S. D. & TURNER, P. 1993. The late Carboniferous Barren Red Bed play of the Silver Pit area, Southern North Sea. *In*: PARKER, J. R. (ed.) *Petroleum Geology of Northwest Europe. Proceedings of the 4th Conference*, The Geological Society, London, 727–740.
BITZER, K. & HARBAUGH, J. W. 1987. Deposim: a Macintosh computer model for two-dimensional simulation of transport, deposition, erosion, and compaction of clastic sediments. *Computers & Geosciences*, **13**, 611–637.
DE BOER, P. L. & SMITH, D. G. (eds) 1994. *Orbital forcing and cyclic sequences*. International Association of Sedimentologists, Special Publications, **19**.
FISCHER, A. G. & BOTTJER, D. J. (eds) 1991. *Orbital forcing and sedimentary sequences. Journal of Sedimentary Petrology, Special Issue*, **61**(7).
——, HERBERT, T. D., NAPOLEONE, G., SILVA, I. P. & RIPEPE, M. 1991. Albian pelagic Rhythms (Piobbico core). *In*: FISCHER, A. G. & BOTTJER, D. J. (eds) *Orbital forcing and sedimentary sequences. Journal of Sedimentary Petrology, Special Issue*, **61**, 1164–1172.

GLENNIE, K. W. 1990. Lower Permian-Rotliegend. *In*: GLENNIE, K. W. (eds) *Introduction to the Petroleum Geology of the North Sea*. Blackwell Scientific Publications, Oxford, 120–152.

HAQ, B. U., HARDENBOL, J. & VAIL, P. R. 1987. Chronology of fluctuating sea levels since the Triassic (250 million years ago to present). *Science*, **235**, 1156–1167.

HAYS, J. D., IMBRIE, J. & SHACKLETON, N. J. 1976. Variations in earth's orbit: pacemaker the ice ages. *Science*, **194**, 1121–1132.

IMBRIE, J. & IMBRIE, J. Z. 1979. Modeling the climate response to orbital variations. *Science*, **207**, 943–953.

KOCUREK, G., HAVHOLM, K. G., DEYNOUX, M. & BLAKEY, R. C. 1991. Amalgamated accumulations resulting from climatic and eustatic changes, Akchar Erg, Mauritania. *Sedimentology*, **38**, 751–772.

MELNYK, D. H. & SMITH, D. G. 1989. Outcrop to subsurface cycle correlation in the Milankovitch frequency band: Middle Cretaceous, central Italy. *Terra Nova*, **1**, 432–436.

NIO, S. D., YANG, C. S., BAUMFALK, Y. A., VAN DEN HURK, J. J., JONKMAN, H., SCHEELE, E., VAN DER VEEN, H. & VAN DE WEERD, A. 1993 Computer analysis of depositional sequences using wireline logs – A new method for determining rates of geologic processes. *In*: ARMENTROUT, J. M., BLOCH, R., OLSON, H. C. & PERKINS, B. F. (eds) *Rates of geologic processes*. Society of Economic Paleontologists and Mineralogists Foundation, Gulf Coast Section, 14th Annual Research Conference, 141–154.

POSAMENTIER, H. W. & JAMES, D. P. 1993. An overview of sequence-stratigraphic concepts: uses and abuses. *In*: POSAMENTIER, H. W. *ET AL*. (eds) *Sequence stratigraphy and facies associations*. International Association of Sedimentologists, Special Publications, **18**, 3–18.

—— & VAIL, P. R. 1988. Eustatic controls on clastic deposition II – sequence and systems tract models. *In*: WILGUS, C. K. *ET AL*. (eds) *Sea-level changes: an integrated approach*. Society of Economic Paleontologists and Mineralogists, Special Publications, **42**, 125–154.

——, JERVEY, M. T. & VAIL, P. R. 1988. Eustatic controls on clastic deposition I – conceptual framework. *In*: WILGUS, C. K. *ET AL*. (eds) *Sea-level changes: an integrated approach*. Society of Economic Paleontologists and Mineralogists, Special Publications, **42**, 109–124.

RIVENÆS, J. C. 1995. 2D computer simulation of the Brent delta development. *Predictive High Resolution Sequence Stratigraphy*, Abstract, Norwegian Petroleum Society, Stavanger, 21A–21B.

SARNTHEIN, M. 1978. Sand deserts during glacial maximum and climatic optimum. *Nature*, **272**, 43–45.

SCHWARZACHER, W. 1991. Milankovitch cycles and the measurement of time. *In*: EINSELE, G., RICKEN, W. & SEILACHER, A. (eds) *Cycles and events in stratigraphy*. Springer-Verlag, 855–863.

——1993. Cyclostratigraphy and the Milankovitch theory. *Developments in Sedimentology*, **52**, Elsevier, 31–35.

SMITH, D. G. 1989. Stratigraphic correlation of presumed Milankovitch cycles in the Blue Lias (Hettangian to earliest Sinemurian), England. *Terra Nova*, **1**, 457–460.

TALBOT, M. R. 1985. Major bounding surfaces in eolian sandstones – a climatic model. *Sedimentology*, **32**, 257–265.

TEN KATE, W. G. H. Z. & SPRENGER, A. 1992. *Rhythmicity in deep-water sediments, documentation and interpretation by pattern and spectral analysis*. Academisch proefschrift, Vrije Universiteit, Amsterdam.

THUNELL, R., RIO, D., SPROVIERI, R. & RAFFI, I. 1991. Limestone-marl couplets: origin of the Early Pliocene Trubi Marls in Calabria, southern Italy. *In*: FISCHER, A. G. & BOTTJER, D. J. (eds) *Orbital forcing and sedimentary sequences. Journal of Sedimentary Petrology, Special Issue*, **61**, 1109–1122.

VAIL, P. R., MITCHUM, R. M., JR. & THOMPSON, S. III 1977*a*. Seismic stratigraphy and global changes of sea level, part four: global cycles of relative changes of sea level. *In*: PAYTON, C. E. (ed) *Seismic stratigraphy – applications to hydrocarbon exploration*. American Association of Petroleum Geologists, Memoirs, **26**, 83–98.

——, TODD, R. G., WIDMIER, J. M., THOMPSON, S. III, SANGREE, J. B., BUBB, J. N. & HATLELID, W. G. 1977*b*. Seismic stratigraphy and global changes of sea level. *In*: PAYTON, C. E. (ed.) *Seismic stratigraphy – applications to hydrocarbon exploration*. American Association of Petroleum Geologists, Memoirs, **26**, 49–212.

VAN WIJHE, D. H., LUTZ, M. & KAASSCHIETER, J. P. H. 1980. The Rotliegend in The Netherlands and its gas accumulations. *Geologie en Mijnbouw*, **59**, 3–24.

WEEDON, G. P. 1991. The spectral analysis of stratigraphic time series. *In*: EINSELE, G., RICKEN, W. & SEILACHER, A. (eds) *Cycles and events in stratigraphy*. Springer-Verlag, 840–854.

WORTHINGTON, P. F. 1990. Sediment cyclicity from well logs. *In*: HURST, A., LOVELL, M. A. & MORTON, A. C. (eds) *Geological Applications of Wire-line Logs*. Geological Society, London, Special Publications, **48**, 123–132.

YANG, C. S. & BAUMFALK, Y. A. 1994. Milankovitch cyclicity in the Upper Rotliegend Group of The Netherlands offshore. *In*: DE BOER, P. L. & SMITH, D. G. (eds) *Orbital forcing and cyclic sequences*. International Association of Sedimentologists, Special Publications, **19**, 47–61.

—— & KOUWE, W. F. P. 1995. Wireline log-cyclicity analysis as a tool for dating and correlating barren strata: an example from the Upper Rotliegend of the Netherlands. *In*: DUNAY, R. E. & HAILWOOD, E. A. (eds.) *Non-biostratigraphical methods of dating and correlation*. Geological Society, London, Special Publications, **89**, 237–259.

—— & NIO, S. D. 1994. Applications of high-resolution sequence stratigraphy to the Upper Rotliegend in The Netherlands offshore. *In*: WEIMER, P. & POSAMENTIER, H. W. (eds) *Siliciclastic sequence stratigraphy: recent developments* and applications. American Association of Petroleum Geologists, Memoirs, **58**, 285–316.

ZIEGLER, P. A. 1989. *Evolution of Laurussia*. Kluwer Academic Publishers, Dordrecht.

# Index